Advanced Bifunctional Electrochemical Catalysts for Metal-Air Batteries

T0172784

Electrochemical Energy Storage and Conversion

Series Editor: Jiujun Zhang
National Research Council Institute for Fuel Cell Innovation
Vancouver, British Columbia, Canada

Recently Published Titles

Advanced Bifunctional Electrochemical Catalysts for Metal-Air Batteries
Yan-Jie Wang, Rusheng Yuan, Anna Ignaszak, David P. Wilkinson, Jiujun Zhang

High-Temperature Electrochemical Energy Conversion and Storage: Fundamentals and Applications
Yixiang Shi, Ningsheng Cai, and Jiujun Zhang

Hydrothermal Reduction of Carbon Dioxide to Low-Carbon Fuels
Fangming Jin

Carbon Nanomaterials for Electrochemical Energy Technologies: Fundamentals and Applications
Shuhui Sun, Xueliang Sun, Zhongwei Chen, Jinli Qiao, David P. Wilkinson, and Jiujun Zhang

Redox Flow Batteries: Fundamentals and Applications
Huamin Zhang, Xianfeng Li, and Jiujun Zhang

Electrochemical Energy: Advanced Materials and Technologies
Pei Kang Shen, Chao-Yang Wang, San Ping Jiang, Xueliang Sun, and Jiujun Zhang

Electrochemical Polymer Electrolyte Membranes
Jianhua Fang, Jinli Qiao, David P. Wilkinson, and Jiujun Zhang

Electrochemical Supercapacitors for Energy Storage and Delivery: Fundamentals and Applications
Aiping Yu, Victor Chabot, and Jiujun Zhang

Photochemical Water Splitting: Materials and Applications
Neelu Chouhan, Ru-Shi Liu, and Jiujun Zhang

Metal–Air and Metal–Sulfur Batteries: Fundamentals and Applications
Vladimir Neburchilov and Jiujun Zhang

Electrochemical Reduction of Carbon Dioxide: Fundamentals and Technologies
Jinli Qiao, Yuyu Liu, and Jiujun Zhang

Electrolytes for Electrochemical Supercapacitors
Cheng Zhong, Yida Deng, Wenbin Hu, Daoming Sun, Xiaopeng Han, Jinli Qiao, and Jiujun Zhang

Solar Energy Conversion and Storage: Photochemical Modes
Suresh C. Ameta and Rakshit Ameta

Lead-Acid Battery Technologies: Fundamentals, Materials, and Applications
Joey Jung, Lei Zhang, and Jiujun Zhang

Lithium-Ion Batteries: Fundamentals and Applications
Yuping Wu

Graphene: Energy Storage and Conversion Applications
Zhaoping Liu and Xufeng Zhou

Proton Exchange Membrane Fuel Cells
Zhigang Qi

Advanced Bifunctional Electrochemical Catalysts for Metal-Air Batteries

Yan-Jie Wang
Rusheng Yuan
Anna Ignaszak
David P. Wilkinson
Jiujun Zhang

CRC Press
Taylor & Francis Group
Boca Raton London New York

CRC Press is an imprint of the
Taylor & Francis Group, an **informa** business

CRC Press
Taylor & Francis Group
6000 Broken Sound Parkway NW, Suite 300
Boca Raton, FL 33487-2742

First issued in paperback 2020

ISBN-13: 978-0-8153-4632-6 (hbk)
ISBN-13: 978-0-367-78050-0 (pbk)

Library of Congress Cataloging-in-Publication Data

Names: Wang, Yan-Jie, author.
Title: Advanced bifunctional electrochemical catalysts for metal-air batteries / Yan-Jie Wang, Rusheng Yuan, Anna Ignaszak, David P. Wilkinson, Jiujun Zhang.
Description: Boca Raton : Taylor & Francis, a CRC title, part of the Taylor & Francis imprint, a member of the Taylor & Francis Group, the academic division of T&F Informa, plc, 2018. | Includes bibliographical references and index.
Identifiers: LCCN 2018031748| ISBN 9780815346326 (hardback : alk. paper) | ISBN 9781351170727 (ebook).
Subjects: LCSH: Catalysts. | Electrocatalysis. | Electric batteries--Materials.
Classification: LCC TP159.C3 W36 2018 | DDC 660/.2995--dc23
LC record available at https://lccn.loc.gov/2018031748

Visit the Taylor & Francis Web site at
http://www.taylorandfrancis.com

and the CRC Press Web site at
http://www.crcpress.com

Contents

Preface

Rechargeable metal-air batteries (RMABs), a type of electrochemical technology for energy storage and conversion, have attracted great interest due to their highly specific energy, low cost, and safety. The development of RMABs has been limited considerably by its relatively low rate capability and lack of efficient and robust air catalysts with long-term stability. The main challenges stem from the sluggish kinetics of the oxygen reduction reaction (ORR) and the oxygen evolution reaction (OER), as well as the corrosion/oxidation of carbon materials in the presence of oxygen and at high electrode potentials. To improve the performance of RMABs, the electrocatalysts must be active for both ORR and OER during discharge and recharge. Therefore, the effective electrocatalysts for oxygen reactions in RMABs need to be "bifunctional." The current developed electrocatalysts are mainly based on composite materials due to the synergistic effect between their major components. The progress made on the development of advanced bi-functional composite electrocatalysts for RMABs is comprehensively reviewed in this book.

The goal of this book is to provide the reader with an appreciation for what bifunctional composite electrocatalysts are capable of, how this field has grown in the past decades, and, in particular, how bifunctional composite electrocatalysts can significantly improve the performance of RMABs. An initial chapter provides the introduction of the current state of bifunctional composite electrocatalysts in RMABs, along with the working mechanisms, design principles, and the synergistic effects between the major components of bifunctional composite electrocatalysts. The following chapters will discuss three types of composite catalysts: carbon-based composite electrocatalysts, doped carbon-based composite electrocatalysts, and noncarbon-based electrocatalysts. The validation of bifunctional composite electrocatalysts in RMABs is also reviewed. Each chapter also provides some suggestions about future research directions for overcoming technical challenges to facilitate further research and development in this area.

We hope this book can be helpful to those readers who are working in electrochemical energy storage areas, including both fundamental research and practical applications. We welcome any constructive comments for further improvement on the quality of this book.

Dr. Yan-Jie Wang
Dongguan University of Technology, Guangdong, China

Dr. Rusheng Yuan
Fuzhou University, Fuzhou, China

Dr. Anna Ignaszak
University of New Brunswick, Fredericton, Canada

Dr. David P. Wilkinson
University of British Columbia, Vancouver, Canada

Dr. Jiujun Zhang
Shanghai University, Shanghai, China

Authors

Dr. Yan-Jie Wang obtained his M.S. in Materials from North University of China in 2002 and his Ph.D. in Materials Science & Engineering from Zhejiang University, China, in 2005. Subsequently, he conducted two and a half years of postdoctoral research at Sungkyunkwan University, Korea, followed by two years as a research scholar at Pennsylvania State University, United States of America, studying advanced functional materials. In April of 2009, he was co-hired by the University of British Columbia, Canada, and the National Research Council of Canada to research advanced materials. During November of 2012 and August of 2017, he worked as a senior research scientist for the University of British Columbia and Vancouver International Clean-Tech Research Institute Inc. (VICTRII), researching core–shell structured materials. In September of 2017, he started as an academic leader in the School of Environment and Civil Engineering at Dongguan University of Technology in China. Also, he is an adjunct professor at Fuzhou University in China. Dr. Wang has published 60 papers in peer-reviewed journals, conference proceedings, and industry reports. His research interests include electrochemistry, electrocatalysis, polymer materials, and nanostructured material synthesis, characterization, and application in energy storage and conversion, biomass engineering, and medical areas.

Dr. Rusheng Yuan is a professor working at the State Key Laboratory of Photocatalysis on Energy and Environment in the College of Chemistry and Chemical Engineering at Fuzhou University, China. He received his M.S. in Chemical Technology from North China Institute of Technology in 2002, and his Ph.D. in Chemical Technology from the Institute of Coal Chemistry, Chinese Academy of Sciences and China University of Petroleum (East China). He then joined the College of Chemistry and Chemical Engineering at Fuzhou University and was promoted to professor in 2014. His research interest focuses on electrocatalysis and photocatalysis, and their applications in energy and the environment.

Dr. Anna Ignaszak is an associate professor in the Department of Chemistry at the University of New Brunswick in Canada and an adjunct assistant professor in the Institute of Organic and Macromolecular Chemistry at the Friedrich-Schiller University (Jena, Germany, Carl-Zeiss junior professorship holder since May 2012). Before 2012, she worked as a research associate at the Clean Energy Research Center (CERC), University of British Columbia (Vancouver, Canada), and National Research Council of Canada Institute for Fuel Cell Innovation (Vancouver, Canada). She has a diverse background in materials for electrochemical energy storage and conversion, electrochemical sensors, functional materials (carbons, composites, metal clusters) and heterogeneous catalysis. The research conducted in her labs in Canada and Germany aims in synthesizing morphology-controlled nano-catalysts and understanding the structure-reactivity interplay for the optimum redox activity, electrochemical characterization, and electrode engineering. She also spent several years working in industrial R&D (ABB Corporate Research, Pliva Pharmaceutical Company, Ballard Power Systems Inc.).

Dr. David P. Wilkinson is a professor and Canada Research Chair in the Department of Chemical and Biological Engineering at the University of British Columbia (UBC). He previously held the positions of Executive Director of the UBC Clean Energy Research Center, Principal Research Officer and Senior Advisor with the National Research Council of Canada Institute for Fuel Cell Innovation, Director and Vice President of Research and Development at Ballard Power Systems, and Group Leader at Moli Energy. His main research interests are in electrochemical and photochemical devices, energy conversion and storage materials, and processes to create clean and sustainable energy and water.

Dr. Jiujun Zhang is a professor at Shanghai University in China and former principal research officer at the National Research Council of Canada (NRC). Dr. Zhang's areas of expertise are electrochemistry, photoelectrochemistry, spectroelectrochemistry, electrocatalysis, fuel cells (PEMFC, SOFC, and DMFC), batteries, and supercapacitors. Dr. Zhang received his B.S. and M.Sc. in Electrochemistry from Peking University (China) in 1982 and 1985, and his Ph.D. in Electrochemistry from Wuhan University (China) in 1988. Starting in 1990, he carried out three terms of postdoctoral research at the California Institute of Technology, York University, and the University of British Columbia. Dr. Zhang holds more than 18 adjunct professorships, including one at the University of Waterloo (Canada) and one at the University of British Columbia.

1 Description of Bifunctional Electrocatalysts for Metal-Air Batteries

1.1 INTRODUCTION

To facilitate the transition from a fossil fuel-based economy to a clean energy one, active research in the field of sustainable energy harvesting, conversion and storage is required. Worldwide development of green energy would result in less CO_2 emissions while benefiting low-carbon economies, sustainable infrastructure, and consciousness in global clean technologies. All of these would greatly promote the progress of sustainable energy development, including electrochemical energy storages and conversions.[1–4] Among different electrochemical energy technologies, metal-air batteries have been considered one of the feasible options for electrical vehicles. When international companies such as *Tesla* produced patents for electric vehicles powered by metal-air batteries to extend the vehicles mileage range,[5] considerable interests in the development and exploration of new metal-air battery energy technologies have been recaptured due to their potential high energy density, rapid refuelability, and long range and ambient temperature operations.[6–8]

Regarding a metal-air battery, it uses a metal such as Li, Na, Zn, Mg, or Al, etc. for the negative electrode, similar to traditional batteries; it is also similar to conventional fuel cells in which continuous and inexhaustible oxygen can be supplied from the surrounding air as the reactant.[1,9,10] Technically, a metal-air battery can be considered to be a cross between a traditional battery and a fuel cell, producing an electrochemical coupling of a metal negative electrode with an air-breathing positive electrode through a suitable electrolyte, resulting in large theoretical energy density, which is roughly 2–10 times higher than those of current lithium-ion batteries (LIBs).[1,6,11]

A metal-air battery (MAB) usually consists of an anode made from a pure metal and an external cathode of ambient air, along with an electrolyte, in which electrochemical charge/discharge reactions occur between a positive air-electrode and a negative metal electrode,[12–14] as shown in Figure 1.1.[14] Currently, the metals that have been developed and used in the fabrication of MABs are Li, Zn, Al, Fe, Na, Ca, Mg, K, Sn, Si, and Ge.[2,10,15,16] It should be noted that when compared to traditional batteries such as $Zn-MnO_2$ (Zn-Mn), rechargeable lead-acid, nickel-metal hydride and LIBs,[10] MABs possess higher theoretical energy densities.[17] Presently, MAB technologies have been and are being considered one of the alternatives and

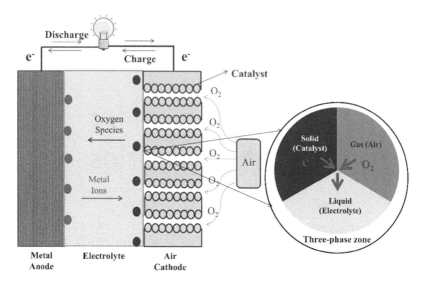

FIGURE 1.1 A basic metal-air battery (MAB) configuration with a simplified solid-liquid-gas three-phase zone. (Modified from Wang, Y.-J. et al. *Electrochem. Energy Rev.* **2018**, *1*, 1–34.[14])

successors for LIBs, especially in the automotive industry, due to their extremely high energy density (e.g., ~3000 Wh kg^{-1} for Li-air battery when considering all cell components).[18,19] Additionally, the calculation of energy densities in literature for metal-air batteries only counts the mass weight of the negative metal electrodes rather than the whole cell's weight. When we see some figures displaying the type of batteries versus energy density, some special attention should be given to this point. Having said this, MABs, whether laboratory scale or large prototype forms, have only been successfully operated with zinc, iron, lithium, and aluminum thus far.[20]

MABs have aqueous MABs and non-aqueous MABs in terms of the electrolyte used. The electrolyte was a conductor for ion conduction of the internal electrical circuits of the battery system,[8,21] which acts as a charge carrier in the internal circuit between the positive and negative electrodes, enabling electrons (i.e., current) to flow through the external circuit. In MAB systems, aqueous electrolytes can be acidic, quasi-neutral, or alkaline. Normally, neutral aqueous electrolytes cannot sustain battery operations due to their insufficient concentrations of either H$^+$ or OH$^-$ ions in positive reactions.[22,23] Recently, non-aqueous MABs have attracted attention due to their advantages in energy density and rechargeability over aqueous ones. It is possible for non-aqueous MABs to replace precious metal-based bifunctional catalysts required in aqueous MABs with non-precious metal-based bifunctional catalysts having low costs, easy operation, and/or good durability.[24]

Additionally, based on the rechargeability of electricity during electrochemical charge/discharge reactions, MABs can be further classified into primary (i.e., non-rechargeable) and secondary (i.e., rechargeable) batteries.[18,25] Compared to the former, the latter should be more efficient and economic. Rechargeable MABs can convert electrical energy directly into chemical energy via reversible electrochemical oxidation-reduction reactions.[26] To date, metals such as Li, V, Fe, Zn, Cd, and Al have been developed for rechargeable MABs with aqueous electrolytes[21] while metals such as Li, Na, K, Mg, Ca, Mo, W, Fe, and Al have been developed for rechargeable MABs with non-aqueous electrolytes. Interestingly, it is reported that the alteration of non-aqueous electrolytes in batteries can turn non-rechargeable (primary) MABs into rechargeable (secondary) ones.[27]

1.2 CURRENT STATUS OF BIFUNCTIONAL CATALYSTS IN METAL-AIR BATTERIES

As discussed above, MABs with atmospheric oxygen as oxidant have much higher theoretical energy densities than those of traditional aqueous batteries and lithium-ion batteries. Both the development and commercialization of MABs have been considerably challenged due to the lack of efficient and robust air electrocatalysts, which can result in relatively low performances such as low rate capability and poor long-term stability.[9,10,28,29] Among the different types of MABs that were developed and marketed, only the primary Zn-air battery can be used in commercial applications such as hearing aids and transportation signals, while Al-air and Mg-air cells are under development as saline systems used in military applications (e.g., underwater propulsion),[11,17,18] whereas their secondary counterparts are still in the early stages of research and development. To improve and enhance the performance of rechargeable MABs, air-electrode electrocatalysts must be active for both the oxygen reduction reaction (ORR) during battery discharge and the oxygen evolution reaction (OER) during battery recharge. Due to this, an effective electrocatalyst in a MAB system must have bifunctionality for ORR/OER. This bifunctionality can directly control the charge and discharge of MABs, determining the power, energy density, and energy efficiency of MABs.

In the past few decades, catalyst research for MABs has been progressing from the study of single functional electrocatalysts (e.g., Pt and its alloys)[30–32] for ORR, into the development of bifunctional electrocatalysts (e.g., Ir-based materials, Ru-based materials, or non-precious metal-based materials)[33,34] for both ORR and OER. Basically, the air-electrode reaction layer for ORR and OER related to O_2 is a matrix structure containing primarily catalyst and/or carbon-supported catalyst particles and ionomer.[35–37] During battery discharge (ORR) and charge (OER) reactions inside this layer, these particles can serve as the reaction sites. However, the presence of O_2 and/or high electrode potentials can result in the corrosion/oxidation of such carbon-based air-electrode catalysts and thereby lower the stability of MABs.[24,38,39] To solve this corrosion issue and improve the stability and performance of MABs, three types of novel catalysts have been explored to replace conventional carbon-based catalysts: carbon-based composites,[10,15,40] modified carbon-based materials,[41,42] and noncarbon-based materials (e.g., transitional metal oxides,[43] nitride,[44] or sulfide[45]). These bifunctional catalyst candidates can be synthesized and fabricated via different nanotechnologies to generate nanostructures with different advanced morphologies and compositions. These nanostructured bifunctional catalysts have high chemical and electrochemical stability as well as good catalytic activity induced by the interaction and synergetic effects of the catalyst components. Typically, graphene-based materials with or without metal and/or metal oxide have been studied as bifunctional catalysts[46–50] and demonstrated the enhanced ORR performance and durability in comparison with conventional carbon black supports. This is because they have unique two-dimensional morphologies with high surface areas, reasonable chemical resistance, and excellent electronic conduction.[46] However, during the fabrication of these catalyst-based electrodes, graphene-based materials are inclined to restack, which considerably degrades their catalytic performance.[46,51] To resolve this issue, doping strategies are used[52–54] to induce heteroatoms such as boron, nitrogen, and sulfur into graphene to form high performance catalysts.

1.3 EXPLORATION AND DEVELOPMENT OF ADVANCED BIFUNCTIONAL CATALYSTS

During the commercial effort of developing rechargeable MABs,[2,55–57] the performance by the slow kinetics of both ORR and OER of the air-electrodes during battery discharge and charge[6,8–10,58,59] can be observed, limiting the battery's performance.[6,8–10,58,59] Thus, the exploration and development of advanced bifunctional catalysts, offering high activity, stable durability, and affordable costs, is absolutely critical for the commercialization of this technology. In addition, due to the low catalyst stability of OER at the positive electrodes, MABs can also be recharged mechanically by providing refuelable units to replace consumed metal anodes. In this case, the catalyst at air-electrode can only perform the function of catalyzing ORR.

Besides conventional carbon materials, carbon-based composite materials have also been used to improve catalytic performances in the research of advanced bifunctional catalysts. This is because the addition of other materials, such as oxides, carbides, sulfides, or hydroxides, can not only prevent carbon corrosion for stability, but also enhance the catalytic activities for ORR and OER via a physiochemical interaction of different components. This interaction can be typically analyzed via structural characterizations such as morphology-related scanning electron microscope (SEM), transmission electron microscopy (TEM), high-resolution TEM (HRTEM), scanning transmission electron microscopy (STEM), and high-angle annular dark field scanning transmission electron microscopy (HAADF-STEM); additionally, X-ray photoelectron spectroscopy (XPS) and electrochemical measurements for MAB performance can be used to validate this interaction. Normally, the hybrid phases exist in the morphology of carbon-based composite catalysts and play an active role in the prevention of corrosion, possibly due to structural changes, which can be observed by XPS spectra for the effects of different elements after their incorporation into carbon. Interestingly, two different morphological carbon materials (carbon nanotubes [CNTs] and graphene[60]) can be used to form new carbon composite materials in the pursuit of advanced catalytic performance.

Regarding single carbon materials, their structural modifications are conducted to form different morphologies. In this regard, many researchers[61] have designed and prepared affordable novel carbon materials with a porous mixture of mesopores and macropores as bifunctional air-electrode catalysts with advanced electrocatalytic properties, particularly a high rechargeability. To increase both ORR and OER activities, single or dual elements have been also utilized in doping strategies to form doped carbon materials in MABs because these dopants, such as B, S, or P,[62–65] can modulate the electronic properties and surface polarities of carbon. To further improve catalytic performance, doped carbon materials have also composited with other noncarbon materials to form modified carbon-based composite bifunctional catalysts for ORR and OER. A typical example of this is the N-doped graphene composited with Co_3O_4, fabricated by Li et al.,[66] which showed highly active and durable catalytic activity in MABs.

In addition to the modifications and compositions mentioned above, the methods of replacing carbon materials with noncarbon materials have also be explored to generate advanced bifunctional catalysts.[67–69] These noncarbon materials (e.g., transitional metal oxides and hydroxides) have exhibited a great potential in the search of highly advanced catalysts.[3] In particular, by using a physical and/or chemical synthetic route, two or three different noncarbon materials can be composited to form a new noncarbon composite material that provides promising ways to obtain highly advanced bifunctionalities for ORR and OER due to the synergistic effects from different noncarbon materials.

For these three advanced bifunctional catalyst candidates (carbon-based composites, modified carbon-based composites, and noncarbon-based composites), it is noted that these catalysts should have advanced nanostructures, morphologies, and compositions to favor catalytic performance and prevent physicochemical corrosion in practice.

1.4 CHAPTER SUMMARY

This chapter gives a brief introduction and discussion about the types and characteristics of metal-air batteries, as well as their associated ORR and OER electrocatalysts. Regarding catalysts, bi- or multi-component catalysts are normally superior to single component catalysts. The composite catalysts not only possess original component characteristics, but also have new characteristics, resulting in the improvement of catalytic activity and stability. In past decades, three types of composite bifunctional air-electrode electrocatalysts (i.e., carbon-based composites, modified carbon-based composites, and noncarbon-based composites) have been extensively explored and developed for rechargeable MABs. Some comprehensive survey, analysis, synthesis, and performance assessment of these catalysts have been carried out, and great progress has been made in the improvement of catalyst activity and stability toward both ORR and OER.

REFERENCES

1. Dell, R. M.; Rand, D. A. J. *Clean Energy.* Cambridge: Royal Society of Chemistry, **2004**.
2. Rahman, Md. A.; Wang, X.; Wen, C. High energy density metal-air batteries: A review. *J. Electrochem. Soc.* **2013**, *160*, A1759–A1771.
3. Wang, Y.-J.; Wilkinson, D. P.; Zhang, J. Noncarbon support materials for polymer electrolyte membrane fuel cell electrocatalysts. *Chem. Rev.* **2011**, *111*, 7625–7651.
4. Wang, Y.-J.; Zhao, N.; Fang, B.; Li, H.; Bi, X.; Wang, H. Carbon-supported Pt-based alloy electrocatalysts for the oxygen reduction reaction in polymer electrolyte membrane fuel cells: Particle size, shape, and composition manipulation and their impact to activity. *Chem. Rev.* **2015**, *115*, 3433–3467.

5. Stewart, S. G.; Kohn, S. I.; Kelty, K. R.; Straubel, J. B. U.S. Patent 20130181511 A1, **2013**.
6. Li, Y.; Dai, H. Recent advances in zinc-air batteries. *Chem. Soc. Rev.* **2014**, *43*, 5257–5275.
7. Luntz, A. C.; McCloskey, B. D. Non-aqueous Li-air batteries: A status report. *Chem. Revs.* **2014**, *114*, 11721–11750.
8. Wang, Z.-L.; Xu, D.; Xu, J.-J.; Zhang, X.-B. Oxygen electrocatalysts in metal-air batteries: From aqueous to non-aqueous electrolytes. *Chem. Soc. Rev.* **2014**, *43*, 7746–7786.
9. Lee, J. S.; Kim, S. T.; Cao, R.; Choi, N. S.; Liu, M.; Lee, K. T.; Cho, J. Metal-air batteries with high energy density: Li-air versus Zn-air. *Adv. Energy Mater.* **2011**, *1*, 34–50.
10. Chen, F. Y.; Chen, J. Metal-air Batteries: From oxygen reduction electrochemistry to cathode catalysts. *Chem. Soc. Rev.* **2012**, *41*, 2172–2192.
11. Lee, D. U.; Xu, P.; Cano, Z. P.; Kashkooli, A. G.; Park, M. G.; Chen, Z. Recent progress and perspectives on bi-functional oxygen electrocatalysts for advanced rechargeable metal-air batteries. *J. Mater. Chem. A* **2016**, *4*, 7107–7134.
12. Kreuer, K. D. Fuel Cells. *Encyclopedia of Sustainability Science and Technology*, Chapter 5. New York: Springer, **2013**.
13. Bagotsky, V. S.; Skundin, A. M.; Volfkovich, Y. M. *Electrochemical Power Sources: Batteries, fuel cells, and supercapacitors*. Hoboken, New Jersey: Wiley & Sons. Inc., **2015**.
14. Wang, Y.-J.; Fang, B.; Zhang, D.; Li, A.; Wilkinson, D. P.; Ignaszak, A.; Zhang, L.; Zhang, J. A review of carbon-composited materials as air-electrode bifunctional electrocatalysts for metal-air batteries. *Electrochem. Energy Rev.* **2018**, *1*, 1–34.
15. Cao, R.; Lee, J.-S.; Liu, M.; Cho, J. Recent progress in non-precious catalysts for metal-air batteries. *Adv. Energy. Mater.* **2012**, *2*, 816–829.
16. Girishkumar, G.; McCloskey, B.; Luntz, A. C.; Swanson, S.; Wilcke, W. Lithium-air battery: Promise and challenges. *J. Phys. Chem. Lett.* **2010**, *1*, 2193–2203.
17. Linden, D.; Reddy, T. B. *Handbook of Batteries*, Third Edition. New York: McGraw-Hill Companies, Inc., **2002**.
18. Sardar, A.; Mubashir, S.; Muttana, S. B.; Prasad, M.; Kumar, K. S. Technology Foresight. *Quest for EV Batteries with High Specific Energy*, **2011**. http://autotechreview.com/component/k2/item/121-quest-for-ev-batteries-with-high-specific-energy.html.
19. Metal-Air Electrochemical Cell. *Wikipedia.* https://en.wikipedia.org/wiki/Metal–air_electrochemical_cell.
20. Winter, M.; Brodd, R. J. What are batteries, fuel cells, and supercapacitors? *Chem. Rev.* **2004**, *104*, 4245–4269.
21. Zhang, Z.; Zhang, S. S. *Rechargeable Batteries: Materials, technologies and new trends*. Switzerland: Springer, **2015**.
22. Zhang, T.; Tao, Z.; Chen, J. Magnesium-air batteries: From principle to application. *Mater. Horiz.* **2014**, *1*, 196–206.
23. Imanishi, N.; Luntz, A. C.; Bruce, P. *The Lithium Air Battery: Fundamentals*. New York: Springer, **2014**.
24. Chen, D.; Chen, C.; Baiyee, Z. M.; Shao, Z.; Ciucci, F. Nonstoichiometric oxides as low-cost and highly-efficient oxygen reduction/evolution catalysts for low-temperature electrochemical devices. *Chem. Rev.* **2015**, *115*, 9869–9921.
25. Palacín, M. E. Recent advances in rechargeable battery materials: A chemist's perspective. *Chem. Soc. Rev.* **2009**, *38*, 2565–2573.
26. Gao, X. P.; Yang, H. X. Multi-electron reaction materials for high energy density batteries. *Energy Environ. Sci.* **2010**, *3*, 174–189.
27. Kraytsberg, A.; Ein-Eli, Y. The impact of nano-scaled materials on advanced metal-air battery systems. *Nano Energy* **2013**, *2*, 468–480.
28. Shao, Y.; Park, S.; Xiao, J.; Zhang, J.-G.; Wang, Y.; Liu, J. Electrocatalysts for non-aqueous lithium–air batteries: Status, challenges, and perspective. *ACS Catal.* **2012**, *2*, 844–857.

29. Christensen, J.; Albertus, P.; Sanchez-Carrera, R. S.; Lohmann, T.; Kozinsky, B.; Liedtke, R.; Ahmed, J.; Kojic, A. A critical review of Li-air batteries. *J. Electrochem. Soc.* **2012**, *159*, R1–R30.

30. Fujiwara, N.; Yao, M.; Siroma, Z.; Senoh, H.; Ioroi, T.; Yasuda, K. Reversible air electrodes integrated with an anion-exchange membrane for secondary air batteries. *J. Power Sources* **2011**, *196*, 808–813.

31. Lu, Y.-C.; Xu, Z.; Gasteiger, H. A.; Chen, S.; Hamad-Schifferili, K.; Shao-Horn, Y. Platinum-gold Nanoparticles: A highly active bifunctional electrocatalyst for rechargeable lithium-air batteries. *J. Am. Chem. Soc.* **2010**, *132*, 12170–12171.

32. Lu, Y.; Wen, Z.; Jin, J.; Cui, Y.; Wu, M.; Sun, S. Mesoporous carbon nitride loaded with Pt nanoparticles as a bifunctional air electrode for rechargeable lithium-air battery. *J. Solid State Electrochem.* **2012**, *16*, 1863–1868.

33. Gorlin, Y.; Jaramillo, T. F. A bifunctional nonprecious metal catalyst for oxygen reduction and water Oxidation. *J. Am. Chem. Soc.* **2010**, *132*, 13612–13614.

34. Lee, Y.; Suntivich, J.; May, K. J.; Perry, E. E.; Shao-Horn, Y. Synthesis and activities of rutile IrO_2 and RuO_2 nanoparticles for oxygen evolution in acid and alkaline solutions. *J. Phys. Chem. Lett.* **2012**, *3*, 399–404.

35. Yang, Y.; Sun, Q.; Li, Y.-S.; Li, H.; Fu, Z.-W. A CoO_x/carbon double-layer thin film air electrode for non-aqueous Li-air batteries. *J. Power Sources* **2013**, *223*, 312–318.

36. Yang, C.-C. Preparation and characterization of electrochemical properties of air cathode electrode. *Inter. J. Hydrogen Energ.* **2004**, *29*, 135–143.

37. Dong, H.; Yu, H.; Wang, X. Catalysis kinetics and porous analysis of rolling activated carbon-PTFE air-cathode in microbial fuel cells. *Environ. Sci. Technol.* **2012**, *46*, 13009–13015.

38. Velraj, S.; Zhu, J. H. Cycle life limit of carbon-based electrodes for rechargeable metal-air battery application. *J. Electroanal. Chem.* **2015**, *736*, 76–82.

39. Wang, J.; Li, Y.; Sun, X. Challenges and opportunities of nanostructured materials for aprotic rechargeable lithium-air batteries. *Nano Energy* **2013**, *2*, 443–467.

40. Chen, Z.; Yu, A.; Ahmed, R.; Wang, H.; Li, H.; Chen, Z. Manganese dioxide nanotube and nitrogen-doped carbon nanotube based composite bifunctional catalyst for rechargeable zinc-air battery. *Electrochim. Acta* **2012**, *69*, 295–300.

41. Sun, B.; Wang, B.; Su, D.; Xiao, L.; Ahn, H.; Wang, G. Graphene nanosheets as cathode catalysts for lithium-air batteries with an enhanced electrochemical performance. *Carbon* **2012**, *50*, 727–733.

42. Chervin, C. N.; Long, J. W.; Brandell, N. L.; Wallace, J. M.; Kucko, N. W.; Rolison, D. R. Redesigning air cathodes for metal-air batteries using MnO_x-functionalized carbon nanofoam architectures. *J. Power Sources* **2012**, *207*, 191–198.

43. Du, G.; Liu, X.; Zong, Y.; Andy Hor, T. S.; Yu, A.; Liu, Z. Co_3O_4 nanoparticle-modified MnO_2 nanotube bifunctional oxygen cathode catalysts for rechargeable zinc-air batteries. *Nanoscale* **2013**, *5*, 4657–4661.

44. Zhang, K.; Zhang, L.; Chen, X.; He, X.; Wang, X.; Dong, S.; Gu, L.; Liu, Z.; Huang, C.; Cui, G. Molybdenum nitride/N-doped carbon nanospheres for lithium-O_2 battery cathode electrocatalyst. *ACS Appl. Mater. Interfaces* **2013**, *5*, 3677–3682.

45. Cao, X.; Zheng, X.; Tian, J.; Jin, C.; Ke, K.; Yang, R. Cobalt sulfide embedded in porous nitrogen-doped carbon as a bifunctional electrocatalyst for oxygen reduction and evolution reactions. *Electrochim. Acta* **2016**, *191*, 776–783.

46. Yu, A.; Park, H. W.; Davies, A.; Higgins, D. C.; Chen, Z.; Xiao, X. Free-standing layer-by-layer hybrid thin film of graphene-MnO_2 nanotube as anode for lithium ion batteries. *J. Phys. Chem. Lett.* **2011**, *2*, 1855–1860.

47. Seger, B.; Kamat, P. V. Electrocatalytically active graphene-platinum nano-composites. Role of 2-D carbon support in PEM fuel cells. *J. Phys. Chem. C* **2009**, *113*, 7990–7995.

48. Liang, Y.; Li, Y.; Wang, H.; Zhou, J.; Wang, J.; Regier, T.; Dai, H. Co_3O_4 nanocrystals on graphene as a synergistic catalyst for oxygen reduction reaction. *Nat. Mater.* **2011**, *10*, 780–786.

49. Seo, M. H.; Choi, S. M.; Kim, H. J.; Kim, W. B. The graphene-supported Pd and Pt catalysts for highly active oxygen reduction in an alkaline condition. *Electrochem. Commun.* **2011**, *13*, 182–185.

50. Lee, D. U.; Kim B. J.; Chen, Z. One-pot synthesis of a mesoporous $NiCo_2O_4$ nanoplatelet and graphene hybrid and its oxygen reduction and evolution activities as an efficient bi-functional electrocatalyst. *J. Mater. Chem. A* **2013**, *1*, 4754–4762.

51. Yoo, E.; Kim, J.; Hosono, E.; Zhou, H.-S.; Kudo, T.; Honma, I. Large reversible Li storage of graphene nanosheet families for use in rechargeable lithium ion batteries. *Nano Lett.* **2008**, *8*, 2277–2282.

52. Sheng, Z.-H.; Gao, H.-L.; Bao, W.-J.; Wang, F.-B.; Xia, X.-H. Synthesis of boron doped graphene for oxygen reduction reaction fuel cells. *J. Mater. Chem.* **2012**; *22*, 390–395.

53. Wang, H.; Maiyalagan, T.; Wang, X. Review on recent progress in nitrogen-doped graphene: Synthesis, characterization, and its potential applications. *ACS Catal.* **2012**, *2*, 781–794.

54. Li, Y.; Wang, J.; Li, X.; Geng, D.; Banis, M. N.; Tang, Y.; Wang, D.; Li, R.; Sham, T.-K.; Sun, X. J. Discharge product morphology and increased charge performance of lithium-oxygen batteries with graphene nanosheet electrodes: The effect of Sulphur doping. *Mater. Chem.* **2012**, *22*, 20170–20174.

55. Xu, Q.; Kobayashi, T. *Advanced Materials for Clean Energy.* Boca Raton, CRC Press, Taylor & Francis Group, **2015**.

56. Lee, D. U.; Park, M. G.; Park, H. W.; Seo, M. H.; Wang, X.; Chen, Z. Highly active and durable nanocrystal-decorated bifunctional electrocatalyst for rechargeable zinc-air batteries. *ChemSusChem* **2015**, *8*, 3129–3138.

57. Gallagher, K. G.; Goebel, S.; Greszler, T.; Mathias, M.; Oelerich, W.; Eroglu, D.; Srinivasan, V. Quantifying the promise of lithium-air batteries for electric vehicles. *Energy Environ. Sci.* **2014**, *7*, 1555–1563.

58. Takeguchi, T.; Yamanaka, T.; Takahashi, H.; Watanabe, H.; Kuroki, T.; Nakanishi, H.; Orikasa, Y. et al. Layered perovskite oxide: A reversible air electrode for oxygen evolution/reduction in rechargeable metal-air batteries. *J. Am. Chem. Soc.* **2013**, *135*, 11125–11130.

59. Shao, Y.; Ding, F.; Xiao, J.; Zhang, J.; Xu, W.; Park, S.; Zhang, J.-G.; Wang, Y.; Liu, J. Making Li-air batteries rechargeable: Material challenges. *Adv. Funct. Mater.* **2013**, *23*, 987–1004.

60. Lu, X.; Chan, H. M.; Sun, C.-L.; Tseng, C.-M.; Zhao, C. Interconnected core-shell carbon nanotube-graphene nanoribbon scaffolds for anchoring cobalt oxides as bifunctional electrocatalysts for oxygen evolution and reduction. *J. Mater. Chem. A* **2015**, *3*, 13371–13376.

61. Sakaushi, K.; Yang, S. J.; Fellinger, T.-P.; Antonietti, M. Impact of large-scale mesor- and macropore structures in adenosine-derived affordable noble carbon on efficient reversible oxygen electrocatalytic redox reactions. *J. Mater. Chem. A* **2015**, *3*, 11720–11724.

62. Wang, S.; Lyyamperumal, E.; Roy, A.; Xue, Y.; Yu, D.; Dai, L. Vertically aligned BCN nanotubes as efficient metal-free electrocatalysts for the oxygen reduction reaction: A synergetic effect by Co-doping with boron and nitrogen. *Angew. Chem. Int. Ed.* **2011**, *50*, 11756–11760.

63. Liang, J.; Jiao, Y.; Jaroniec, M.; Qiao, S. Z. Sulfur and nitrogen dual-doped mesoporous graphene electrocatalyst for oxygen reduction with synergistically enhanced performance. *Angew. Chem. Int. Ed.* **2012**, *51*, 11496–11500.

64. Xue, Y.; Yu, D.; Dai, L.; Wang, R.; Li, D.; Roy, A.; Lu, F.; Chen, H.; Liu, Y.; Qu, J. Three-dimensional B, N-doped graphene foam as a metal-free catalyst for oxygen reduction reaction. *Phys. Chem. Chem. Phys.* **2013**, *15*, 12220–12226.

65. Jiao, Y.; Zheng, Y.; Jaroniec, M.; Qiao, S. Z. Origin of the electrocatalytic oxygen reduction activity of graphene-based catalysts: A roadmap to achieve the best performance. *J. Am. Chem. Soc.* **2014**, *136*, 4394–4403.

66. Li, G.; Wang, X.; Fu, J.; Li, J.; Park, M. G.; Zhang, Y.; Lui, G.; Chen, Z. Pomegranate-inspired design of highly active and durable bifunctional electrocatalysts for rechargeable metal-air batteries. *Angew. Chem. Int. Ed.* **2016**, *55*, 4977–4982.

67. Park, M.-S.; Kim, J.; Kim, K. J.; Lee, J.-W.; Kim, J. H.; Yamauchi, Y. Porous nanoarchitectures of spinel-type transition metal oxides for electrochemical energy storage systems. *Phys. Chem. Chem. Phys.* **2015**, *17*, 30963–30977.

68. Gupta, S.; Kellogg, W.; Xu, H.; Liu, X.; Cho, J.; Wu, G. Bifunctional perovskite oxide catalysts for oxygen reduction and evolution in alkaline media. *Chem. Asian J.* **2016**, *11*, 10–21.

69. Zhang, K.; Zhang, L.; Chen, X.; He, X.; Wang, X.; Dong, S.; Han, P. et al. Mesoporous cobalt molybdenum nitride: A highly active bifunctional electrocatalyst and its application in lithium-O_2 batteries. *J. Phys. Chem. C* **2013**, *117*, 858–865.

2 Reaction Mechanisms of Bifunctional Composite Electrocatalysts of Metal-Air Batteries

2.1 INTRODUCTION

This book's main focus is the rechargeable metal-air batteries (RMABs), the catalysts used at the air-electrodes of RMABs should have bifunctionality for catalyzing both oxygen reduction reaction (ORR) and oxygen evolution reaction (OER).[1] Figure 2.1 shows the basic configuration of a RMAB. It is evident when the battery is discharged, the negative metal electrode is oxidized and releases electrons to the external circuit while the metal ions diffuse through the electrolyte into the air-electrode coated with a bifunctional catalyst. The electrons from the negative electrode pass through the external circuit to the air-electrode where oxygen takes on electrons and is reduced to oxygen-containing species. When the RMAB is charged, the process above will be reversed, leading to metal plating at the negative electrode and oxygen evolution at the air-electrode, respectively. The bifunctional catalyst at the air-electrode, coupled with the electrolyte and metal electrode, determines the formation of different electrochemical reactions and overall products in various metal dependent RMABs.[1–6]

2.1.1 MECHANISMS IN AQUEOUS ELECTROLYTE-BASED MABS

In general, aqueous electrolytes for MABs can be classified as acidic, quasi-neutral, or alkaline. Among these three aqueous electrolytes, neutral aqueous electrolytes are sparsely studied in the development of advanced bifunctional cathode catalysts in MABs. This is because the electrochemical reactions in such electrolytes often result in larger overpotentials and high polarizations, and thereby poor battery performances, imposing large hindrances in the search for highly efficient bifunctional catalysts.[7] According to the ORR at the liquid-gas-solid interface,[8] a possible reaction mechanism in neutral solutions can be proposed as below:

$$O_2 + 2H_2O + 4e^- \rightarrow 4OH^- \tag{2.1}$$

$$O_2 + H_2O + 2e^- \rightarrow HO_2^- + OH^- \tag{2.2}$$

$$HO_2^- + H_2O + 2e^- \rightarrow 3OH^- \tag{2.3}$$

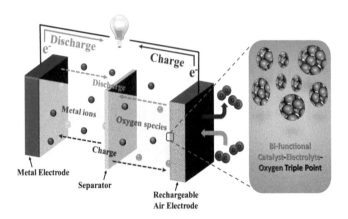

FIGURE 2.1   A rechargeable metal-air battery (MAB) configuration with an air-electrode solid-liquid-gas interface. (From Li, D. U. et al. *J Mater. Chem. A.* **2016**, *4*, 7107–7134.[2])

Equation 2.1 is a direct four-electron ORR process, which is heavily related to the reaction processes in the three-phase interface of the catalyst layer. Equations 2.2 and 2.3 are the ORR process to produce water through a 2 + 2-electron pathway. Normally, a bifunctional catalyst is required to have a direct four-electron catalytic ability to solve the issues of overpotential and polarization. Sometimes, a catalyst that works well in weak alkaline solutions also does well for neutral solutions[9] and is seen as a potential bifunctional catalyst for neutral solutions.

Compared to alkaline aqueous electrolytes, acidic aqueous electrolytes often result in severe corrosion at the metal negative electrode because metals such as Li, Mg, Al, and Zn will violently react with acidic media to generate hydrogen along with a great release of heat.[4,10,11] As generally observed, few bifunctional catalyst materials become stable under strong acidic and aggressive oxidizing environments at the air-electrodes. This suggests that, with acidic aqueous electrolytes in MABs, it is important to develop not only suitable bifunctional catalysts with highly efficient catalytic performance, but also to solve the two large issues of anode corrosion and complex thermal management.[4,12,13] According to literature,[14–16] four-electron direct reduction (Equation 2.4) and two-electron indirect pathways (Equations 2.5 and 2.6) are the two oxygen reduction mechanisms found in MABs in acidic solution, as shown below:

$$O_2 + 4H^+ + 4e^- \rightarrow 2H_2O \tag{2.4}$$

$$O_2 + 2H^+ + 2e^- \rightarrow 2H_2O_2 \tag{2.5}$$

$$HO_2^- + H_2O + 2e^- \rightarrow 3OH^- \tag{2.6}$$

The four-electron reduction can directly produce water while the two-electron pathway has an intermediate reaction to produce hydrogen peroxide which is then further reduced to water.

The ORR at the air-electrode occurs at the triple-phase interface of a gas (oxygen from the air), a liquid (e.g., the acidic aqueous electrolyte), and a solid (the catalyst). Owing to that the active O_2 is diluted in the air, high surface area electrodes are favorable to the reduction of diffusion limitations at higher currents. In comparison with alkaline electrolytes, acidic electrolytes in MABs are only suitable for limited types of bifunctional catalysts (e.g., Pt- or Pt alloy-based catalysts). In practical MABs, the ORR reaction pathway is strongly dependent on the type of bifunctional catalysts.[17–19]

Aside from acidic and neutral aqueous electrolytes, alkaline aqueous electrolytes have become more popular in the research and development of advanced MABs, particularly in recent decades; this is due to lower metal corrosion, wide oxygen catalyst material selection, and cost-effectiveness. However, alkaline media can cause an accumulation of carbonate ions at the supply of air, resulting in the clogging of electrode pores and/or blocking of electrolyte channels, thus decreasing battery performance.[4,13] Similar to the ORR mechanism in acidic and neutral aqueous electrolytes, the ORR at the air-electrode in an alkaline electrolyte is also related to the complicated multi-step electron-transfer reactions involving different oxygen species such as O, OH, O_2^-, and HO_2.[1,13,20,21] There are three typical ORR mechanisms at the triple-phase interface of the gas, alkaline aqueous electrolyte, and catalyst: a four-electron pathway, a two-electron pathway, and a charge-compensated electron pathway. Among these three ways, the former two are associated with the ORR on the metal catalysts while the latter one is associated with the ORR on the surface of the metal oxides. Considering the effects of the catalyst structure on ORR, many researchers have proposed that the four-oxygen adsorption on the catalyst surface (i.e., on top end-on, bridge end-on, bridge side-on one site, and bridge side-on two sites) is closely related to the distribution of active sites, surface geometry, and the binding energy of the catalyst, as illustrated in Figure 2.2.[4] The electrochemical catalysis process of the ORR on the surface of metal catalysts can be simplified as two oxygen adsorptions: bidentate O_2 adsorption (two O atoms coordination with the

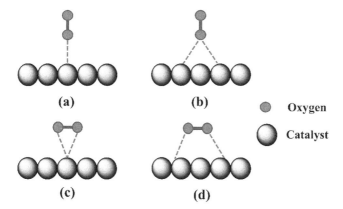

FIGURE 2.2 Four-oxygen adsorption modes on catalyst surface: on top end-on (a), bridge end-on (b), bridge side-on one site (c), and bridge side-on two sites (d). (From Chen, F. Y.; Chen, J. *Chem. Soc. Rev.* **2012**, *41*, 2172–2192.[4])

metal) and end-on O_2 adsorption (one O atom coordination perpendicularly to the surface) corresponding to a four-electron pathway and a two-electron pathway,[1,20] respectively. The four-electron pathway of the bidentate adsorption can be described below:

$$O_2 + 2H_2O + 2e^- \rightarrow 2OH + 2OH^- \tag{2.7}$$

$$2OH + 2e^- \rightarrow 2OH^- \tag{2.8}$$

$$\text{Overall: } O_2 + 2H_2O + 4e^- \rightarrow 4OH^- \tag{2.9}$$

while the two-electron pathway of the end-on adsorption can be presented as:

$$O_2 + H_2O + e^- \rightarrow HO_2 + OH^- \tag{2.10}$$

$$HO_2 + e^- \rightarrow HO_2^- \tag{2.11}$$

$$\text{Overall: } O_2 + H_2O + 2e^- \rightarrow HO_2^- + OH^- \tag{2.12}$$

Considering the different charge distributions on the metal oxide surface from the metal surface, the anion coordination on the metal oxides has to be compensated by the oxygen of H_2O in alkaline aqueous solutions to complete the full reduction of surface cations.[4] This charge-compensation can be done after the protonation of an adsorbed oxygen ligand, as indicated in the four-step catalytic mechanism (see Equations 2.13–2.16) for ORR on the surfaces of transition-metal oxide catalysts.[13,22] This mechanism includes the sequential generation of surface hydroxide displacement, surface peroxide formation, surface oxide formation, and surface hydroxide regeneration. In this mechanism, the competition between the O_2^{2-}/OH^- displacement and OH^- regeneration is considered to be the determinable limiting factor of the ORR rate for metal oxides in alkaline solution.[22,23]

Interestingly, based on experimental studies and mathematical modellings, other research on carbon-supported Pt nanoparticles[24] revealed that at high potential intervals, direct four-electron pathway was the dominating one while series $2 + 2$-electron pathway (see Equations 2.13) and 2.14) occurred in the reduction reaction at low potential using the peroxide-mediated pathway. This research clearly showed that for carbon-supported Pt catalysts, four-electron and two-electron mechanisms might exist concurrently. It was reported that in electrochemical reactions, the desorption and re-adsorption of intermediates and the non-covalent association of hydrated alkali-metals with adsorbed species could effectively determine the ORR kinetics and pathway.[25,26] However, insufficient analysis techniques in the experimental characterization and theoretical approach limited the determination of more ORR pathways on the catalyst surfaces.

It has been demonstrated that, in most cases, noble metal catalysts such as Pt-based ones catalyze the four-electron ORR, while the two-electron pathway is primarily associated with carbonaceous materials.[2,8] Various ORR pathways may be further found in transition-metal oxides and metal macrocycles[8,27] if the specific crystal structure, molecular composition, or experimental parameters can be better understood and analyzed using new technologies and approaches.

$$HO_2^- + H_2O + 2e^- \rightarrow 3OH^- \qquad (2.13)$$

$$2HO_2^- \rightarrow 2OH^- + O_2 \qquad (2.14)$$

For a RMAB system, the OER must be considered in the investigation and development of bifunctional catalysts. In general, compared to the ORR, the OER on the catalyst surface is even more complex and sensitive because oxygen evolution is rationally linked to the oxide surface rather than the metal surface.[26,28–30] The site geometries of metal(s) and the chemical valence(s) of metal ion(s) play an important role in the OER mechanism due to the interactions of metal(s) with oxygen-containing intermediates.[30] So far, the mostly proposed OER mechanisms for transition metal oxides are expressed using alkaline media rather than an acidic or neutral one. This OER mechanism, alternatively referred to an acid-base mechanism,[31,32] has a series of acid-base steps (Equations 2.15–2.17)[2] in which OH^- and an oxygen nucleophile (Lewis acid) can attack a metal-bound electrophile oxygen surface species (Lewis base). The existence of active transition metal cations can participate in the catalysis process by reacting with oxygen species under decreased adsorption energy. Interestingly, although RuO_2 and IrO_2 as typical noble metal oxides that can exhibit highly OER activities owing to their low redox potential and high electrical conduction, their poor ORR activities as well as high-cost lead to their low usage as bifunctional catalysts in RMABs.[33] Regarding non-noble metal-based catalysts, Co-based and/or Ni-based oxides (e.g., spinel Co_3O_4[34] and $NiCo_2O_4$[35]) have been explored as promising bifunctional catalysts in alkaline solutions.

$$M^{m+} - O^{2-} + OH^- \rightarrow M^{(m-1)+} - O - OH^- + -O^- + e^- \qquad (2.15)$$

$$M^{(m-1)+} - O - OH^- + OH^- \rightarrow M^{m+} - O - O^{2-} + H_2O + e^- \qquad (2.16)$$

$$2M^{m+} - O - O^{2-} \rightarrow 2M^{m+} - O^{2-} + O_2 \qquad (2.17)$$

2.1.2 Mechanisms in Non-Aqueous Electrolytes-Based MABs

Thus far, the research and development of RMABs have focused more frequently on non-aqueous and aprotic electrolytes than on aqueous electrolytes due to the unstable metal electrodes in water solutions[36] as well as the limited working potential range in which H_2O is stable against the evolution of hydrogen or oxygen.[4] After the early research on a non-aqueous Li-air battery (LAB) in the late 1990s,[37] increasing attention has been given to non-aqueous MABs.[38,39–41] For example, since studies demonstrated the reduction of molecular oxygen could generate superoxide (O^{2-}) in non-aqueous environments,[42,43] organic solvents have become feasible in bifunctional catalyst fabricated rechargeable Li-air batteries (RLABs). The aprotic electrolyte could prevent the disproportionation of superoxide anion radicals into oxygen and hydro-peroxide anion through a hydrogen abstraction mechanism.[44]

Similar to the aqueous solution-related oxygen electrochemical mechanism, the oxygen electrochemistry related to both the ORR and OER processes in RLABs is reported to be complex and sensitive to the choice of types of non-aqueous electrolytes, catalyst materials, oxygen pressures, and electrochemical stressings.[45–48]

In the exploration and development of advanced novel bifunctional catalysts, it is important to fundamentally understand the working electrochemical reaction mechanisms of the catalyst in the presence of non-aqueous electrolytes. By investigating the intrinsic ORR behavior on glass carbon (GC), polycrystalline Au and Pt electrodes in aprotic organic electrolytes, Lu et al.[49] proposed an ORR mechanism (see Equations 2.18–2.21) with a one-electron reduction of oxygen[43,50] in which the reduction mechanism and the discharge product were related to different catalyst surfaces. The first step (see Equation 2.18) of ORR is to form superoxide radicals as weakly adsorbed species that can be solvated by different cations and solvents to diffuse into the bulk of the electrolyte. Then, these superoxide radicals can react with Li^+ to form surface-adsorbed LiO_2 (see Equation 2.19), which has also been demonstrated by the density functional theory (DFT) study carried out by Hummelshøj et al.[51] In a further relatively weak oxygen chemisorption step (e.g., on the surface of carbon),[51] the surface-adsorbed LiO_2 can be then reduced to Li_2O_2 (see Equation 2.20), as evident by ex-situ Raman spectroscopy which monitors the surfaces of the carbon and carbon/MnO_2 composites.[37,52] However, when the catalyst is changed to a polycrystalline metal like Pt[53] to produce strong bonding with O_2, the surface-adsorbed Li_2O_2 is reduced to Li_2O (see Equation 2.21). It is reported that in the O_2 reduction reaction, the generated intermediates and products (e.g., Li_2O_2 or carbonate species) can accumulate to clog the pore of the air-electrode, block the catalyst surface, and decrease the charge transfer and mass transport, resulting in degradation of battery performance.[2] Therefore, the selection and utilization of bifunctional catalysts are critical factors in obtaining high performance MABs because the ORR and the reaction products are closely related to the type of catalysts used.

$$O_2 + e^- \rightarrow O_2^{*-} \tag{2.18}$$

$$O_2^{*-} + Li^+ \rightarrow LiO_2 \tag{2.19}$$

$$LiO_2 + e^- + Li^+ \rightarrow Li_2O_2 \tag{2.20}$$

$$Li_2O_2 + 2e^- + O_2 + 2Li^+ \rightarrow 2(Li_2O_2) \tag{2.21}$$

$$O_2 + e^- \rightarrow O_2^- \tag{2.22}$$

$$O_2^- + Li^+ \rightarrow LiO_2 \tag{2.23}$$

$$2LiO_2 \rightarrow Li_2O_2 + O_2 \tag{2.24}$$

$$Li_2O_2 \rightarrow 2Li^+ + O_2 + 2e^- \tag{2.25}$$

Apart from the ORR, the OER under non-aqueous electrolytes is also a necessary electrochemical process in rechargeable MABs, especially in RMABs. Peng et al.[54] probed the OER mechanism using in-situ surface-enhanced Raman spectroscopy (SERS), and demonstrated that, as an intermediate of oxygen reduction in non-aqueous solution, LiO_2 could disproportionate to Li_2O_2 (see Equations 2.22–2.24). However, the spectroscopy data indicated that the Li_2O_2 oxidation did not follow the reverse pathway to reduction but directly decomposed in a one-step reaction to evolve O_2 (see Equation

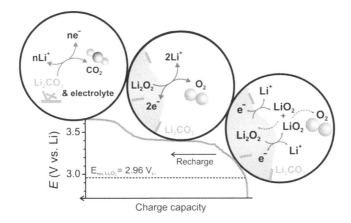

FIGURE 2.3 Two-stage OER reaction mechanism of Li-O$_2$ recharge. The OER process associated with Stage I (sloping and catalyst-insensitive) is attributed to a deintercalation process via solid-solution route from the outer part of Li$_2$O$_2$ to form LiO$_2$-like species on the surface in which LiO$_2^-$-like species disproportionate to evolve O$_2$, yielding an overall 2e$^-$/O$_2$ OER process. The OER process at the flat potential plateau is attributed to the oxidation of bulk Li$_2$O$_2$ particles to form Li$^+$ ions and O$_2$ via a two-phase transition. Lastly, a rising charge plateau after Stage II is assigned to the decomposition of carbonate-type byproducts and electrolytes. (From Lu, Y. C.; Yang, S. H. *J. Phys. Chem. Lett.* **2013**, *4*, 93–99.[58])

2.25). In particular, based on the Nernst Equation,[55,56] the thermodynamic potential for Li-O$_2$ batteries with non-aqueous solutions should be 2.96 V, but the practical Li-O$_2$ showed a much higher overpotential (~1.0–1.5 V) for charging, resulting in smaller cell voltage.[57] To further find suitable OER mechanisms, Lu et al.[58] proposed a reasonable two-stage OER reaction process (see Equations 2.26–2.29 and Figure 2.3) based on their experimental measurements using Vulcan carbon (VC) air electrodes.

They believed the OER mechanism was sensitive to discharge/charge rates and catalysts in the tested batteries. As shown in Figure 2.3, at the first OER stage, a slopping voltage profile, corresponding to a process insensitive to charge rates and catalysts, occurs at low overpotentials (<400 mV) in which a delithiation of the outer part of Li$_2$O$_2$ results in the formation of LiO$_2$-like species (see Equation 2.26) following a solid-solution route. These LiO$_2$-like species can disproportionate to evolve O$_2$ (see Equation 2.27), yielding an overall 2e$^-$/O$_2$ OER process (see Equation 2.28). This proposed OER reaction is consistent with the research by McCloskey et al.[6,59] At the second stage, with a high overpotential ranging from 400 to 1200 mV, bulk Li$_2$O$_2$ can be oxidized to form Li$^+$ and O$_2$ (see Equation 2.29) via a two-phase transition in which the kinetics are strongly sensitive to charge rates and catalysts, as supported by the previous results.[60–62] In this situation, a rising charge after the second stage is resulted from the decomposition of carbonate-type byproducts and electrolytes, possibly during discharge.

$$Li_2O_2 \rightarrow LiO_2 + Li^* + e^- \qquad (2.26)$$

$$LiO_2 + LiO_2 \rightarrow Li_2O_2 + O_2 \qquad (2.27)$$

$$Li_2O_2 \rightarrow 2Li^+ + O_2 + 2e^- \tag{2.28}$$

$$Li_2O_2 \rightarrow 2Li^+ + 2e^- + O_2 \tag{2.29}$$

2.1.3 MECHANISMS IN HYBRID ELECTROLYTES-BASED MABs

To alleviate issues in rechargeable non-aqueous Li-O$_2$ batteries,[2] a typical hybrid LAB was designed to combine aqueous and non-aqueous electrolytes[63,64] in which a lithium-electrode in non-aqueous electrolyte and an air catalytic cathode in aqueous electrolyte was separated by LISICON (a water-stable, solid-state lithium-ion conducting ceramic electrolyte).[63] A significant benefit from using an aqueous electrolyte at the air-electrode is the prevention of lithium oxide products being generated, which can clog electrode pores, block catalyst surfaces, and decrease battery performances. Moreover, the use of incombustible aqueous electrolytes can reduce safety issues that occur in non-aqueous electrolytes, such as organic electrolytes.[63,65]

Based on the fabrication of hybrid LAB, bifunctional catalyst-fabricated air-cathodes can display sensitive electrochemical reactions depending on the alkalinity of the electrolyte used. When basic electrolytes were used at the air-electrode, the ORR is described below:

$$O_2 + 2H_2O + 4e^- \rightarrow 4OH^- \tag{2.30}$$

Whereas, the ORR of acidic electrolytes is presented as follows:

$$O_2 + 4H^+ + 4e^- \rightarrow 2H_2O \tag{2.31}$$

The ORR mechanisms at air-electrode in the proposed hybrid LAB should be similar to those at the cathodes of H$_2$–O$_2$ fuel cells, as they both have continuous OH$^-$ from the reduction of supplied O$_2$.[66] This hybrid LAB, however, produced a lower volumetric capacity compared to non-aqueous batteries, as reported in literature.[67] The solid LISICON ceramic was also found to be unstable in strong acidic and strong basic solutions, resulting in limitations of long-term applications.[68] Limitations were especially apparent for the acidic electrolyte because acidic solutions could inevitably become alkaline due to the continuous generation of OH$^-$ after long-term use, causing serious challenges to the development of bifunctional catalysts. At the same time, the use of expensive noble metals in the fabrication of highly active bifunctional catalysts is also limited due to their high cost. To overcome these disadvantages, careful design and engineering is vital in the development of hybrid LAB.[69]

However, little research has been focused on the OER of hybrid RLABs. Wang et al.[67] examined a graphene supported spinel CoMn$_2$O$_4$ catalyst for the OER in a hybrid Li-O$_2$ cell, along with the ORR. However, the OER mechanism was not explored in their report. Wang and Zhou[66] discussed the electrode reactions for the ORR and OER after they investigated a Mn$_3$O$_4$/activated carbon composite as a metal-free composite bifunctional catalyst in a hybrid RLAB. They summarized the ORR and OER reactions (see Equations 2.32 and 2.33) as follows:

$$O_2 + 2H_2O + 4e^- \Leftrightarrow 4OH^- \tag{2.32}$$

$$4Li \Leftrightarrow 4Li^+ + 4e^- \tag{2.33}$$

where the forward and backward rows indicate the discharge and charge processes, respectively. It was claimed that during discharge, O_2 from the supplied air could continuously diffuse into the porous catalytic electrode, at which ORR occurred (see Equation 2.32) while metal lithium at negative electrode was oxidized to form Li^+ (Equation 2.33). The Li^+ ions were transported from the non-aqueous solution to the aqueous solution through the LISICON film. During charge, the OER happened at air-electrode, according to the backward reaction direction of Eq. 32, and Li^+ would be transferred from aqueous solution back to non-aqueous solution to be reduced, forming metal Li at the negative electrode.

2.1.4 Mechanisms in Solid-State MABs

Solid-state electrolytes have been developed for decades, and used in solid-state LIBs, including solid organic electrolytes[37,70] and solid inorganic electrolytes.[71,72] In recent years, this types of solid-state RLAB has been extensively studied from prototype to real configurations. Based on a design by Kitaura and Zhou,[73] a typical configuration of LABs (see Figure 2.4) composing of a Li metal anode, a Li-ion conductive solid-state electrolyte and an air-electrode was constructed.

Using a PEO-based polymer electrolyte, Hassoun et al.[70] fabricated a solid-state cell of Li/PEO–LiCF$_3$SO$_3$–ZrO$_2$/Super-P to investigate the electrochemical behavior during discharge/charge, where Super-PA, acting as an air-electrode, is a super high surface area carbon material. Base on the analysis using potentiodynamic cycling with galvanostatic acceleration and cyclic voltammetry (CV) measurements, they proposed

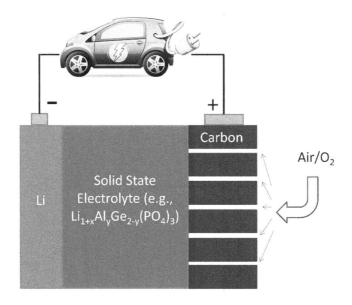

FIGURE 2.4　Schematic configuration for a solid-state Li-air battery (LAB).

a mechanism for the ORR and OER similar to that in non-aqueous (i.e., organic liquid electrolyte) RLABs. Compared to other research on the electrochemical mechanism of solid-state RMABs,[73–75] their proposed mechanism seemed to be more specific and reasonable. For ORR process, three electrochemical oxygen reduction reactions were proposed to take place (see Equations 2.34–2.36) during discharge, corresponding to the three tested overlapping peaks at 2.7, 2.65, and 2.63 V versus Li, respectively. In the discharge process, LiO_2 tended to form Li_2O_2 due to its low thermodynamic stability. The third reduction peak at 2.63 V was attributed to the transformation of Li_2O_2 into Li_2O (see Equation 2.36). The presence of the assumed products (e.g., Li_2O_2 and LiOH) in the ORR was confirmed by their ex-situ X-ray analysis and supported by previous literature.[76,77] For the OER process, their results indicated that two overlapping oxidation peaks at 3.13 and 3.15 V versus Li could be associated with two oxidation reactions (i.e., Equations 2.37 and 2.38, respectively). Although this proposed mechanism seemed reasonable, two intermediate products, LiO_2 and Li_2O were not detected by X-ray diffraction or other characterization techniques in the solid-state electrolyte. Moreover, the $1 - e^-/O_2$ and $4 - e^-/O_2$ processes were not detected in the charging process by differentiation electrochemical mass spectrometry.[78]

$$4O_2 + 4e^- + 4Li^+ \rightarrow 4LiO_2 \tag{2.34}$$

$$4LiO_2 + 4e^- + 4Li^+ \rightarrow 4Li_2O_2 \tag{2.35}$$

$$Li_2O_2 + 2e^- + 2Li^+ \rightarrow 2Li_2O \tag{2.36}$$

$$Li_2O_2 \rightarrow LiO_2 + e^- + Li^+ \tag{2.37}$$

$$LiO_2 \rightarrow O_2 + e^- + Li^+ \tag{2.38}$$

2.2 DESIGN PRINCIPLES OF BIFUNCTIONAL COMPOSITE CATALYSTS FOR MABS

As one type of the most promising energy conversion and storage systems, MABs are considered to be the important candidates in replacing the popular LIBs in highly energy-intensive applications, such as electric vehicles,[79,80] due to its extremely high theoretical energy densities.[81] To be commercially viable as a next generation energy system, RMABs have to address several technical challenges; in particular, the sluggish reaction kinetics in both ORR and OER, resulting in high overpotentials and, therefore, limited energy efficiency. Due to this, bifunctional electrocatalysts for air-electrodes need to be first addressed as it is required to be able to expedite the kinetics of oxygen reactions in the commercial application of MABs. Under RMAB operations, the bifunctional catalysts will take part in the discharge process as well as in the charge process.[82,83] Therefore, it is important that after the OER is efficiently conducted, the catalyst can recover its active sites for the ORR in the next discharge-charge cycle, thus producing cycle efficiency. Hence, bifunctional ORR/OER catalysts should be designed to offer both high activity and stability/durability in RMABs.[59,84,85]

When bi-components or multi-components are physically and/or chemically combined to form novel composite materials, the new efficient properties should be expected for matching the more rigorous requirements of practical applications. As observed, among conventional materials used as bifunctional catalysts, single-component materials, particularly carbon materials, show serious corrosion/oxidation problems in the presence of O_2 at high electrode potentials even though they have desirable properties such as large surface areas, high electrical conductivity, well-developed pore structure and low costs.[86] This corrosion/oxidation problem also is often observed for carbon-based catalyst materials.

Meanwhile, the use of noble metals (e.g., Pt, Ru, and Ir)[87] have large limitations due to their high cost. There are several strategies to overcome this limitation, such as (1) using low or ultra-low loading of noble metal in catalyst materials,[88] (2) developing non-noble metal-based catalyst materials,[89] and (3) exploring metal-free catalyst materials.[90] Recent research suggests these three strategies can significantly benefit from the introduction of composite materials because different components can exhibit strong synergistic effects on the composite catalyst materials, resulting in improvement of both catalytic activity and stability/durability.[86,91,92]

With increasing focus on MABs as promising energy sources to power automobiles and portable devices, there has been rapid growth in the research and development of novel bifunctional electrocatalysts, especially composite bifunctional catalysts. These composite catalysts and their novel structures can provide improved density power, cycling capability, and round-trip energy efficiency for RMABs[52,84] It is believed that the design and selection of material(s) in advanced composite catalysts is vital to catalyst properties, performance, longevity, and cost-effectiveness of RMABs.[4,93] The basic principles are categorically proposed in the design and fabrication of advanced bifunctional composite catalysts, based on the understanding of carbon-based and noncarbon-based materials with various physical and/or chemical properties, as well as on the synergistic effects that exist in the supported and unsupported bifunctional catalysts in various MABs, as follows:

1. In supported catalysts, novel support materials (i.e., modified carbon-based composite supports and noncarbon-based composite supports) should match some basic requirements:[86,91,92] (i) large surface area, (ii) high electrical conductivity, (iii) hierarchical pore and porous structure, (iv) high resistance to electrochemical corrosion, (v) stability in acidic or alkaline media, and (vi) favorable interaction with catalyst particles. Support materials that possess these properties can serve as suitable substrates to match the distribution and utilization of active catalyst particles for the electrochemical process.[96,91,94]

2. Based on the first principle theory and calculation, studies on the ORR activity of doped carbon materials,[95,96] coupled with experimental measurements,[97] the optimal design for doped carbon catalysts can be proposed, which should have the best ORR and OER activities that are strongly related to the doping structure. When dual element-doped carbon-based catalysts are researched as high active metal-free bifunctional catalysts, the p-electrons between co-dopants can result in significant synergistic interactions[30] that reduce overpotentials, stabilize adsorbates, and speed up both ORR and OER kinetics.

3. For the invention and fabrication of novel noble metal-based composite catalysts and non-noble metal-based composite catalysts, it is important to incorporate bi- or multi-metals or alloys that exhibit different absorption and desorption energies for oxygen.[98] For instance, Ag, as a non-noble metal, has strong oxygen desorption energy while Cu has a strong absorption energy. Their incorporation can produce synergistic effects on the catalytic activity according to the d-band center theory.[99,100]

4. When new catalytic materials are researched and developed as bifunctional catalysts, the details of nanostructure, such as controlled composition, optimized size, designed shape, and/or porous structure, should be tailored and tuned to create more active catalytic sites along with high stability/ durability. Examples of materials with mesoporous[101] or hollow[102] structure are oxide-based composites that display more active sites, facilitating oxygen diffusion and electrolyte impregnation throughout the electrode.

5. Apart from the covalence of cation-oxygen bonds and the unity e_g-filling of B site transition metals, the positions of oxygen p-band and metal d-band in relation to Fermi levels should be considered in obtaining highly active perovskite-based catalyst materials using molecular orbital theory (MOT) and band theory (BT).[103]

6. Strong interactions can lead to the synergistic effects between components in the catalysts made from bi- or multi-components, which should be created by novel synthesis method strategies. The synergistic effect can not only benefit both the ORR and OER activities, but also be favorable to the enhancement of material structure stability/durability.[86]

In terms of electrode design and fabrication, bifunctional composite catalysts tend to be fabricated into the electrode with a favorite formation of the three-phase boundary between the supplied gas, the electrolyte, and the catalyst surface. In addition, the cost-effectiveness also needs to be considered in the commercial application of MABs catalysts. Although these principles above are selectively proposed in the design and innovation of bifunctional composite catalysts, it is not easy to immediately obtain excellent composite materials as the optimal catalysts in RMABs, in terms of one or two design principle selections, because the preparation and fabrication of highly active bifunctional catalysts is also influenced by other factors and experimental conditions, including temperature, pH, technical procedure, solvent type, and concentration.[2,4,94,95]

2.3 SYNERGISTIC EFFECT AND ITS CHARACTERIZATION FOR MAJOR COMPONENTS OF MABs

Over the past ten to twenty years, synergistic effect has been greatly emphasized in the enhancement of catalytic performance. This is because the effect of the interaction arising from two or more components in a composite catalyst is greater than the sum of their individual effects. For instance, the synergistic effect between carbon and Pt or Pt-alloy[86,91,92] has been found to produce a strong physicochemical interaction, resulting in effective charge transfers and/or electronic effects, thereby improving catalytic performance. In the research and development of advanced bifunctional composite catalysts, the synergistic effect on catalytic activities, including ORR, OER, and stability/durability, have been studied by experimental and/or theoretical approaches in terms of their physical and/or chemical structures such as the electronic structure, chemical bonds, functional groups, doping, nanostructured morphology, and so on.

Using the first-principle theory studies, Ren et al.[104] conducted an investigation into B, P co-doped graphene as a carbon-based bifunctional composite electrocatalyst in RMAB. In the comparison of B, P co-doped graphene, B-doped graphene and P-doped graphene, the thermodynamic calculations indicated that by combining the active ORR activity of the P-doped graphene and the high OER activity of the B-doped graphene, the B, P co-doped graphene could produce an effective synergistic effect and reduce the rate-determining oxygen evolution barrier as well as the charge voltage, resulting in the synergistic enhancement of both ORR and OER. This effect was due to the electron transfer in Li-sited adsorption and the structural effect on O-sited adsorption, corresponding to the combination of B-doping and P-doping. According to the exponential relationship of the charge rate $(i(U) = \Upsilon e^{[-\Delta G^r / k_b T]})$,[104] B,P co-doping could improve the equilibrium current density by 10–12 times in comparison with that of pure graphene. This synergistic catalytic effect of co-doped elements was also observed in other experiments.[97,105] DFT calculations in Zheng et al.[105] demonstrated that the synergistic effect could be resulted from the changes of spin density and charge density, leading to a large charge delocalization effect in the improvement of ORR.[105] As for the effects of p-block elements in co-doped carbon materials, Zhao and Xia[97] used the products of electronegativity and electron affinity as a descriptor to examine the distance between two dopants and its synergistic effects. They theoretically predicted that the synergistic effect of co-doping on ORR and OER activities should occur within a certain distance between the co-dopants.

As evidenced in X-ray photoelectron spectroscopy (XPS), transitional metal oxides play an important interaction between carbon and Pt-alloy to produce active ORR.[106–108] As a special oxide for bifunctional catalysts, spinel $MnCo_2O_4$ examined by XPS[109] was shown to present a synergistic effect in the establishment of covalent interfacial Mn–O–C interactions in which electron sharing between C and O could reduce the electron cloud density around Mn. Accordingly, the intensities of C–O peaks in $MnCo_2O_4$/CNT and $MnCo_2O_4$/graphene were found higher than those in CNT and graphene alone while the binding energy of both C–C and C–H bonds were decreased with the addition of $MnCo_2O_4$. It was confirmed that in the presence of $MnCo_2O_4$, a stronger oxidized C environment existed in the oxide-nanocarbon interaction. Parallel to the research of spinel $MnCo_2O_4$ in its carbon composites, a strong synergistic effect between a non-spinel cobalt-manganese oxide ($Co_xMn_{1-x}O$) and nitrogen-doped carbon nanotubes (N–CNTs) was also found using X-ray absorption near edge structure (XANES) spectra and the extended X-ray absorption fine structure (EXAFS) spectra.[110] It was demonstrated that the synergistic effect of Co and Mn could result in higher OER and ORR activities than those of the two single metal-based catalysts of CoO/N–CNT and MnO/N–CNT. This was confirmed by its Co and Mn K-edge X-ray absorption spectra. The results of XANES indicated that similar to CoO and MnO, the Co K-edge shifted to a high energy band and the Mn K-edge reflected a purely electric dipole-allowed transition from the 1 s core level to the 4 p in the fabricated $Co_xMn_{1-x}O/N–CNT$ composite. The results of EXAFS further indicated that the Co K-edge in the $Co_xMn_{1-x}O/N–CNT$ showed two peaks split in the range of 1.0–2.0 Å, corresponding to Co−O bonding. This was clearly different from that of CoO/N–CNT with one peak at 1.69 Å. Moreover, the Mn K-edge of $Co_xMn_{1-x}O/N–CNT$ showed a peak shifting toward a larger interatomic distance when compared to that of MnO/N–CNT while no obvious well-defined structural peaks and valleys could be observed in the range of 4.0–8.0 Å for the Co K-edge of $Co_xMn_{1-x}O/N–CNT$, indicating the distinct local neighbor environment of Co atoms influenced by Mn.

In addition to the oxides mentioned above, two cobalt sulfide materials, Co_9S_8[111] and CoS_2[112] were found to exhibit synergistic effects on both the ORR and OER activities in their composites with doped carbon materials. To understand the synergistic interaction between Co_9S_8 and N–doped carbon (N–C) material, the compared XPS spectra of Co_9S_8, N-doped carbon, and $Co_9S_8/N–C$ showed that the introduction of Co_9S_8 could provide S as a dopant that formed S-doped C in $Co_9S_8/N–C$ composites, along with the formation of N-doped C.[111] The independent peak at a low binding energy of 158.9 eV in the spectrum of S 2p suggested that the coupling between Co_9S_8 and N–C[113] existed in the C–S–Co combination with sulfur species doped into carbon. This indicated that the electrons could transfer from N–C to Co_9S_8 and thus favored the improvement of catalytic activity and durability. Interestingly, Ganesan et al.[112] also demonstrated a strong coupling between CoS_2 and N,S co-doped carbon materials for the improvement of oxygen electrode performance by using pyrolysis of graphene (G) oxide to produce the carbon material (N,S–G). To confirm the strong coupling between CoS_2 and N, S–G, UV-visible spectra, Fourier transform infrared spectroscopy (FT–IR), Raman spectra, and XPS were used in their measurements to examine the interaction of CoS_2 particles with N,S–G. Based on the Kubelka-Munk function,[114–116] it was found that compared to pure CoS_2 products such as CoS_2(commercial) and CoS_2(400), the typical CoS_2(400)/N,S–G composite prepared through a heat treatment route at 400°C displayed a lower band gap (3.25 eV) and thus better ORR and OER performance due to the in-situ incorporation of CoS_2 nanoparticles. The IR and Raman results suggested the presence of C–$(S–S)_n$ bonds, showing stronger bonding of C and S for CoS_2(400)/N,S–G than the two CoS_2 materials, confirming the electronic transfer from CoS_2(400) to N,S–G. The XPS results showed the peak of S_{2p} at 163.4 eV resulted from polysulfide and carbon, suggesting the strong coupling of CoS_2 with N,S–G for the improvement of bifunctional activity.

Regarding the "strong metal-support interaction" (SMSI) studied for the synergistic effect of Pt and oxide-based support materials[91] on the enhancement of ORR electrocatalysis, Han et al.[117] performed XPS to characterize the SMSI of calcium-manganese oxides supported Pt bifunctional catalysts (e.g., $Pt/CaMnO_3$), where calcium-manganese oxides were promising materials for oxygen electrocatalysis.[118–120] In their XPS analysis, the Pt 4f slightly was shifted to higher binding energies relative to bulk Pt after the addition of $CaMnO_3$, resulted from the modification of the electronic structure of Pt. Furthermore, the slightly more positive peak position of Mn $2p_{3/2}$ in $Pt/CaMnO_3$ suggested a higher oxidation state of Mn, especially, the analysis and comparison of two deconvolution peaks at 528.9 and 531.0 eV for O1s spectra indicated a high surface oxidation of Pt clusters and $CaMnO_3$ nanoparticles as well as the stronger interaction between them. The interaction was believed to play an important role in the improvement of bifunctional activity and durability for $Pt/CaMnO_3$. Aside from $Pt/CaMnO_3$, three other metal-oxide composites, such as Au/MnO_2, Ag/MnO_2, and $Ag/LaMnO_3$, were also found to possess synergistic effects that could enhance both ORR and OER activities through physical and/or chemical interactions between the metal and oxide. In comparison with the XPS analysis of the electronic structures in both Au/MnO_2 and Ag/MnO_2, the combination of XPS and electron energy loss spectroscopy (EELS), coupled with DFT calculations, were conducted to examine the synergistic effects in $Ag/LaMnO_3$ in which $LaMnO_3$ acted as an active perovskite support for ORR and OER.[121] In a detailed analysis in which the electronic structure was an important factor determining the ORR activity, the electrons were found to transfer between catalyst metals and other components in co-catalysts or supports, showing as a ligand effect. Specifically, the electrons could transfer from Mn to Ag in the $Ag/LaMnO_3$ as evidenced by the behavior of the binding energies of Mn 2p and Ag 3d in XPS, resulted from the incorporation of Ag to $LaMnO_3$. Importantly, this charge transfer was further demonstrated by the difference in total charges when the electron population for each orbital was calculated by the DFT method with the CASTEP code[122] based on a plane-wave expansion of the wave functions and a Vanderbilt-type ultra-soft pseudopotential formalism.[123] Therefore, after Ag was incorporated into $LaMnO_3$, the charge transfer as a ligand effect could result in a beneficial effect to both the ORR of Ag and the OER of $Ag/LaMnO_3$.

2.4   CHAPTER SUMMARY

RMABs have attracted considerable research and development interest because they demonstrate high specific energy densities with a great potential to meet the ever-increasing demand for electrical energy storage and conversion in emerging applications, such as electric vehicles and smart grids. Depending on the types of metal electrode and electrolyte, RMABs have significantly exhibited their diversity in research and development. ORR, at the air-electrode, includes a series of complex electrochemical reaction steps, which are associated with multiple electron-transfer processes and complicated oxygen-containing species. According to the literature, the ORR process may occur by a four-electron pathway or a two-electron pathway, depending on the surface properties of the air-electrode catalysts. Different from the ORR, the OER on the surface of air-electrode catalyst is more complex and sensitive due to OER process occurs in the oxide surface, which is related to the interaction between metal and oxygen-containing intermediates. So far, the OER mechanism is still in continuous studies for various types of RMAB, which is associated with the site geometries of metal and the chemical valence of metal ion in the process.

REFERENCES

1. Jörissen, L. Bifunctional oxygen/air electrodes. *J. Power Sources.* **2006**, *155*, 23–32.
2. Li, D. U.; Xu, P.; Cano, Z. P.; Kashlooli, A. G.; Park, M. G.; Chen, Z. W. Recent progress and perspectives on bi-functional oxygen electrocatalysts for advanced rechargeable metal–air batteries. *J Mater Chem A.* **2016**, *4*, 7107–7134.

3. Padbury, R.; Zhang, X. Lithium-oxygen batteries—limiting factors that affect performance. *J. Power Sources.* **2011**, *196*, 4436–4444.
4. Chen, F. Y.; Chen, J. Metal-air Batteries, From oxygen reduction electrochemistry to cathode catalysts. *Chem. Soc. Rev.* **2012**, *41*, 2172–2192.
5. Kowalczk, I.; Read, J.; Salomon, M. Li-air batteries: A classic example of limitations owing to solubilities. *Pure Appl. Chem.* **2007**, *79*, 851–860.
6. McCloskey, B. D.; Bethune, D. S.; Shelby, R. M.; Girishkumar, G.; Luntz, A. C. Solvents' critical role in non-aqueous lithium-oxygen battery electrochemistry. *J. Phys. Chem. Lett.* **2011**, *2*, 1161–1166.
7. Zhang, T.; Tao, Z.; Chen, J. Magnesium-air batteries: From principle to application. *Mater. Horiz.* **2014**, *1*, 196–206.
8. Morozan, A.; Jousselme, B.; Palacin, S. Low-platinum and platinum-free catalysts for the oxygen reduction reaction at fuel cell cathodes. *Energy Environ. Sci.* **2011**, *4*, 1238–1254.
9. Cheng, F.; Shen, J.; Peng, B.; Pan, Y.; Tao, Z.; Chen, J. Rapid room-temperature synthesis of nanocrystalline spinels as oxygen reduction and evolution electrocatalysts. *Nature Chem.* **2011**, *3*, 79–84.
10. Wang, Z.-L.; Xu, D.; Xu, J.-J.; Zhang, X.-B. Oxygen electrocatalysts in metal-air batteries: From aqueous to non-aqueous electrolytes. *Chem. Soc. Rev.* **2014**, *43*, 7746–7786.
11. Kraytsberg, A.; Ein-Eli, Y. Review on Li-air batteries—opportunities, limitations and perspective. *J. Power Sources.* **2011**, *196*, 886–893.
12. Linden, D.; Reddy, T. B. *Handbook of Batteries*, Third Edition. New York: McGraw-Hill Companies, Inc., **2002**.
13. Vielstich, W.; Lamm, A.; Gasteiger, H. A. Metal-air batteries: The zinc-air case. Fundamentals and survey of systems. *Handbook of Fuel Cells: Fundamentals, Technology and Applications.* United Kingdom: John Wiley & Sons, Ltd., **2010**.
14. Zhang, J. *PEM Fuel Cell Electrocatalysts and Catalyst Layers: Fundamentals and Applications.* London: Springer, **2008**.
15. Harnisch, F.; Schröder, U. From MFC to MXC: Chemical and biological cathodes and their potential for microbial bioelectrochemical systems. *Chem. Soc. Rev.* **2010**, *39*, 4433–4448.
16. He, P.; Zhang, T.; Jiang, J.; Zhou, H. Lithium-air batteries with hybrid electrolytes. *J. Phys. Chem. Lett.* **2016**, *7*, 1267–1280.
17. Zhang, Z.; Zhang, S. S. *Rechargeable Batteries: Materials, Technologies and New Trends.* Switzerland: Springer, **2015**.
18. Kinoshita, K. *Electrochemical Oxygen Technology.* New York: John Wiley & Sons Inc., **1992**.
19. Feng, X. *Nanocarbons for Advanced Energy Conversion*, (First Edition). Weinheim, Germany: Wiley-VCH Verlag GmbH & Co, **2015**.
20. Spendelow, J. S.; Wieckowshi, A. Electrocatalysis of oxygen reduction and small alcohol oxidation in alkaline media. *Phys. Chem. Chem. Phys.* **2007**, *9*, 2654–2675.
21. Christensen, P. A.; Hamnett, A.; Linares-Moya, D. Oxygen reduction and fuel oxidation in alkaline solution. *Phys. Chem. Chem. Phys.* **2011**, *13*, 5206–5214.
22. Suntivich, J.; Gasteiger, H. A.; Yabuuchi, N.; Nakanishi, H.; Goodenough, J. B.; Shao-Horn, Y. Design principles for oxygen-reduction activity on perovskite oxide catalysts for fuel cells and metal-air batteries. *Nat. Chem.* **2011**, *3*, 546–550.
23. Débart, A.; Bao, J.; Armstrong, G.; Bruce, P. G. An O_2 Cathode for rechargeable lithium batteries: The effect of a catalyst. *J. Power Sources.* **2007**, *174*, 1177–1182.
24. Ruvinskiy, P. S.; Bonnefont, A.; Pham-Huu, C.; Savinova, E. R. Using ordered carbon nanomaterials for shedding light on the mechanism of the cathodic oxygen reduction reaction. *Langmuir* **2011**, *27*, 9018–9027.
25. Schneider, A.; Colmenares, L.; Seidel, Y. E.; Jusys, Z.; Wickman, B.; Kasemo, B.; Behm, R. J. Transport effect in the oxygen reduction reaction on nanostructured, planar glassy carbon supported Pt/GC model electrodes. *Phys. Chem. Chem. Phys.* **2008**, *10*, 1931–1943.

26. Strmcnik, D.; Kodama, K.; van der Vliet, D.; Greeley, J.; Stamenkovic, V. R.; Markovićm, N. M. The role of non-covalent interactions in electrocatalytic fuel-cell reactions on platinum. *Nat. Chem.* **2009**, *1*, 466–472.

27. Feng, Y. J.; Alonso-Vante, N. Nonprecious metal catalysts for the molecular oxygen-reduction reaction. *Phys. Status Solidi B.* **2008**, *245*, 1792–1806.

28. Rossmeisl, J.; Logadottir, A.; Nørskov, J. K. Electrolysis of water on (oxidized) metal surfaces. *Chem. Phys.* **2005**, *319*, 178–184.

29. Nørskov, J. K.; Bligaard, T.; Logadottir, A.; Bahn, S.; Hansen, L. B.; Bollinger, M.; Bengaard, H.; Hammer, B.; Sljivancanin, Z.; Mavrikakis, M. Universality in heterogeneous catalysis. *J. Catalysis* **2002**, *209*, 275–278.

30. Man, I. C.; Su, H. Y.; Calle-Vallejo, F.; Hansen, H. A.; Martínez, J. I.; Inoglu, N. G.; Kitchin, J.; Jaramillo, T. F.; Nørskov, J. K.; Rossmeisl, J. Universality in oxygen evolution electrocatalysis on oxide surfaces. *ChemCatChem* **2011**, *3*, 1159–1165.

31. Mavros, M. G.; Tsuchimochi, T.; Kowalczyk, T.; McIsaac, A.; Wang, L.-P.; Voorhis, T. V. What can density functional theory tell us about artificial catalytic water splitting? *Inorg. Chem.* **2014**, *53*, 6386–6397.

32. Betley, T. A.; Wu, Q.; Van Voorhis, T.; Nocera, D. G. Electronic design criteria for O–O bond formation via metal-oxo complexes. *Inorg. Chem.* **2008**, *47*, 1849–1861.

33. Rasiyah, P.; Tseung, A. C. C. The role of the lower metal oxide/higher metal oxide couple in oxygen evolution reactions. *J. Electrochem. Soc.* **1984**, *131*, 803–808.

34. Gao, R.; Zhu, J.; Xiao, X.; Hu, Z.; Liu, J.; Liu, X. Facet-dependent electrocatalytic performance of Co_3O_4 for rechargeable Li-O_2 battery. *J. Phys. Chem. C* **2015**, *119*, 4516–4523.

35. Prabu, M.; Ketpang, K.; Shanmugam, S. Hierarchical nanostructured $NiCo_2O_4$ as an efficient bifunctional non-precious metal catalyst for rechargeable zinc-air batteries. *Nanoscale* **2014**, *6*, 3173–3181.

36. Liu, X.; Cui, B.; Liu, S.; Chen, Y. Progress of non-aqueous electrolyte for Li-air batteries. *J. Mater. Sci. Chem. Eng.* **2015**, *3*, 1–8.

37. Abraham, K. M.; Jiang, Z. A Polymer electrolyte–based rechargeable lithium/oxygen battery. *J. Electrochem. Soc.* **1996**, *143*, 1–5.

38. Li, Y.; Dai, H. Recent advances in zinc-air batteries. *Chem. Soc. Rev.* **2014**, *43*, 5257–5275.

39. Zhang, L.-L.; Wang, Z.-L.; Xu, D.; Zhang, X.-B.; Wang, L.-M. The development and challenges of rechargeable non-aqueous lithium-air batteries. *Int. J. Smart Nano Mater.* **2013**, *4*, 27–46.

40. Tan, P.; Wei, Z.; Shyy, W.; Zhao, T. S. Prediction of the theoretical capacity of non-aqueous lithium-air batteries. *Appl. Energ.* **2013**, *109*, 275–282.

41. Wang, Y.; Liang, Z.; Zou, Q.; Cong, G.; Lu, Y.-C. Mechanistic insights into catalyst-assisted non-aqueous oxygen evolution reaction in lithium-oxygen batteries. *J. Phys. Chem. C* **2016**, *120*, 6459–6466.

42. Hayyan, M.; Hashim, M. A.; AlNashef, I. M. Superoxide ion: Generation and chemical implications. *Chem. Rev.* **2016**, *116*, 3029–3085.

43. Laoire, C. O.; Mukerjee, S.; Abraham, K. M.; Plichta, E. J.; Hendrickson, M. A. Influence of non-aqueous solvents on the electrochemistry of oxygen in the rechargeable lithium-air battery. *J. Phys. Chem. C* **2010**, *114*, 9178–9186.

44. Balaish, M.; Kraytsberg, A.; Ein-Eli, Y. A critical review on lithium-air battery electrolytes. *Phys. Chem. Chem. Phys.* **2014**, *16*, 2801–2822.

45. Lau, K. C.; Curtiss, L. A.; Greeley, J. Density functional investigation of the thermodynamic stability of lithium oxide bulk crystalline structures as a function of oxygen pressure. *J. Phys. Chem. C* **2011**, *115*, 23625–23633.

46. Yang, X. H.; Xia, Y. Y. The effect of oxygen pressures on the electrochemical profile of lithium/oxygen battery. *J. Solid State Electrochem.* **2010**, *14*, 109–114.

47. Zhang, S. S.; Foster, D.; Read, J. Discharge characteristic of a non-aqueous electrolyte Li/O$_2$ battery. *J. Power Sources* **2010**, *195*, 1235–1240.
48. Laoire, C. O.; Mukerjee, S.; Plichta, E. J.; Hendrickson, M. A.; Abraham, K. M. Rechargeable lithium/TEGDME–LiPF$_6$/O$_2$ battery. *J. Electrochem. Soc.* **2011**, *158*, A302–A308.
49. Lu, Y.-C.; Gasteiger, H. A.; Crumlin, E.; McGuire, Jr., R.; Yang, S. Electrocatalytic activity studies of select metal surfaces and implications in Li-air batteries. *J. Electrochem. Soc.* **2010**, *157*, A1016–A1025.
50. Laoire, C. O.; Mukerjee, S.; Abraham, K. M.; Plichta, E. J.; Hendrickson, M. A. Elucidating the mechanism of oxygen reduction for lithium-air battery applications. *J. Phys. Chem. C* **2009**, *113*, 20127–20134.
51. Hummelshøj, J. S.; Blomqvist, J.; Datta, S.; Vegge, T.; Rossmeisl, J.; Thygesen, K. S.; Luntz, A. C.; Jacobsen, K. W.; Nørskov, J. K. Elementary oxygen electrode relations in the aprotic Li-air battery. *J. Chem. Phys.* **2010**, *132*, 071101.
52. Débart, A.; Paterson, A. J.; Bao, J.; Bruce, P. G. α-MnO$_2$ nanowires: A catalyst for the O$_2$ electrode in rechargeable lithium batteries. *Angew. Chem. Int. Ed.* **2008**, *47*, 4521–4524.
53. Nørskov, J. K.; Bligaard, T.; Rossmeisl, J.; Christensen, C. H. Towards the computational design of solid catalysts. *Nat. Chem.* **2009**, *1*, 37–46.
54. Peng, Z. Q.; Freunberger, S. A.; Hardwick, L. J.; Chen, Y. H.; Giordani, V.; Barde, F.; Novak, P.; Graham, D.; Tarascon, J. M.; Bruce, P. G. Oxygen reactions in a non-aqueous Li$^+$ electrolyte. *Angew. Chem. Int. Ed.* **2011**, *50*, 6351–6355.
55. Zhang, S.; Read, J. Partially fluorinated solvent as a co-solvent for the non-aqueous electrolyte of Li/air battery. *J. Power Sources* **2011**, *196*, 2867–2870.
56. Lu, Y. C.; Gasteiger, H. A. Parent, M. C.; Chiloyan, V.; Shao-Horn, Y. The influence of catalysts on discharge and charge voltages of rechargeable Li-oxygen batteries. *Electrochem. Solid-State Lett.* **2010**, *13*, A69–A72.
57. Li, F. J.; Zhang, T.; Zhou, H. S. Challenges of non-aqueous Li-O$_2$ batteries: Electrolytes, catalysts, and anodes. *Energy Environ. Sci.* **2013**, *6*, 1125–1141.
58. Lu, Y. C.; Yang, S. H. Probing the reaction kinetics of the charge reaction of non-aqueous Li-O$_2$ batteries. *J. Phys. Chem. Lett.* **2013**, *4*, 93–99.
59. McCloskey, B. D.; Bethune, D. S.; Shelby, R. M.; Mori, T.; Scheffler, R.; Speidel, A.; Sherwood, M.; Luntz, A. C. Limitations in rechargeability of Li-O$_2$ batteries and possible origins. *J. Phys. Chem. Lett.* **2012**, *3*, 3043–3047.
60. Lee, J.-H.; Black, R.; Popov, G.; Pomerantseva, E.; Nan, F.; Botton, G. A.; Nazar, L. F. The role of vacancies and defects in Na$_{0.44}$MnO$_2$ nanowire catalysts for lithium/oxygen batteries. *Energy Environ. Sci.* **2012**, *5*, 9558–9565.
61. Oh, S. H.; Nazar, L. F. Oxide catalysts for rechargeable high-capacity Li-O$_2$ batteries. *Adv. Energy Mater.* **2012**, *2*, 903–910.
62. Harding, J. R.; Lu, Y. C.; Tsukada, Y.; Yang, S. H. Evidence of catalyzed oxidation of Li$_2$O$_2$ for rechargeable Li-air battery applications. *Phys. Chem. Chem. Phys.* **2012**, *14*, 10540–10546.
63. Wang, Y.; He, P.; Zhou, H. A Lithium-air capacitor battery based on a hybrid electrolyte. *Energy Environ. Sci.* **2011**, *4*, 4994–4999.
64. He, P.; Wang, Y.; Zhou, H. Titanium nitride catalyst cathode in Li-air fuel cell with an acidic aqueous solution. *Chem. Commun.* **2011**, *47*, 10701–10703.
65. Christensen, J.; Albertus, P.; Sanchez-Carrera, R. S.; Lohmann, T.; Kozinsky, B.; Liedtke, R.; Ahmed, J.; Kojic, A. A critical review of Li-air batteries. *J. Electrochem. Soc.* **2012**, *159*, R1–R30.
66. Wang, Y., Zhou, H. A Lithium-air battery with a potential to continuously reduce O$_2$ from air for delivering energy. *J. Power Sources* **2010**, *195*, 358–361.
67. Wang, L.; Zhao, X.; Lu, Y.; Xu, M.; Zhang, D.; Ruoff, R. S.; Stevenson, K. J.; Goodenough, J. B. CoMn$_2$O$_4$ spinel nanoparticles grown on graphene as bifunctional catalyst for lithium-air batteries. *J. Electrochem. Soc.* **2011**, *158*, A1379–A1382.

68. Hasegawa, S.; Imanishi, N.; Zhang, T.; Xie, J.; Hirano, A.; Takeda, Y.; Yamamoto, O. Study on lithium/air secondary batteries-stability of NASICON-type lithium ion conducting glass-ceramics with water. *J. Power Sources.* **2009**, *189*, 371–377.
69. Black, R.; Adams, B.; Nazar, L. F. Non-aqueous and hybrid Li-O_2 batteries. *Adv. Energy Mater.* **2012**, *2*, 801–815.
70. Hassoun, J.; Croce, F.; Armand, M.; Scrosati, B. Investigation of the O_2 electrochemistry in a polymer electrolyte solid-state cell. *Angew. Chem. Int. Ed.* **2011**, *50*, 2999–3002.
71. Kumar, B.; Kumar, J.; Leese, R.; Fellner, J. P.; Rodrigues, S. J.; Abraham, K. M. A Solid-state, rechargeable, long cycle life lithium-air battery. *J. Electrochem. Soc.* **2010**, *157*, A50–A54.
72. Liu, Y.; Li, B.; Kitaura, H.; Zhang, X.; Han, M.; He, P.; Zhou, H. Fabrication and performance of all-solid-state Li-air battery with SWCNTs/LAGP cathode. *ACS Appl. Mater. Interfaces.* **2015**, *7*, 17307–17310.
73. Kitaura, H.; Zhou, H. Electrochemical and reaction mechanism of all-solid-state lithium-air batteries composed of lithium, $Li_{1+x}Al_yGe_{2-y}(PO_4)_3$ solid electrolyte and carbon nanotube air electrode. *Energy Environ. Sci.* **2012**, *5*, 9077–9084.
74. Suzuki, Y.; Kami, K.; Watanabe, K.; Imanishi, N. Characteristics of discharge products in all-solid-state Li-air batteries. *Solid State Ionics.* **2015**, *278*, 222–227.
75. Park, J.; Park, M.; Nam, G.; Lee, J.-S.; Cho, J. All-solid-state cable-type flexible zinc-air battery. *Adv. Mater.* **2015**, *27*, 1396–1401.
76. Visco, S. J.; Nimon, E.; De Jonghe, L. C. *Encyclopedia of Electrochemical Power Sources*, pp. 376–383. New York: Elsevier, **2009**.
77. Freunberger, S.; Hardwick, L. J.; Peng, Z.; Giordani, V.; Chen, Y.; Maire, P.; Novák, P.; Tarascon, J.-M.; Bruce, P. G. *Fundamental Mechanism of the Lithium-air Battery.* Abstract #830, IMLB 2010, The Electrochemical Society, **2010**.
78. Meini, S.; Piana, M.; Tsiouvaras, N.; Garsuch, A., Gasteiger, H. A. The effect of water on the discharge capacity of a non-catalyzed carbon cathode for Li-O_2 batteries. *Electrochem. Solid-State Lett.* **2012**, *15*, A45–A48.
79. Hwang, H. J.; Koo, J.; Park, M.; Park, N.; Kwon, Y.; Lee, H. Multilayer graphenes for lithium ion battery anode. *J. Phys. Chem. C* **2013**, *117*, 6919–6923.
80. Park, H. W.; Lee, D. U.; Nazar, L. F.; Chen, Z. Oxygen reduction reaction using MnO_2 nanotubes/nitrogen-doped exfoliated graphene hybrid catalyst for Li-O_2 battery applications. *J. Electrochem. Soc.* **2013**, *160*, A344–A350.
81. Bruce, P. G.; Freunberger, S. A.; Hardwick, L. J.; Tarascon, J.-M. Li-O_2 and Li-S batteries with high energy storage. *Nat. Mater.* **2012**, *11*, 19–29.
82. Wu, J.; Park, H. W.; Yu, A.; Higgins, D.; Chen, Z. Facile synthesis and evaluation of nanofibrous iron–carbon based non-precious oxygen reduction reaction catalysts for Li-O_2 battery applications. *J. Phys. Chem. C* **2012**, *116*, 9427–9432.
83. Park, H. W.; Lee, D. U.; Liu, Y.; Wu, J.; Nazar, L. F.; Chen, Z. Bifunctional N-doped CNT/graphene composite as highly active and durable electrocatalyst for metal air battery applications. *J. Electrochem. Soc.* **2013**, *160*, A2244–A2250.
84. Lu, Y.-C.; Xu, Z.; Gasteiger, H. A.; Chen, S.; Hamad-Schifferili, K.; Shao-Horn, Y. Platinum-gold nanoparticles: A highly active bifunctional electrocatalyst for rechargeable lithium-air batteries. *J. Am. Chem. Soc.* **2010**, *132*, 12170–12171.
85. Armand, M.; Tarascon, J. M. Building better batteries. *Nature* **2008**, *451*, 652–657.
86. Wang, Y.-J.; Fang, B.; Li, H.; Bi, X. T.; Wang, H. Progress in modified carbon support materials for Pt and Pt-alloy cathode catalysts in polymer electrolyte membrane fuel cells. *Prog. Mater. Sci.* **2016**, *82*, 445–498.
87. Lee, D. U.; Xu, P.; Cano, Z. P.; Kashkooli, A. G.; Park, M. G.; Chen, Z. Recent progress and perspectives on bifunctional oxygen electrocatalysts for advanced rechargeable metal-air batteries. *J. Mater. Chem. A* **2016**, *4*, 7107–7134.

88. Kim, B. G.; Kim, H. J.; Back, S.; Nam, K. W.; Jung, Y.; Han, Y. K.; Choi, J. W. Improved reversibility in lithium-oxygen battery: Understanding elementary reactions and surface charge engineering of metal alloy catalyst. *Sci. Rep.* **2014**, *4*, 4225.

89. Wang, H.; Lee, H.-W.; Deng, Y.; Lu, Z.; Hsu, P.-C.; Liu, Y.; Lin, D.; Cui, Y. Bifunctional non-noble metal oxide nanoparticle electrocatalysts through lithium-induced conversion for overall water splitting. *Nat. Commun.* **2015**, *6*, 7261.

90. Yang, H. B.; Miao, J.; Hung, S.-F.; Chen, J.; Tao, H. B.; Wang, X.; Zhang, L. et al. Identification of catalytic sites for oxygen reduction and oxygen evolution in N-doped graphene materials: Development of highly efficient metal-free bifunctional electrocatalyst. *Sci. Adv.* **2016**, *2*, e1501122.

91. Wang, Y.-J.; Wilkinson, D. P.; Zhang, J. Noncarbon support materials for polymer electrolyte membrane fuel cell electrocatalysts. *Chem. Rev.* **2011**, *111*, 7625–7651.

92. Wang, Y.-J.; Zhao, N.; Fang, B.; Li, H.; Bi, X.; Wang, H. Carbon-supported Pt-based alloy electrocatalysts for the oxygen reduction reaction in polymer electrolyte membrane fuel cells: Particle size, shape, and composition manipulation and their impact to activity. *Chem. Rev.* **2015**, *115*, 3433–3467.

93. Li, Q.; Cao, R.; Cho, J.; Wu, G. Nanostructured carbon-based cathode for non-aqueous lithium-oxygen batteries. *Phys. Chem. Chem. Phys.* **2014**, *16*, 13568–13582.

94. Caramia, V.; Bozzini, B. Materials science aspects of zinc-air batteries: A review. *Mater Renew Sustain Energy* **2014**, *3*, 28.

95. Li, M.; Zhang, L.; Xu, Q.; Niu, J.; Xia, Z. N-doped graphene as catalysts for oxygen reduction and oxygen evolution reactions theoretical considerations. *J. Catal.* **2014**, *314*, 66–72.

96. Zhao, Z.; Li, M.; Zhang, L.; Dai, L.; Xia, Z. Design principles for heteroatom-doped carbon nanomaterials as highly efficient catalysts for fuel cells and metal-air batteries. *Adv. Mater.* **2015**, *27*, 6834–6840.

97. Zhao, Z.; Xia, Z. Design principles for dual-element-doped carbon nanomaterials as efficient bifunctional catalysts for oxygen reduction and evolution reactions. *ACS Catal.* **2016**, *6*, 1553–1558.

98. Jackson, A.; Viswanathan, V.; Forman, A. J.; Larsen, A. H.; Nørskov, J. K.; Jaramillo, T. F. Climbing the activity volcano: Core–shell Ru@Pt electrocatalysts for oxygen reduction. *ChemElectroChem.* **2014**, *1*, 67–71.

99. Lima, F. H. B.; Zhang, J.; Shao, M. H.; Sasaki, K.; Vukmirovic, M. B.; Ticianelli, E. A.; Adzic, R. R. Catalytic activity-d-band center correlation for the O_2 reduction reaction on platinum in alkaline solutions. *J. Phys. Chem. C.* **2006**, *111*, 404–410.

100. Nørskov, J. K.; Rossmeisl, J.; Logadottir, A.; Lindqvist, L.; Kitchin, J. R.; Bligaard, T.; Jónsson, H. Origin of the overpotential for oxygen reduction at a fuel-cell cathode. *J. Phys. Chem. B.* **2004**, *108*, 17886–17892.

101. Sun, B.; Huang, X.; Chen, S.; Zhao, Y.; Zhang, J.; Munroe, P.; Wang, G. Hierarchical macroporous/mesoporous $NiCo_2O_4$ nanosheets as cathode catalysts for rechargeable Li-O_2 batteries. *J. Mater. Chem. A* **2014**, *2*, 12053–12059.

102. Yoon, K. R.; Lee, G. Y.; Jung, J.-W.; Kim, N. H.; Kim, S. O.; Kim, I.-D. One-dimensional RuO_2/Mn_2O_3 hollow architectures as efficient bifunctional catalysts for lithium-oxygen batteries. *Nano Lett.* **2016**, *16*, 2076–2083.

103. Gupta, S.; Kellogg, W.; Xu, H.; Liu, X.; Cho, J.; Wu, G. Bifunctional perovskite oxide catalysts for oxygen reduction and evolution in alkaline media. *Chem. Asian J.* **2016**, *11*, 10–21.

104. Ren, X.; Wang, B.; Zhu, J.; Liu, J.; Zhang, W.; Wen, Z. The doping effect on the catalytic activity of graphene for oxygen evolution reaction in a lithium-air battery: A first-principles study. *Phys. Chem. Chem. Phys.* **2015**, *17*, 14605–14612.

105. Zheng, Y.; Jiao, Y.; Ge, L.; Jaroniec, M.; Qiao, S. Z. Two-step boron and nitrogen doping in graphene for enhanced synergistic catalysis. *Angew. Chem. Int. Ed.* **2013**, *52*, 3110–3116.

106. Akalework, N. G.; Pan, C. J.; Su, W.-N.; Rick, J.; Tsai, M.-C.; Lee, J.-F.; Lin, J.-M.; Tsai, L.-D.; Hwang, D.-J. Ultrathin TiO$_2$-coated MWCNTs with excellent conductivity and SMSI nature as Pt catalyst support for oxygen reduction reaction in PEMFCs. *J. Mater. Chem.* **2012**, *22*, 20977–20985.

107. Rigdon, W. A.; Larrabee, D.; Huang, X. Composite carbon nanotube and titanium catalyst supports for enhanced activity and durability. *ECS Trans.* **2013**, *58*, 1809–1821.

108. Wang, C.; Gao, H.; Chen, X.; Yuan, W. Z.; Zhang, Y. Enabling carbon nanofibers with significantly improved graphitization and homogeneous catalyst deposition for high performance electrocatalysts. *Electrochim. Acta.* **2015**, *152*, 383–390.

109. Ge, X.; Liu, Y.; Thomas Goh, F. W.; Andy Hor, T. S.; Zong, Y.; Xiao, P.; Zhang, Z. et al. Dual-phase spinel MnCo$_2$O$_4$ and spinel MnCo$_2$O$_4$/Nanocarbon hybrids for electrocatalytic oxygen reduction and evolution. *ACS Appl. Mater. Interfaces.* **2014**, *6*, 12684–12691.

110. Liu, X.; Park, M.; Kim, M. G.; Gupta, S.; Wang, X.; Wu, G.; Cho, J. High-performance non-spinel cobalt-manganese mixed oxide-based bifunctional electrocatalysts for rechargeable zinc-air batteries. *Nano Energy* **2016**, *20*, 315–325.

111. Cao, X.; Zheng, X.; Tian, J.; Jin, C.; Ke, K.; Yang, R. Cobalt sulfide embedded in porous nitrogen-doped carbon as a bifunctional electrocatalyst for oxygen reduction and evolution reactions. *Electrochim. Acta.* **2016**, *191*, 776–783.

112. Ganesan, P.; Prabu, M.; Sanetuntikul, J.; Shanmugam, S. Cobalt sulfide nanoparticles grown on nitrogen and sulfur co-doped graphene oxide: An efficient electrocatalyst for oxygen reduction and evolution reactions. *ACS Catal.* **2015**, *5*, 3625–3637.

113. Feng, L.-L.; Li, G.-D.; Liu, Y. P.; Wu, Y. Y.; Chen, H.; Wang, Y.; Zou, Y.-C.; Wang, D. J.; Zou, X. X. Carbon-armored Co$_9$S$_8$ nanoparticles as all-ph efficient and durable H$_2$-evolving electrocatalysts. *ACS Appl. Mater. Interface* **2015**, *7*, 980–988.

114. Chakraborty, L.; Malik, P. K.; Moulik, S. P. Preparation and characterization of CoS$_2$ nanomaterial in aqueous cationic surfactant medium of cetyltrimethylammonium bromide (CTAB). *J. Nanopart. Res.* **2006**, *8*, 889–897.

115. Klaas, J.; Schulz-Ekloff, G.; Jaeger, N. I. UV-visible diffuse reflectance spectroscopy of zeolite-hosted mononuclear titanium oxide species. *J. Phys. Chem. B* **1997**, *101*, 1305–1311.

116. Dang, Y.; Meng, X.; Jiang, K.; Zhong, C.; Chen, X.; Qin, J. A promising nonlinear optical material in the mid-IR region new results on synthesis, crystal structure and properties of noncentrosymmetric β–HgBrCl. *Dalton Trans.* **2013**, *42*, 9893–9897.

117. Han, X.; Cheng, F.; Zhang, T.; Yang, J.; Hu, Y.; Chen, J. Hydrogenated uniform Pt clusters supported on porous CaMnO$_3$ as a bifunctional electrocatalyst for enhanced oxygen reduction and evolution. *Adv. Mater.* **2014**, *26*, 2047–2051.

118. Najafpour, M. M.; Ehrenberg, T.; Wiechen, M.; Kurz, P. Calcium manganese (III) oxides (CaMn$_2$O$_4$·xH$_2$O) as biomimetic oxygen-evolving catalysts. *Angew. Chem. Int. Ed.* **2010**, *49*, 2233–2237.

119. Wiechen, M.; Zaharieva, I.; Dau, H.; Kurz, P. Layered manganese oxides for water-oxidation: Alkaline earth cations influence catalytic activity in a photosystem II-like fashion. *Chem. Sci.* **2012**, *3*, 2330–2339.

120. Han, X.; Zhang, T.; Du, J.; Cheng, F.; Chen, J. Porous calcium-manganese oxide microspheres for electrocatalytic oxygen reduction with high activity. *Chem. Sci.* **2013**, *4*, 368–376.

121. Lu, M.; Qu, J.; Yao, Q.; Xu, C.; Zhan, Y.; Xie, J.; Lee, J. Y. Exploring metal nanoclusters for lithium-oxygen batteries. *ACS Appl. Mater. Interface.* **2015**, *7*, 5488–5496.

122. Segall, M. D.; Lindan, P. J. D.; Probert, M. J.; Pickard, C. J.; Hasnip, P. J.; Clark, S. J.; Payne, M. C. First-principles simulation: Ideas, illustrations and the CASTEP code. *J. Phys. Condens. Ma.* **2002**, *14*, 2717–2744.

123. Vanderbilt, D. Soft self-consistent pseudopotentials in a generalized eigenvalue formalism. *Phys. Rev. B* **1990**, *41*, 7892–7895.

3 Carbon-Based Bifunctional Composite Electrocatalysts for Metal-Air Batteries

3.1 INTRODUCTION

3.1.1 SINGLE CARBON CATALYST MATERIALS FOR MABS

Due to significant advantages such as high electronic conductivity, tunable porosity, lightweight, and low-cost, carbon materials (e.g., porous carbons, carbon sheets, carbon fibers, graphene, carbon nanotubes, carbon aerogel, etc.) have been extensively explored in the research of air-electrode (or air-cathode) materials for rechargeable metal-air batteries (RMABs).[1–3] In the exploration and development of high-performance, durable MAB cathodes, single carbon materials are increasingly replaced by carbon-based composite materials. Although single carbon materials are not our focus in this book, it is necessary to discuss single carbon material-related electrode electrocatalysts for comparison with other carbon-based composite electrocatalysts.

In the search for non-precious metal materials to address issues of Li-O_2 batteries (LOBs) such as poor cycling stability, poor rate capability, short lifetime, and low round-trip efficiency,[4] nanostructured porous metal-free carbon catalysts have been extensively investigated as promising air-electrode materials due to their analogous functions and comparable reactivity to those of conventional noble metals.[5] Considering that carbon microspheres (CMs) have a high surface area and unique porous structure, Meng et al.[2] used a reflux-calcination method to obtain as-prepared CMs as an advanced air-electrode material in LOBs. The morphological characterization by TEM and SEM showed that the nanosized pores could uniformly distributed over the CM sample particles, corresponding to a pore-size distribution of around 0–5 nm and a relatively lower proportion at 9 nm shown by the N_2 adsorption-desorption isotherms. The existence of these nanosized pores not only led to the sample of a high surface area ($\sim$626.919 m^2 g^{-1}) but also provided favorable environments for oxygen diffusion and electrode/electrolyte contact. When the sample was tested as the air-electrode, without adding any metal particles in a non-aqueous electrolyte containing lithium bis(trifluoromethanesulfonyl) imide (LITFSI) and tetraethylene glycol dimethyl ether (TEGDME), this as-prepared carbon microspheres were found to deliver a high initial discharge capacity of 12081 mA h g^{-1} at a current density of 200 mA g^{-1} and allowed for 60 cycles at a cut off capacity of 1000 mA h g^{-1}

at 200 mA g^{-1}. Moreover, the capacity loss was insignificant after 20 full charge-discharge cycles under a current density of 300 mA g^{-1}. Interestingly, when compared with a pure commercial carbon black, such as acetylene black (AB), the CM sample could result in more pronounced ORR and OER currents, suggesting its higher catalytic activity for the Li-O$_2$ battery. Their results indicated that with high surface areas and novel structures, the air-electrode material could favor both the ORR and OER activities.

Similar to CM samples, 2D carbon sheets have been studied as low-cost air-electrode materials for rechargeable Zn-air batteries (RZABs) due to their microporous structure and high surface area. Li et al.[6] used eggplants as a renewable, high yield, and natural starting material for the production of additive-, metal-free, and highly active electrocatalysts for RZABs via simple carbonization and activation steps. This might be because eggplants could be easily converted to a low-cost unique 2D eggplant carbon (EPC) sheet after carbonization due to its high hydrocarbon content and its sponge-like microstructure from its interconnected framework of thin cell walls.[7] In their study, the carbonization temperature (Tc) was varied from 600°C to 1000°C for the preparation of different EPC-Tc samples for RZABs. For their EPC-900 sample, a further activation process following KOH method[8,9] was carried out to form an A-EPC-900 sample allowing minerals (e.g., K, Ca, Mg, Na, etc.) to be removed to produce ultra-small voids and thus high surface areas with a high population of active sites for facilitating electrochemical reactions. The morphological characterization showed that there was no significant change for the size and shape of the samples regardless of carbonization temperature or whether the as-prepared sample was activated or not. The similar morphology exhibited in the 2D sheets possessed an in-plane size of a few to tens of micrometers and a thickness of 100–200 nm for all samples. The activated A-EPC-900 sample, however, looked much rougher due to numerous distinguishable pores on the surface and edges, compared to the EPC-900 sample. The Brunauer-Emmett-Teller (BET) measurements showed a much higher specific surface area of 1051.5 m^2 g^{-1} of the A-EPC-900 than that (9.5 m^2 g^{-1}) of the pristine EPC-900. The N$_2$ adsorption-desorption isotherms, as well as the Barrett-Joyner-Halenda (BJH) plots demonstrated a sharp peak centered at 1.9 nm with a tail into a pore size of ~4 nm for the A-EPC-900, indicating a large number of microspores with the co-existence of mesopores. Small BET surface areas and low populations of pores were found for all EPC samples prepared at 600–1000°C without activation. When the electrocatalytic activity was investigated, using cyclic voltammetry (CV) and linear sweep voltammetry (LSV) with a rotating disk electrode (RDE), it was found that, at a rotation speed of 1600 rpm, the A-EPC-900 showed superior ORR activity to that of EPC samples obtained by the carbonization process and even better than that of commercial 20 wt% Pt/C. Moreover, the calculated electron transfer number (n) per oxygen molecule in the ORR catalyzed by A-EPC-900 was approximately 4.0 at a potential of −0.4 to −0.6 V, indicating a four-electron transfer behavior to produce water. The investigated OER results also showed that the A-EPC-900 had notably less positive onset potential and a larger current density than those of commercial Pt/C and other EPC samples obtained through the carbonization process, suggesting higher OER activity of the A-EPC-900. XRD, Raman spectra, TGA, and energy dispersive X-ray spectroscopy (EDS) measurements were also used to analyze the origin of the enhanced catalytic activity of the A-EPC-900 sample. The graphitic structure of A-EPC-900 was found to exist with a higher content than that of EPC-900, resulted from the introduction of activation. Multiple minerals such as K and Ca could be removed by the activation process. The measurements demonstrated that this activation process could further result in high performance of the A-EPC-900 sample. When A-EPC-900 was compared with 20 wt% Pt/C using the air-electrode measurements in practical RZABs, it was found that the specific capacity per gram of the consumed Zn was ~669 mA h g^{-1} for the A-EPC-900 based RZAB. This was comparable to RZABs built with highly efficient CoO/N-CNT[10] or nanostructured NiCo$_2$O$_4$[11] catalysts. The cycling tests demonstrated insignificant changes in the discharge or charge voltages for A-EPC-900 based RZABs over 160 hours, while the voltage gap remained at ~0.85 V. This is significantly better than the 20 wt% Pt/C based RZAB. The results of 2D carbon sheets based of air electrode show nature-derived carbon materials could play an important role in the development of advanced air-electrodes in large RZABs; however, these carbon materials require further improvements in both their durability and catalytic activity in the application of RZAB air-electrodes.

Differing from the porous structures in both carbon microspheres and carbon sheets discussed above, the porous structures of carbon aerogels are composed of mesopores and micropores. The mesopores tends to favor mass (e.g., oxygen) transport and enlarge reaction active sites, facilitating reaction kinetics and possibilities; while micropores with little or non-catalytic active centers can be filled with highly conductive Li$^+$ electrolytes due to the coalition of hydrophilic carbon surfaces and capillary pressure, favoring the Li$^+$ transfer. To build a reasonable electrode design to get rid of pore blockage, electrode passivation and degradation, as well as to provide a free path for oxygen, electrons, and Li$^+$ transports, Wang et al.[3] developed a novel highly conductive dual pore carbon aerogel based air-electrode for RLAB using a resorcinol-formaldehyde (R-F) polycondensation and carbonization route.[12] The employed route consisted of a series of complex physical and chemical processes including sol-gel reaction, solvent exchange, and high-temperature sintering. The BET surface area and BJH pore size distribution were measured to reveal that the typical carbon aerogels CRF1000-30 and CRF 1500-40 possessed characteristic single and dual pore size distributions, respectively, having extremely similar average meso-pore diameters of around 8.0 nm. The CRF 1500-40 sample exhibited a dual-pore structure, centering at a larger mesopore size of 17.95 nm and a smaller micropore size of 3.85 nm. The CRF1000-30 only had a mesopore size of 16.00 nm. The BET surface areas were 669.17 and 680.22 m^2 g^{-1} for the CRF 1500-40 and the CRF 1000-30 respectively, which are very similar. In the electrochemical measurements of the carbon aerogel samples as the air-electrode in a non-aqueous RLAB with a lithium disk and a separator dipped in 1 M LITFSI/sulfolane electrolyte, it was found that, compared to the single porous structure of CRF1000-30, the dual porous structure of CRF1500-40 could give higher galvanostatic charge/discharge cycling numbers and lower voltage gaps during cycling. This suggested that CRF1500-40 possessed better battery performance than that of CRF1000-30 because the dual pore of CRF1500-40 could create a more suitable tri-phase boundary reaction zone, facilitating oxygen and Li$^+$ transport with two independent pore channels, allowing for relatively higher discharge rate capability, lower pore blockage and cathode passivation. At current densities varying from 1.00 to 0.05 mA cm^{-2}, the CRF1500-40 electrode showed 18–525 cycles, as well as a high energy efficiency of 78.32%. The above results demonstrate that such nano-porous carbon aerogels can provide a novel dual pore structure and a promising material for RLAB air-electrodes even if carbon aerogels are single carbon materials.

3.1.2   COMPOSITES OF DIFFERENT CARBONS FOR MABS

In literature, the combination of two different carbon materials has been explored as the modified carbon support materials for Pt-based catalysts in PEM fuel cells.[13] Based on the advantages of different carbon materials, the formed Pt-based catalysts supported on carbon-carbon composite materials can give enhanced catalytic activities towards ORR. Similar combinations of two or more different carbon materials have also been investigated as noble metal-free candidates to replace commercial noble metal-based electrocatalysts for MABs.

To obtain a hybrid phase and the synergetic advantage of different materials, two different carbons, which have different physical properties and morphologies, have been used to fabricate all-carbon hybrid electrocatalysts with improved electrochemical properties such as mass transport, ORR or OER in the application of RMABs. Graphene quantum dots (GQDs), regarded as zero dimensional carbon nanostructures with atomic sizes below 100 nm, have shown satisfactory ORR performance due to their strong quantum confinement, edge effects, and oxygen-rich functional groups on the surface.[14,15] GQDs have been used to composite with other carbon materials (e.g., graphene, multi-walled carbon nanotubes, or graphene hydrogel) to create efficient metal-free catalysts for RMABs.[16–18] The existence of GQDs on carbon substrates was found to notably improve the ORR performance by cutting off the electrical neutrality of carbon materials to produce rich active sites for oxygen adsorption/desorption. By encapsulating GQDs in a graphene hydrogel structure, Wang et al.[18] prepared a carbon-carbon composite electrocatalyst in which the GQDs could offer more active sites while the graphene hydrogel served as a conductive substrate to fix and sputter GQDs to prevent the agglomeration of GQDs and facilitate electron transfer. According to a proposed synthesis mechanism (see Figure 3.1a), graphene quantum dot/graphene hydrogel (GH-GQD) composite samples were prepared and labeled as GH, GH-GQD-45, GH-GQD-90, GH-GQD-180, corresponding to GQD amounts of 0, 45, 90, and 180 mg, respectively.

In the structural characterization of all GH-GQD samples, the typical SEM and TEM images (see Figure 3.1b–e) can not only demonstrate the porous structure of the graphene hydrogel but also show the existence of GQDs with different sizes from 2 to 20 nm in the porous structure. In particular; for GQDs, the HRTEM images (see Figure 3.1f) prove the graphitization of carbon dots while no peaks can be found at ~270 nm in the tested UV-vis absorption spectra. This confirms the presence of rich edges in GQDs.[15,19] When electrochemical characterizations were run at room temperature, the RDE measurements revealed that the GH-GQD composite samples exhibited better ORR activities than GH and graphene oxide due to the synergistic

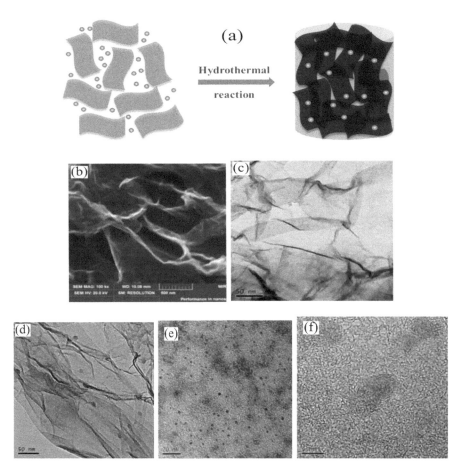

FIGURE 3.1 (a) Illustration of the synthesis of GQDs, (b) SEM image of GQD-GH-90 with a scale bar of 500 nm, and TEM images of (c) GH, (d) GQD-GH-90, and (e) GQD, and (f) HRTEM image (scale bar = 5 nm) of GQDs. (Modified from Wang, M. et al. *Nanoscale* **2016**, *8*, 11398–11402.[18])

effect between GH and GQDs in which the GQDs could offer more active sites while the GH served not only as a conductive substrate to enhance the charge transfer of the electrodes, but also as a good support material to realize the high dispersion of GQDs and their active sites. The GH-GQD-90 sample demonstrated the highest ORR onset potential of −0.13 V among all GH-GQD composite samples, indicating that the optimized GQD amount was 90 mg. When linear sweep voltammetry (LSV) curves were run in O_2-saturated 0.1 M KOH to evaluate the ORR mechanism, it was found that the enhanced electrocatalytic activity resulted from the rich edge defects of the GH-GQD-90 composite sample was related to ORR through both the two-electron and four-electron transfer pathways,[20] evidenced by the tested peroxide percentages and electronic transfer numbers. Moreover, after 1000 CV cycles was run for the GH-GQD-90 with a scan rate of 100 mV s^{-1}, the 1st CV cycle generally matched the

1000th CV cycle, suggesting that the GH-GQD-90 had excellent durability in alkaline solutions, and it should be a superior catalyst for alkaline RMABs. However, the GH-GQD-90 only had a comparable discharge property to commercial 20 wt% Pt/C at higher current densities ($\sim$20 mA cm^{-2}) and no OER was discussed in the study.

To further demonstrate the advantages of hybrid phases in the development of advanced carbon-carbon composite electrocatalysts in RMABs, the Ketjenblack-carbon paper (KB/CP) and Ketjenblack-Super P (KB/SP) composite catalysts were explored for rechargeable Li-O2 battery (RLOB) air-electrodes.[21,22] In their studies, the KB was the same carbon black, called EC-600JD. The ultra-large specific discharge capacities of the KB/CP air-electrode in a RLOB were much greater than those of KB and CP electrode.[21] After the incorporation of SP to KB, the KB/SP could increase O$_2$ transport, improving the usage of the whole electrode volume, resulting in an enhanced weight mass specific capacity ($\sim$1219 mAh g^{-1}).[22] All these results suggested the synergetic effect between two different carbon components, even if the two components have different structural morphologies and physical properties. Cheon et al.[23] also prepared and studied a graphitic nanoshell/mesoporous carbon (GNSL/MC) nanohybrid composite as a bifunctional oxygen electrocatalyst using mesoporous silica (i.e., SBA-15) as a template. Their rechargeable aqueous Na-air battery tests indicated that due to the hybrid effect, GNSL/MC as an air-electrode could exhibit a superior performance to Ir/C and Pt/C catalysts.

Core-shell structures have also been explored in carbon-carbon composite catalysts for RMAB air-electrodes in which one carbon material is the core while the other acts as the shell. This advanced core-shell structure is believed to play a significant role in optimizing and maximizing the performance of two different carbon materials to obtain optimal carbon-carbon composite catalysts. For example, a carbon nanofiber@mesoporous carbon (CNF@mesoCs) core-shell composite catalyst was designed and fabricated by Song and Shin[24] to improve the electrochemical reactions of RLAB electrodes in which mesoporous carbon was coated onto carbon nanofibers (CNFs). This catalyst not only had an enlarged surface area, but also provided a highly conductive graphitized surface to increase electrical conductivity. In a typical experimental route, an electrospinning technique was first used to produce CNFs as a core, which entangled with one another to form a self-standing three-dimensional cross-linked web structure. Nanocasting was then carried out in the second step in which the mesoporous carbon was coated on to the CNFs to form the final CNF@mesoCs core-shell composite material. In the structural characterization, the collected Raman spectra of the CNF@mesoCs and the pristine CNFs demonstrated that the former showed a lower peak ratio of D to G bands than the latter, indicating that the surface of the CNF@mesoCs had a more graphitized structure than that of the pristine CNFs. The investigation of TEM images showed that the CNF@mesoCs exhibited a different structure with an aligned carbon layer ($<$20 nm) due to the coating of mesoporous carbon from pristine CNFs. The CNF@mesoCs displayed a higher electrical conductivity of $\sim$4.638 S cm^{-1} and a larger surface area of 2194 m^2 g^{-1} than the pristine CNFs ($\sim$3.0759 S cm^{-1} and 708 m^2 g^{-1}, respectively). This confirmed the effect of mesoporous carbon onto CNFs. When the electrochemical performance of the RLAB was examined under oxygen atmosphere, it was found in a discharging test using a fabricated RLAB with non-aqueous electrolyte that as the air-electrode, the CNF@mesoCs could give a higher discharge capacity (4000 mA h g^{-1}) than the pristine CNFs (2750 mA h g^{-1}) due to its higher electrical conductivity and larger mesopore size and volume. This research provided a significant new design route to create advanced RLAB air-electrodes.

3.2 COMPOSITE CATALYSTS OF CARBON AND NONMETAL FOR MABs

3.2.1 COMPOSITES OF CARBON AND SINGLE ELEMENT

The composites of carbon and single elements are mainly defined as a pure carbon material composited with a non-metal heteroatom such as nitrogen, boron, or phosphorus. These non-metal heteroatoms can be introduced into pure carbon materials to develop advanced RMAB electrocatalysts via doping strategies because the dopant can act as a secondary phase, improving the electrocatalytic activities of ORR and/or OER.[25,26] Recently, various carbon materials (e.g., porous carbon materials, CNT, graphenes, carbon fibers, carbon xerogel, carbon aerogel, nano-cage carbons, carbon nano-onions, etc.) have been researched in the development of advanced heteroatom-doped carbon materials for RMABs because these carbon materials can exhibit desirable characteristics such as appropriate pore size/volume, large surface area, and high stability.

3.2.1.1 N-Doped Carbons

The doping of heteroatoms is often used to modify the nature and chemical properties of pure carbon materials for generating advanced metal-free carbon catalysts for RMABs. Among these heteroatoms, nitrogen (N) has become the most popular dopant for carbon materials because nitrogen has a larger electronegativity than carbon, a comparable atomic size to carbon, and five valence electrons for bonding with carbon atoms.[27,28] Specifically, N-doped carbon materials can not only generate structural defects and withdraw electrons from carbon atoms and improve mass/ion transfer, but also enhance the conductivity of the resulting carbon materials and improve battery performances when compared with pure carbon materials.[29–32] With respect to this, many carbon materials have been studied using different synthesis techniques in the exploration of heteroatom-doped carbon catalysts as RMAB air-electrodes.

To find affordable, highly active carbon-based RLOB air-electrode catalysts and to design favorable mesoporous structure to locate active sites,[33,34] Sakaushi et al.[35] developed ionic liquid (IL)-based, N-doped mesoporous carbons (meso-NDCs) in which the designed mesoporous structure was used to produce active sites while the composition of carbon and nitrogen was changed and compared in terms of their electrocatalytic activity. In their silica template assisted (Ludox HS40) carbonization method, ionic liquid (N-butyl-3-methylpyridinium dicyanamide) was used as sources of both N and C[33,36] and carbonization temperatures of 900°C and 1000°C were used to prepare two samples: meso-NDC 900 and meso-NDC 1000. In structural characterizations for these two samples, the broad XRD patterns indicated typical characteristics of highly porous materials, and confirmed by TEM images (see Figure 3.2a,b). The pore distribution exhibits an average pore diameter of ca.

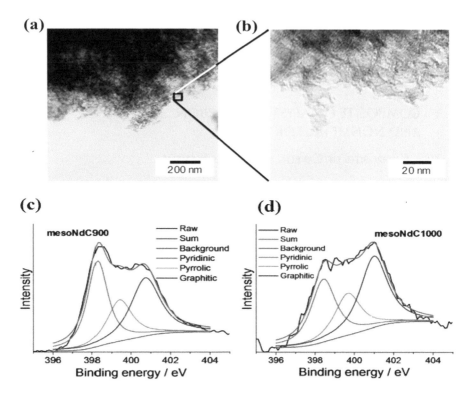

FIGURE 3.2 (a, b) TEM images of meso-NdCs samples. (c, d) X-ray photoemission spectra of mesoNDC900 and mesoNDC1000. The peaks at 398.3, 399.4, and 401 eV are corresponding to pyridinic, pyrrolic, and graphitic nitrogen, respectively. (Modified from Sakaushi, K. et al. *ChemSusChem* **2015**, *8*, 1156–1160.[35])

8 nm, due to defects at the carbon walls or irregular fragments.[37] Importantly, in the analysis of the three N species; pyridinic (398.3 eV), pyrrolic (399.4 eV), and graphitic (401 eV) nitrogen, in N1s XPS spectra (see Figure 3.2c,d), it was found that increasing carbonizing temperatures could result in higher ratios of graphitic nitrogen, confirming the results of XRD patterns in which mesoNDC 1000 showed two typical peaks at (002) and (101) associated with a higher degree of graphitization as compared to mesoNDC 900.

Elemental analysis showed that mesoNDC 1000 had a higher nitrogen content of ca. 12 wt% than mesoNDC 900 ($\sim$6 wt%). When a RLOB cell was assembled using 0.5 M LITFSI + TEGDME (Aldrich) as an electrolyte, the electrochemical characteristics of mesoNDC 1000 were examined by charge-discharge curves in a potential range of 2.0–4.0 V versus Li/Li$^+$ at current densities of 100 and 200 mA g^{-1}. Based on the combination of Ex-situ XRD measurements and charge-discharge tests, mesoNDC1000 was found to have a lower discharge overpotential of only 0.3 V for ORR, while its lowest recharge overpotential was 0.45 V (see Table 3.1) for OER. At current densities of 100 and 200 mA g^{-1}, mesoNDC 1000 also had specific discharge capacities of ca. 1750 and 1280 mAh g^{-1}, respectively, indicating a good rate capability

TABLE 3.1
Comparative Data on the Electrochemical Properties of Several Air-Electrodes for RLOB

Catalyst	η_{OER}[a] (V)	Q_{disc} (mAh g^{-1})	I (mA g^{-1})	Electrolyte
Noble carbon	0.45	1750	100	1 M LiTFSI/TEGDME
Porous Au	0.6	325	500	0.1 M LiClO$_4$/DMSO
Co$_3$O$_4$/RGO	0.6	14,000[c]	140	1 M LiPF$_6$/TEGDME
TiC[b]	0.8	350	250	0.5 M LiClO$_4$/DMSO
Co$_3$O$_4$/carbon	0.9	2000[c]	70	1 M LiPF$_6$/PC
α-MnO$_2$/carbon	1.0	2500[c]	70	1 M LiPF$_6$/PC
Carbon (Super-S)	1.8	850	70	1 M LiPF$_6$/PC

Source: Modified from Sakaushi, K. et al. *ChemSusChem* **2015**, *8*, 1156–1160.[35]

[a] A comparison of average overpotentials (η_{OER}) for several catalysts to fully recharge a RLOB cell. Here, η_{OER} (V) = E_{OER}-2.96. E_{OER} (V vs. Li/Li$^+$) is the average reaction potential of a RLOB cell at recharging, that is, an OER reaction. The value 2.96 (V vs. Li/Li$^+$) is the electromotive force of the following reaction: $O_2 + 2Li^+ + 2e^- \leftrightarrow Li_2O_2$. The η_{OER} data were selected from reports showing full discharge-recharge measurements.

[b] LiFePO$_4$/carbon was substituted for Li metal anode for this measurement to supply Li$^+$. Other measurements used Li metal-based negative electrode.

[c] Q_{disc} was calculated based on the mass of carbon.

of the RLOB. Moreover, using mesoNDC 1000 as an air electrode, a continuous 15th charge-discharge cycling in the tested RLOB under harsh conditions demonstrated better electrocatalytic behavior and cycling performance. This is possibly due to the increased ratio of graphitic nitrogen at higher temperatures, leading to a favorable structural formation of doped nitrogen in mesoporous carbon and thus enhancing electrochemical properties.

In addition to N-doped mesoporous carbon materials designed from ionic liquids, other N-doped carbon materials with mesoporous structures[34,38–41,43,44] were also obtained for RMAB air-electrodes. For instance, a novel hierarchical N-doped carbon ORR catalyst (N:C-MgNTA) with graphitic shells was prepared by using an in-situ templating method, coupled with etching and pyrolysis, and the microporous, mesoporous, and macroporous structures were obtained. It was proposed that the electron-withdrawing ability of N-doping could result in its neighboring carbon being active for O_2 adsorption, facilitating the nucleation of discharge products during ORR.[45] In addition, the active sites tended to stay in the mesoporous structure while the macroporous structure could promote mass transport from these active sites. Importantly, the co-existence of microporous, mesoporous and macroporous structures demonstrated to produce short electron and ion transport paths and enlarge the active surface, benefitting catalytic activity.[46,47] When the surface structure was characterized, the XPS revealed high N-doping (~6 wt%) in which three nitrogen moieties, pyridinic N (~35%), graphitic (~37%), and oxidized environments (~28%), were found. Both pyridinic and graphitic nitrogen atoms had a total 4 wt% in the carbon material. These nitrogen moieties are generally believed to be ORR catalytic active sites for ORR. At the same time, TEM measurements showed two porous size ranges of 0.3–0.6 nm and 5–100 nm, corresponding to micropores, mesopores, and macropores, with the mesopores being lined with highly graphitic shells of 0.33 nm interlayer spacing. These graphitic shells could possibly improve electrical conductivity in the carbon material. Additionally, the tested surface area was very high ($\sim$1320 m^2 g^{-1}), which should favor the ORR air-electrode catalysts for MRAB. Based on the investigation of electrocatalytic ORR in O$_2$- and N$_2$-saturated 0.1 M KOH solutions, it was found that with a 4e$^-$ transfer mechanism, the synthesized N:C-MgNTA exhibited more effective ORR activity and durability than the commercial 20 wt% Pt/C sample or the un-doped carbon reference sample. This result suggests that hierarchical porous carbon materials are an exciting avenue of research as ORR catalysts for RMABs.

Aside from the hierarchical N-doped carbon ORR catalyst with three types of porous structures above, a spherical N-doped hollow mesoporous carbon (HMC) material was also obtained as an efficient RZAB air-electrode by Hadidi et al.,[40] which combined hollow and mesoporous structures with N-doping to obtain a novel carbon bifunctional electrocatalyst via polymerization and carbonization of dopamine on a sacrificial spherical SiO$_2$ template that was then removed by hydrofluoric acid. In their experiment, as-synthesized stöber silica particles were coated with dopamine and triblock copolymer PEO-PPO-PEO (F127). After the polymerization of dopamine, carbonization in Ar at 400°C for two hours and 800°C for three hours was conducted to form the mesoporous carbon coated on silica (silica@mC). An aqueous 48% HF solution was then used to remove the silica, forming the hollow structured HMC material. Both polydopamine (PDA) and F127 acted as a carbon source while PDA afforded the doping of N. The characterization by SEM, TEM and high resolution TEM (HR-TEM) confirmed the formation of hollow mesoporous spheres. Moreover, carbon shells (ca. 21 nm ± 28%) were formed after carbonization, revealing the graphitic and amorphous domains. D band and G band in the Raman spectra further confirmed a mixture of graphitic and amorphous carbon in all carbonized samples. Fourier transform infrared (FTIR) spectra (see Figure 3.3a) also clearly revealed that with the removal of silica, the Si-O-Si stretching feature at 1103 cm^{-1} disappeared from the HMC while a feature at 1570 cm^{-1} could be associated with C-N bending for the silica@mC and HMC samples. To evaluate both ORR and OER, electrochemical measurements were performed in a three-electrode cell configuration with a catalyst coated glassy carbon working electrode, a Pt coil counter electrode, and a Hg/HgO (1 M NaOH, 0.098 V vs. NHE at 25°C) reference electrode. The tested linear sweep voltammograms (LSV), as seen in Figure 3.3b scanning from −0.8 to 0.05 V vs. Hg/HgO in O$_2$-saturated 0.1 M KOH solution at 1600 rpm shows that N-HMC exhibits 54 mV more positive ORR onset potential than that of Pt/C. A further ORR test at different rotation speeds indicated that HMC followed the four-electron transfer in its ORR mechanism. When OER was evaluated, the obtained LSV curves in 0.1 M KOH solution at 1600 rpm (see Figure 3.3c) clearly shows the OER activity of HMC is superior to those of other carbon-based materials as well as 40 wt% Pt/C. When the difference between ORR and OER onset potentials was calculated and compared to assess the overall oxygen catalytic activity, HMC presented a smaller difference (see Table 3.2), indicating a better catalytic behavior. Recently, some biomasses have

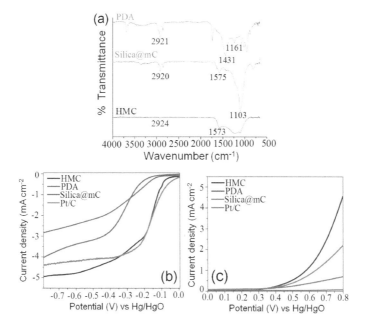

FIGURE 3.3 (a) FTIR spectra for polydopamine (PDA) beads, silica@mC, and HMC, (b) ORR LSV curves for HMC, silica@mC, PDA beads, and Pt/C in an O_2-saturated 0.1 M KOH solution at 1600 rpm, (c) OER LSV curves for HMC, silica@mC, PDA beads, and Pt/C in 0.1 M KOH solution at 1600 rpm. (Modified from Hadidi, L. et al. *Nanoscale* **2015**, *7*, 20547–20556.[40])

been explored to create porous carbon materials for RMABs because biomass are widely accessible, scalable, and recyclable. Wang et al.[48] and Zeng et al.[49] prepared two novel metal-free N-doped porous carbon materials derived from tapioca and nori, respectively. The tapioca-derived N-doped carbon acted as an aluminum-air battery (AAB) air-electrode catalyst while the nori-derived N-doped carbon was used as a

TABLE 3.2

Catalytic Activities as Function of BET Surface Area and N Content for Several Different Oxygen Electrode Catalysts

Catalyst	N Content (at%)	BET Surface Area (m² g⁻¹)	ORR Onset (V)	j_l (mA cm⁻²)	n at −0.7 V	OER Onset (V)	j_l (mA cm⁻²)	$E_{OER}-E_{ORR}$ (V)
HMC	7.08	340	−0.055	−4.95	3.97	0.365	4.53	0.420
Silica@mC	4.34	22	−0.112	−2.85	2.83	0.373	0.68	0.485
PDA beads	4.90	45	−0.165	−4.05	3.47	0.463	0.08	0.628
Pt/C	–	–	0.001	−4.39	4.00	0.430	2.18	0.429

Source: Modified from Hadidi, L. et al. *Nanoscale* **2015**, *7*, 20547–20556.[40]

j_l: Limiting current density. n: number of electrons transferred.

RLOB air-electrode catalyst. Both exhibited decent electrocatalytic activities and stability as well as satisfactory battery performance.

Aside from porous carbon materials in the research of metal-free N-doped carbon electrode for RMABs, other carbon materials such as graphene,[50–54] CNTs,[31,55–58] CNFs,[59–62] and carbon papers[63] have also been explored using the strategy of N-doping to enhance RMAB electrochemical properties, including ORR, OER and battery performance. As an important N-doped carbon material, N-doped graphene (NG) has been studied in the development of advanced metal-free carbon-based bifunctional catalysts for RMABs because graphene possesses intriguing properties such as superior electrical conductivity, large surface area, and excellent mechanical flexibility.[13,64] Among N-doped graphene materials, a special N-doped three-dimensional graphene (N-3DG) material was designed and fabricated as a free-standing air-electrode for RLOBs.[53] The 3DG air-electrode could form a conductive framework with large specific surface and porous structure. Melamine was selected as the nitrogen source due to its high nitrogen content and strong interactions with graphene oxide (GO), thus avoiding severe stacking in the structure of graphene nanosheets (GNS). Based on a typical hydrothermal self-assembly and subsequent annealing process, N-3DG and 3DG materials were obtained. The 3DG was an un-doped G material and was used as a reference sample in the characterization of structures and properties. To determine the effects of melamine in the synthesis, TEM and EDX mapping images of N-3DG, coupled with SEM images revealed a highly porous two-dimensional structure of N-doped graphene nanosheet with a homogeneous distribution of carbon, nitrogen and oxygen elements. Further comparisons of N-3DG and 3DG in N_2 adsorption-desorption analysis indicated that due to the relatively severe stacking of GNSs in 3DG, more mesopores and micropores were formed in N-3DG while 3DG had less. This suggested that N-3DG possibly had more appropriate pore distributions due to the co-existence of mesopores and macropores, favoring fast O_2 diffusion and electrolyte infiltration in comparison with 3DG. XPS measurements were also used to investigate the state of surfaced C and N atoms in N-3DG. The measurements demonstrated that the apparent $C = N$ and C-N peaks reflected from the spectra of C1 s clearly indicated successful N-doping with pyridinic N, pyrrolic N, and graphitic N found in the spectra of N1 s. Here, pyridinic N ($\sim$54%) was regarded as an effective N type to improve ORR in the graphene plane or edge because it could donate more available lone electron pairs for effective oxygen activation.[65] Moreover, according to the first-principle computation,[66] carbon sheets with pyridine N was found to be more thermodynamics favorable in attracting Li^+ and the lithiated pyridinic N could provide excellent active sites for O_2 adsorption and activation in the discharge process of RLOBs. When N-3DG and 3DG were fabricated as free-standing air-electrodes in RLOB tests, N-3DG demonstrated more positive ORR onset potential and cathodic peak current, implying higher ORR activity than 3DG. At three different current densities of 50, 100, 200 mA g^{-1}, the tested discharge capacities of N-3DG were 7300, 5110, and 3900 mAh g^{-1}, corresponding to average operating voltages at 2.67, 2.64, and 2.56 V, respectively, all of which were higher than those of 3DG. Cycling performance measurements with a cut-off capacity of 500 mAh g^{-1} at a current density of 100 mA g^{-1} showed that, when compared to the 3DG, the N-3DG air-electrode exhibited a better cycling performance over 21 cycles with a more stable

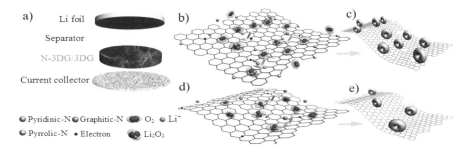

FIGURE 3.4 (a) Schematic illustration for the fabrication components of Li-O$_2$ batteries with free-standing N-3DG and 3DG electrodes. The corresponding ORR mechanism for (b, c) N-3DG and (d, e) electrodes. (From He, M. et al. *Electrochim. Acta* **2016**, *191*, 90–97.[53])

reversible capacity and a higher initial round-trip efficiency of 65.9%. 3DG only performed for about 8 cycles with an initial cycle round trip efficiency of 60.7%. To detect the working mechanisms for N-3DG, a morphological comparison of N-3DG and 3DG electrodes after the first discharge and charge was carried out using a combination of SEM, XRD, and FTIR. It was shown that after the first discharge, the deposited Li$_2$O$_2$ tended to stay on the surface of the N-3DG electrode with a typical toroidal structure,[67] while after charge, the toroidal Li$_2$O$_2$ completely decomposed and the pristine porous structure recovered, confirming the reversible formation and decomposition of Li$_2$O$_2$ in the N-3DG electrode. This finding demonstrated that N-doping was an effective approach not only to enhance the affiliation of Li$^+$ and O$_2$ adsorption but also increase the active sites for promoting the formation of Li$_2$O$_2$ and thus resulted in improvements of discharge specific capacity of RLOB. (see Figure 3.4). It is believed the combination of N-doping and tailored porous structure should be used in the development and exploration of advanced carbon-based cathode materials in various RMABs.

Compared with porous carbon and graphene, one-dimensional carbon nanotubes (CNTs) have been found to be not easily agglomerated and could provide their full surface in electrode applications. This has led to much focus from researchers in developing N-doped CNT (N-CNT) materials as advanced air-electrodes for RMABs. To probe the effects and mechanisms of N-doping in both carbonate-based and ether-based electrolytes in which Li$_2$CO$_3$ and Li$_2$O$_2$ were produced as the main discharge products, Mi et al.[31] synthesized one-dimensional N-CNTs and CNTs samples via a floating catalyst chemical vapor deposition (FCCVD) method in which ethylene and melamine were used as the carbon and nitrogen sources. When both carbonate-based (propylene, carbonate/ethylene, carbonate, PC/EC) and ether-based (1,3-dioxolane/ethylene glycol dimethyl ether) electrolytes were used in a discharge capacity test at a current density of 100 mA g^{-1}, it was found that, compared to CNTs, N-CNTs showed a higher discharge capacity in both electrolytes due to the better dispersion of N-doping and more available sites for O$_2$ adsorption and reduction, as well as a better electrical conductivity facilitating the reduction of kinetics during discharge. Additionally, CV curves in an O$_2$-saturated solution showed a positive shift in the peak potential of N-CNTs in both electrolytes, demonstrating the superior electrocatalytic activity of N-CNTs to that of CNTs. To further evaluate the influences of N-doping on the increase of discharge capacity and ORR kinetics, the porous structure, which resulted in a decreasing surface area from 59.62 to 13.65 m^2 g^{-1} after nitrogen-doping due to the diameter (180–220 nm) of N-CNTs being much larger than that (40–60 nm) of CNTs, was found to have insignificant effect on the increase of discharge capacity for N-CNTs in both electrolytes above. SEM and TEM revealed that a denser population of discharge products on N-CNTs in both electrolytes, especially in the ether-based electrolyte, could result in a better discharge capacity of N-CNTs, while CNTs exhibited some aggregated discharge products on the surface. Therefore, the enhancement of ORR activity after N-doping could be attributed to three main reasons: the improved conductivity, more nucleation sites around nitrogen, and less aggregation of discharge products. To check the structural characteristics of N-CNTs after N-doping, XPS and Raman spectra were also collected. Of the two revealed N species (i.e., pyridinic nitrogen and graphitic nitrogen) observed in the N1 s of the XPS spectra, the pyridinic nitrogen was responsible for the wall roughness and interlinked morphologies of N-CNTs. While in the Raman spectra, an increasing intensity ratio of D band to G band indicated an improved defectiveness of the graphite after N-doping, which matched with the observations from SEM, TEM and XPS spectra. The combined measurements of XRD, Raman spectra and FTIR showed that Li_2O_2 dominated the discharge products when ether-based (1,3-dioxolane/ethylene glycol dimethyl ether) electrolytes were used in the test battery, while Li_2CO_3 was the main discharge products in the case of carbonate-based (PC/EC) electrolytes. Compared to Li_2O_2, Li_2CO_3 was harder to decompose and therefore will cover nucleation sites of the air-electrode in the charge process, resulting in more capacity fading in carbonate-based electrolytes as compared to ether-based electrolytes. This can be seen from Figure 3.5 when cycling performances of CNTs and N-CNTs electrodes were tested. Nevertheless, the capacity fading of N-CNTs was slower than that of CNTs in both electrolytes due to the uniform distribution of discharge products, as well as the improved OER activity, favoring the charge performance of RLABs.[68,69] Corresponding to the tested cycling number, the retained discharge capacities are summarized in Table 3.3, confirming that in N-doped CNTs, N-doping can provide more nucleation sites, promoting a higher dispersion of discharge products, thus improving battery performances in terms of slower capacity fading during cycling.

The studies on the fabrication of N-doped carbon nanofibers (CNFs)[59–62] and N-doped carbon papers (CPs)[63] materials for RMABs showed that CNFs also provide sufficiently large surface areas (50–1000 m^2 g^{-1}) and high electrical conductivity (>200 S cm^{-1}), along with promising porous structure and morphology, while carbon paper could provide a unique texture that benefitted the formation of a 3D open structure with preferential alignment and high porosity. When nitrogen-doped CNF films were used as the air-electrodes in a ZAB operating in ambient air,[59] a high open-circuit voltage (1.48 V) was obtained with a maximum power density of 185 mW cm^{-2} and a maximum energy density of ~776 wh kg^{-1}. The corresponding RZAB showed a small charge-discharge voltage gap (0.73 V at 10 mA cm^{-2}), high reversibility (initial round-trip efficiency of 62%), and stability (voltage gap was increased ~0.13 V after 500 cycles). Moreover, using N-doped CNF films as the air-electrode, flexible all-solid-state rechargeable Zn-air batteries displayed excellent mechanical and cycling

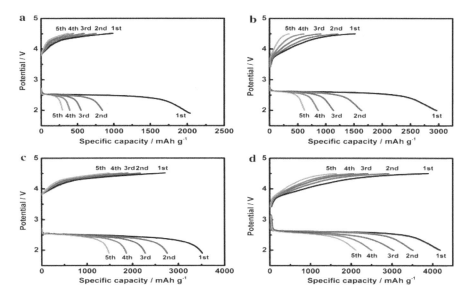

FIGURE 3.5 Cycling performance of CNTs (a, c) and N-CNTs (b, d) electrodes in carbonate-based (a, b) and ether-based (c, d) electrolytes. (From Mi, R. et al. *Carbon* **2014**, *67*, 744–752.[31])

stabilities with a low overpotential (a high discharge and low charge voltage of ~1.0 and 1.78 V at 2 mA cm,$^{-2}$ respectively) as well as a long cycle life (6 hours, which can be recharged by the mechanical replacement of the Zn anode) even under repeated bending conditions. Parallel to this research of nitrogen-doped CNF films as an air-electrode in ZABs and RZABs, N-dope CP air-electrode tested in an assembled RLOB with a non-aqueous electrolyte (1 M lithium perchlorate, LiClO$_4$, in dimethyl sulfoxide, DMSO) displayed a cyclability of more than 30 cycles at a constant current density of 0.1 mA cm^{-2}. The 1st discharge capacity reached 8040 mAh g^{-1} with a cell

TABLE 3.3

Discharge Capacities of CNTs and N-CNTs Air-Electrodes in Carbonate-Based and Ether-Based Electrolytes at Different Cycles

Cycle Number	CNTs in Ether-Based Electrolyte (mAh g⁻¹)	N-CNTs in Ether-Based Electrolyte (mAh g⁻¹)	CNTs in Carbonate-Based Electrolyte (mAh g⁻¹)	N-CNTs in Carbonate-Based Electrolyte (mAh g⁻¹)
1	3516 (100%)	4187 (100%)	2039 (100%)	2963 (100%)
2	2757 (78%)	3517 (84%)	877 (43%)	1636 (55%)
3	2280 (65%)	3042 (73%)	571 (28%)	1141 (39%)
4	1861 (52%)	2507 (60%)	407 (20%)	868 (29%)
5	1486 (42%)	2106 (50%)	305 (15%)	622 (21%)

Source: Mi, R. et al. *Carbon* **2014**, *67*, 744–752.[31]

voltage of around 2.81 V at a current density of 0.1 mA cm^{-2}, whereas the coulombic efficiency was 81% on the 1st cycle at a current density of 0.2 mA cm^{-2}. These research results indicate that N-doped CP materials are promising in the development of low-cost, versatile, paper-based O_2 electrodes for RLOBs.

3.2.1.2   Other Heteroatom-Doped Carbons

As discussed above, nitrogen has been explored to form N-doped carbon electrodes in MABs because the N-doping has the ability to act as an electron donor, increasing the conductivity of carbon and creating ORR active sites by changing the electronic structure of carbon and exposing more carbon edges. Besides N, other heteroatoms such as boron (B), phosphorus (P), sulfur (S), and fluorine (F), have also been selectively used in doped carbon materials for RMABs, as they can also produce different physicochemical properties.

B has a lower electronegativity ($\sim$2.04) than C ($\sim$2.55).[70] and when B is doped to carbon, it can become an electronic acceptor, inducing a certain amount of positive charge and resulting in charged sites (B$^+$) that favor O_2 adsorption and thus ORR activity.[71,72] To study the important effects of B-doping and to create sufficient active sites, the researchers[70,73,74] have focused their attention on the combination of mesoporous carbon structures and B-doping in exploring advanced metal-free carbon catalysts. For example, Shu et al.[73] prepared mesoporous boron-doped onion-like carbon (B-OLC) microspheres for rechargeable sodium-oxygen batteries (RNOBs) using nanodiamond and boric acid as C and B sources. Un-doped onion-like carbon (OLC) and B-doped super P (B-Super P) prepared with the same annealing method, without boric acid and with a replacement of nanodiamond with Super P, respectively, were used as the reference samples. To characterize and examine B-doping and the transformation of the carbon structure, comparisons of HRTEM images and the selected area electron diffraction (SEAD) patterns of B-OLC and OLC samples (see Figure 3.6a,b) reveals the existence of multilayered sp^2 fullerene shells forming a quasi-spherical nanoparticle-like onion with particle sizes of 5–8 nm. Moreover, the appearance of the graphite ring (002), as observed in the SEAD patterns, confirmed the effective conversion of nanodiamond into the sp^2 carbon phase[75,76] with an identified interlayer spacing ($\sim$0.34 nm) that did not change after B-doping. EELS results indicated that the well-defined π^* ($\sim$284.5 eV) and σ^*($\sim$291.9 eV) fine structure features of the carbon K-edge could correspond to the characteristics of sp^2-bonded well-graphitized hexagonal networks[77] while the π^* peak ($\sim$189.5 eV) in the boron K-edge was associated with sp^2 bonding between carbon and boron in a hexagonal carbon/boron structure (BC$_3$).[78,79] Based on B 1 s spectra in XPS, six deconvoluted peaks were obtained at 185.5, 188.8, 190.2, 191.5, 192.9 and 194.8 eV, corresponding to B$_4$C, BC$_3$, BC$_2$O, BCO$_2$ and B$_2$O$_3$,[80,81] respectively. These six species of B were also found in the XPS spectra for the B-Super P sample, confirming the successful doping of boron, while B-super P had a lower content ($\sim$0.83 at%) than B-OLC ($\sim$5.47 at%). Raman spectra demonstrated the effects of B-doping on the improvement of structural defects, and BJH porous analysis showed that both the OLC and the B-OLC had a mesoporous size ($\sim$20 nm) while BET tests showed a small difference in the surface areas between the OLC ($\sim$358 m^2 g^{-1}) and the B-OLC ($\sim$327 m^2 g^{-1}), indicating that B-doping did not affect the mesoporous structure.

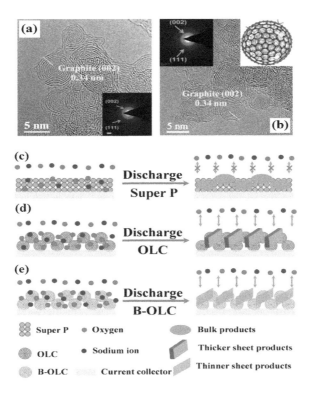

FIGURE 3.6 HRTEM image and SAED pattern (inset) of (a) OLC and (b) B–OLC illustration of the proposed formation mechanism of discharge products on different electrodes (Note: Schematic illustration of the B-OLC structure also shown in inset of (b)). (c) The formation of bulk products on the Super P air-electrode; (d) The formation of thicker sheet products on the OLC air-electrode; and (e) The formation of thinner sheet products on the B-OLC electrode. (Modified from Shu, C. et al. *J. Mater. Chem. A* **2016**, *4*, 6610–6619.[73])

When an assembled RNOBs with B-OLC, OLC, or B-Super P as the air-electrodes cathode were tested, the obtained discharge/charge curves at 0.15 mA cm^{-2} showed that, among the three electrodes, B-OLC could deliver the highest discharge capacity ($\sim$10,200 mA h g^{-1}) and the lowest overpotential, demonstrating higher bifunctional electrocatalytic activity towards ORR and OER than those of OLC and Super P. In the investigation of rate capabilities, it was found that at a high current density of 0.6 mA cm^{-2}, B-OLC presented a discharge capacity of $\sim$7455 mA h g^{-1}. This was nearly 40 times higher than that of Super P ($\sim$160 mA h g^{-1}) and about twice that of OLC ($\sim$3558 mA h g^{-1}). The enhanced rate capabilities could be resulted from the synergistic effects between B-doping, mesoporous structure, and particle size of OLC. Here, B-doping provided more oxygen adsorption sites, benefiting the rate capability, while the mesoporous structure of B-OLC offered substantial electrolyte and oxygen transportation paths, favoring the uniform distribution of electrolyte and O$_2$ inside the electrode and thus improving the rate capability of the battery. However, B-Super P had similar battery performances to Super P because the low B-doping content (~0.83 at%) played no effect on the Super P. In a further examination of the reversibility, cycling performance tests demonstrated that at 0.3 mA cm^{-2}, B-OLC had a better cyclability (up to 125 cycles) than OLC (~56 cycles) and Super P (~6 cycles), indicating the combined effects of B-doping and mesoporous structures. According to the relationship between the morphology and composition of the discharge products being monitored by SEM, the low surface area of Super P could result in the formation of particles with large sizes that limited the nucleation sites. Particles with small sizes were produced and observed on the surface of OLC and B-OLC; however, due to their higher surface area, with a greater density of nucleation sites. Although the OLC and B-OLC samples had similar surface areas, the sheet-shaped products on B-OLC were thinner than those on OLC because of their different surface chemistries. Generally, B-doping inevitably causes the loss of electroneutrality in OLC to create charged sites for O_2 adsorption, acting as nucleation sites for product formation. During O_2 adsorption, the nucleation rate for B-OLC become faster than the growth rate, resulting in a dominant yield of thin sheet-shaped products. As shown in the proposed mechanism (see Figure 3.6c–e), the morphology of the discharge product can significantly influence charge potential,[82] and the uniform dispersion of thin sheets of the discharge products can lead to a low charge voltage due to the increased interactions between the discharge products and the electrode. Therefore, the effective combination of B-doping and mesoporous structure in carbon-based materials can affect the surface chemistry of the air-electrode to control the morphologies of discharge products, resulting in the improvement of RNOB performance.

Although the electronegativity of P (~2.19) is also lower than C (~2.55), P is also an electron donor that can be incorporated into carbon by P-doping easier than N or B due to a much larger covalent radium (107 ± 3 pm) than C (71 ± 1 pm) and B (84 ± 3 pm). According to the studies of density functional theory (DFT)[83,84] on P-doped single walled carbon nanotubes (SWCNTs), P-doping could modify electron transport properties and display affinity towards acceptor molecules (such as O_2), resulting in high electrocatalytic activities. Similar to the combination of B-doping (or N-doping) and porous structures, P-doping was also used in combination with porous structures in the development of metal-free carbon-based bifunctional catalysts for RMABs.[85-87] For example, using a sol-gel polymerization method followed by pyrolysis and P-doping, Wu et al.[85] developed a low-cost and highly active P-doped carbon xerogel electrocatalyst to examine the effect of P-doping and porous structures on ORR. In their synthesis, resorcinol and formaldehyde were used as the carbon sources while H_3PO_4 was used as the phosphorus source. The as-prepared samples were labeled as P-C-1, P-C-2, P-C-3, P-C-4, and P-C-5, corresponding to the weight ratio of 1:10, 2:10, 3:10, 4:10, and 5:10 phosphoric acid to carbonized sample (Co-C), respectively. In the structure characterizations, the combination of XRD and Raman spectra confirmed the existence of increased defects and disordered carbon after P-doping, corresponding to increasing P content from 0.78 at% to 3.56 at% while the P contents (see Table 3.4) tested in inductively coupled plasma-atomic emission spectroscopy (ICP-MS) were 0.78 at%, 1.41 at%, 1.64 at%, 2.77 at%, and 3.56 at% for P-C-1, P-C-2, P-C-3, P-C-4, and P-C-5 samples, respectively. In the P 2p spectra of XPS, two deconvolved contributions at 132.5

TABLE 3.4

P Content, BET Specific Surface Area, Total Pore Volume and Average Pore Size of Carbon Xerogel, and Different P-Doped Carbon Xerogels

Samples	P Content[a] (%)	BET Surface Area (m² g⁻¹)	Total Pore Volume (m³ g⁻¹)	Average Size for Mesopore/Micropore (nm/nm)
Pure carbon	–	332.8	0.13	-/0.45
P-C-1	0.78	695.8	0.36	3.51/0.52
P-C-2	1.41	813.3	0.42	3.66/0.54
P-C-3	1.64	906.5	0.49	3.79/0.55
P-C-4	2.77	972.4	0.58	3.89/0.56
P-C-5	3.56	1166.9	0.77	4.12/0.58

Source: Modified fromWu, J. et al. *Electrochim. Acta.* **2014**, *127*, 53–60.[85]

[a] P content is obtained from ICP-AES.

and 134.5 eV could be assigned to P-C bonding and P-O bonding,[88,89] respectively. The presence of P-O bonding confirmed the formation of P-O-C presented in C_3PO, C_2PO_2, and CPO_3 forms, confirming the successful doping of P.[90] With increasing P content (see Table 3.4), the tested BET specific surface area (see Figure 3.7) was increased from 69.8 to 1166.9 m² g⁻¹ while the average mesopore size was

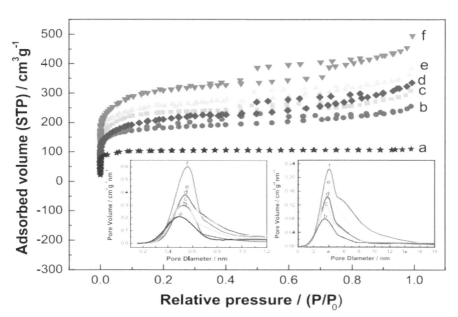

FIGURE 3.7 BET specific surface area and pore size distribution (inset) of carbon xerogel and P-doped carbon xerogels: (a) pure carbon, (b) P-C-1, (c) P-C-2, (d) P-C-3, (e) P-C-4, and (f) P-C-5. (From Wu, J. et al. *Electrochim. Acta* **2014**, *127*, 53–60.[85])

increased from 3.51 to 4.12 nm, along with the increased micropore sizes from 0.52 to 0.58 nm. The increases in surface area and average pore size were attributed to the activation of phosphoric acid in carbon.[91,92] When ORR activity was examined in a standard three-electrode electrochemical cell at room temperature via a rotating ring-disc electrode (RRDE) technique, it was found that after the comparison of onset potential and diffusion limiting current density, the catalytic activity was significantly increased in an order: pure carbon < P-C-1 < P-C-2 < P-C-5 < P-C-4 < P-C-3. Moreover, P-C-3 had a negative shift of ~70 mV in the half-wave potential when compared to 20 wt% Pt/C. The tested LSVs in O_2-saturated 0.1 M KOH with a scan rate of 10 mV s^{-1} at different electrode rotating speeds demonstrated that P-C-3 had an electron transfer number of ~3.81, suggesting that P-C-3 had a similar four-electron reaction pathway in ORR mechanism to commercial 20 wt% Pt/C. Interestingly, when CV was continuously run in Ar-saturated 0.1 M KOH, the half-wave potential loss of P-C-3 after 2000 cycles was 36 mV; lower than that of Pt/C (82 mV), demonstrating the better long-term performance of P-C-3 than Pt/C. This could be possibly due to the strong covalent bond between C and P, as well as the absence of metal agglomeration and migration. According to the studies of density function theory DFT,[93,94] P-doping induced charged sites P$^+$ and asymmetric spin density in carbon atoms should be the most likely active sites for ORR. Moreover, P-doping resulted in high edge exposure and active sites on the high specific surface area of P-doped carbon. Therefore, the amount of P-doping should be critical for the improvement of ORR in P-doped carbon xerogels, as confirmed by that the ORR activity was increased when the P content was increased from 0.78 at% (P-C-1) to 1.64 at% (P-C-2) but decreased when the P content continued to increase to 2.77 at% (P-C-4) and 3.56 at% (P-C-5). In this research, the optimal P content was found to be1.64 at%, resulting in maximum ORR activity.

To obtain the best electrocatalytic activity and stability/durability of the bifunctional catalysts for RMAB, a comparison of different types of heteroatom-doping is crucial in developing heteroatom-doped carbon-based catalysts. Su et al.[74] compared the activities of N, B, and P single heteroatom-doped ordered mesoporous carbons (OMCs) and the effects of doping. They found that improvements to charge transfer kinetics strongly depended on the nature of the heteroatom. At a doping level below 1 at%, the ORR activity tested in alkaline solution increased in the order of: N-OMCs < P-OMCs < B-OMCs, while no data was reported for their stability. There was also research conducted on heteroatom(s) doped graphdiyne[95] in which N, S, B, and F doped-graphdiyne were synthesized. This research focused on the comparison between single heteroatom and dual atoms-doped graphdiyne electrode materials, which will be discussed in the next section.

3.2.2 COMPOSITES OF CARBON AND DUAL ELEMENTS

Dual heteroatoms, introduced into carbon materials to form dual heteroatom-doped carbon composite materials, have become attractive in the development of advanced metal-free carbon-based air-electrode in RMABs. Compared with single heteroatoms, two different heteroatoms can produce a larger asymmetrical spin and charge density, resulting in the further enhancement of electrocatalytic activity and stability/durability.[96,97] Compared to single heteroatom-doped carbon materials whose ORR activities are still inferior to that of conventional Pt/C catalysts, especially in acidic media and neutral solutions. Doped carbon materials with two or more selected heteroatoms have been predicted to further improve the ORR catalytic activity due to the synergetic effects arising from the co-doping of two or more heteroatoms in carbon.[98]

For example, co-doping carbons with two heteroatoms, such as N-doped carbon doped with B, S, or P, was found to be able to modulate electronic properties and surface polarities to further increase ORR activity.[99–102] In this regard, co-doped N and P into carbon materials[103–105] were easily prepared to form porous co-doped carbon air-electrode for MABs due to the same number of valence electrons and similar chemical properties between P and N. To investigate the OER of the doped-carbon materials, Zhang et al.[103] conducted a scalable fabrication of three-dimensional N and P co-doped mesoporous nanocarbons (NPMC foams) using a template-free method (see Figure 3.8a) with a pyrolysis process of polyaniline (PANI) aerogels synthesized in the presence of phytic acid. According to the template-free method, three N, P co-doped mesoporous carbon (NPMC) foams, NPMC-900, NPMC-1000, and NPMC-1100, were prepared with corresponding to annealing temperatures of 900, 1000, and 1100°C, respectively, while a referenced N-doped mesoporous carbon foam (NMC-1000) was prepared via the same method with an annealing temperature of 1000°C. N, P co-doped mesoporous carbon (NPC-1000) prepared at an annealing temperature of 1000°C through the same template-free method without the middle process of the formation of PANI aerogel was also used as the further reference samples. TEM images coupled with elemental mapping for a typical NPMC-1000 sample revealed the uniform distribution of C, N and P while a combination of XRD, Raman spectra, and TEM indicated that pyrolysis resulted in the majority of the thermally stable domains of the graphitic carbon being occupied with the co-dopants of N and P from the PANI and phytic acid. Particularly, many edge-like graphitic structures were found in the HRTEM images of the examined NPMC-1000 sample, which oversaw active sites and thus catalytic activity, while the BET showed surface areas of 635.6, 1548, and 1663 m^2 g^{-1} for the NPMC-900, NPMC-1000, and NPMC-1100 samples, respectively. Further, BJH analysis indicated the presence of mesopores with diameters below 10 nm and significantly enhanced pore volumes ranging from 0.17 cm^3 g^{-1} for pure PANI aerogel to 0.50, 1.10 and 1.42 cm^3 g^{-1} for NPMC-900, NPMC-1000 and NPMC-1100, respectively. This confirmed that the NPMC samples possessed three-dimensional mesoporous structures with large surface areas, high pore volumes, and suitable pore sizes for electrocatalytic activity. In further examining the presence of N and P after doping, the deconvolution of N1 s in XPS spectra of NPMC samples revealed pyridinic, pyrrolic N, graphitic, and oxidized pyridinic N species, confirming the existence of N and indicating a conversion from pyrrolic to graphitic nitrogen with increasing temperatures. When compared to the two deconvoluted peaks in the P 2p of pure PANI aerogel, the two deconvoluted peaks of all the NPMC samples showed a slight binding shift to lower energy due to the gradual dehydration and condensation of phosphoric groups into polyphosphates in the synthesis, corresponding to P-C and P-O bonds.[92,106] Both N and P results revealed by XPS confirmed the successful doping of N and P heteroatoms into the carbon network through thermal pyrolysis. To evaluate the ORR and OER of the NPMC samples, NMC-1000 and NPC-1000 samples were used with RuO_2 and commercial Pt/C (20 wt%, ETEK) as references in electrochemical measurements. When CV tests were run in O_2-saturated 0.1 M KOH solution, NPMC-1000 showed comparable electrocatalytic activity to commercial Pt/C and higher than the other NPMC samples. Results also indicated that 1000°C was the temperature limit because overheating temperatures at 1100°C could result in decomposition of the dopants and the decrease of electrocatalytic activity. Based on further LSV curves (see Figure 3.8b) obtained at 1600 rpm in O_2-saturated 0.1 M KOH solution with a scan rate of 5 mV s^{-1}, it was confirmed that among all carbon-based samples, NPMC-1000 had the highest ORR activity with a comparable onset potential ($\sim$0.94 V), half-wave potential ($\sim$0.85 V), and limiting current density ($\sim$10^{-6} mA cm^{-2}) to those of Pt/C. Although NPMC-900 has a higher N and P content than NPMC-1000, the relatively low pyrolysis temperature possibly led to a high charge-transfer resistance and relatively poor electrocatalytic activity, while for NPMC-1100, the removal of doped heteroatoms at higher temperature ($\sim$1100°C) resulted in the decrease of active sites and then overall electrocatalytic activity. After an analysis of the peroxide species using a RRDE technique, NPMC-1100 demonstrated an electron transfer number above $\sim$3.85, suggesting a four-electron pathway in its ORR. With the use of alkaline (e.g., 1 M or 6 M KOH solution) and acidic electrolytes (e.g., 0.1 M HClO$_4$), similar RRDE tests confirmed that the NPMCs possessed ORR activities comparable to, or even better than Pt/C in alkaline media, and that the NPMCs showed could give comparable ORR activity in acidic environments. Additionally, as seen in the LSVs of Figure 3.8c in which the anodic current is rapidly increased above $\sim$1.30 V and associated with OER, the NPMC-1000 and NPMC1100 samples display lower onset potentials and higher currents, suggesting better OER activities than Pt/C. When compared to RuO_2 nanoparticles, the NPMC-1000 electrode exhibits lower onset potentials, although its current densities are lower at higher potentials. After their ORR and OER activities were evaluated, the NPMC samples were tested as the air-electrode catalyst in both primary and rechargeable ZABs. The primary ZAB (see the design in Figure 3.8d) demonstrates an open-circuit potential of 1.48 V, a specific capacity of 735 mAh g$_{Zn}$$^{-1}$ (corresponding to an energy density of 835 Wh kg$_{Zn}$$^{-1}$), a peak power density of 55 mW cm^{-2} (at a current density of $\sim$70 mA cm^{-2}). Figure 3.8e shows a stable operation of 240 hours after mechanical recharging, while the rechargeable ZAB could be cycled stably for 180 discharge/charge cycles over a period of 30 hours at 2 mA cm^{-2}. As revealed by DFT calculations, these tested battery performances confirmed the synergistic effects of N, P co-doping and graphene edge on the improvement of bifunctional electrocatalytic activity towards both OER and ORR.[107] This was due to the combination of tailored and optimized co-doping of N and P, and the active sites at the edge of the graphene resulted in the optimal NPMC-1000 sample in this research. Additionally, active sites were reported to be more inclined to be located at the graphene edge than at the mesoporous structure while the mesoporous structure could benefit the ionic and electronic transport,

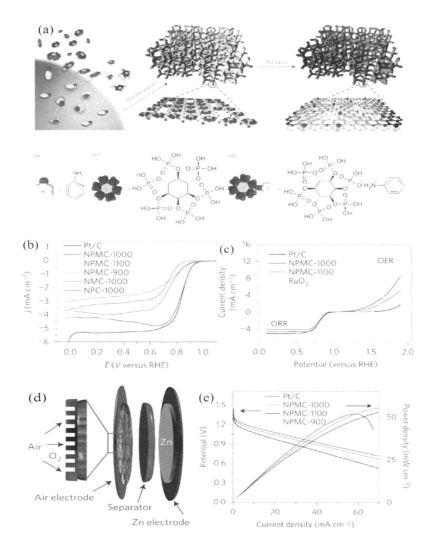

FIGURE 3.8 (a) Schematic illustration of the preparation process for NPMC foams with aniline (i), phytic acid (ii), and aniline-phytic acid complex (iii); (b) linear scan voltammogram (LSV) curves for NPMC-900, NPMC-1000, NPMC-1100, NMC-1000, NPC-1000, and commercial Pt/C catalyst at an RDE (1600 rpm) in O_2-saturated 0.1 M KOH solution with a scan rate of 5 mV s^{-1}; (c) LSV curves of NPMC-1000, NPMC-1100, RuO$_2$, and commercial Pt/C catalysts on an RDE (1600 rpm) in 0.1 M KOH with a scan rate of 5 mV s^{-1}, showing the electrocatalytic activities towards both ORR and OER; (d) schematic illustration of the basic configuration of a primary Zn-air battery in which a carbon paper pre-coated with NPMC is used as an air cathode and is coupled with a Zn anode and a glassy fiber membrane soaked with aqueous KOH electrolyte as the separator, the enlarged part illustrates the porous air electrode loaded with electrocatalyst, which is permeable to air; (e) polarization and power density curves of primary Zn-air batteries using Pt/C, NPMC-900, NPMC-1000, NPMC-1100 as ORR catalyst (mass loading of 0.5 mg cm^{-2}) and 6 M KOH electrolyte (scan rate, 5 mV s^{-1}). (Modified from Zhang, J. et al. *Nat. Nanotech.* **2015**, *10*, 444–452.[103])

providing shorter conducting paths when compared to other porous structures such as macropores and micropores.

Compared to N, P co-doping, N, B co-doping into carbon materials seems more difficult because, during doping, the formation of byproducts (e.g., hexagonal boron nitride, h-BN) can act as inert agents to reduce or stop the electrochemical activity of the catalyst.[108] Therefore, there are a few researches into the preparation of N, B co-doped carbon materials for MAB cathodes. To incorporate heteroatoms at selected sites of the graphene framework and to prevent the formation of inactive byproducts, Zheng et al.[5] prepared a B, N co-doped graphene (B,N-graphene) catalyst using a two-step doping strategy in which graphene was selected as the carbon matrix because co-doped graphene could display stronger doping effects and become more catalytically active than the single-doped graphene.[99,109,110] In their two-step experiment, N-doping was first carried out with NH_3 as the N source. B-doping was then carried out with H_3BO_3 as the B source. To examine the effects of co-doping, structural characterizations using a combination of TEM, SEM, EELS, and elemental mapping confirmed the uniform distribution of B and N in the preserved nanosheets while FTIR confirmed the existence of B and N coupled to C, indicating that byproducts such as h-BN did not exist in the two-step synthesized B, N-graphene sample, which was further confirmed by XPS. A one-step synthesized h-BN/graphene reference sample clearly showed the h-BN phase, suggesting a h-BN/graphene hybrid phase. The analysis of D and G bands in Raman spectra revealed that, compared to one-step synthesized N-graphene, the two-step synthesized B, N-graphene had a higher intensity ratio of D band to G band (i.e., I_D/I_G), indicating higher defects after the co-doping of B and N, while the hybrid h-BN phase in the composite resulted in a higher ratio of one-step synthesized h-BN/graphene than that of B, N-graphene. When the comparison of CV curves in an O_2-saturated 0.1 M KOH was carried out, the B, N-graphene showed a higher cathodic current density, indicating a better ORR activity of B, N-graphene than that of h-BN/graphene. As a series of LSVs tested in O_2-saturated 0.1 M KOH solution were collected using a RDE technique, it was found that the onset potential of B, N-graphene was closer to commercial Pt/C and higher than one-step synthesized single heteroatom-doped graphenes (such as N-graphene and B-graphene). But N-graphene possessed higher onset potentials than B-graphene, suggesting that the co-doping of B and N could result in higher synergistic activities than single-heteroatom-doped graphene, and that N-doping in N-graphene was more effective in the improvement of activity than B-doping in B-graphene. RDE experiments at different rotation speeds ranging from 500 to 2500 rpm revealed that B, N-graphene had a similar ORR electron transfer number ($\sim$3.97) to commercial Pt/C ($\sim$3.98), confirmed by RRDE measurements. At a potential range from -0.4 to -0.8 V, RRDE monitored the yield of HO_2^- (i.e., the intermediate peroxide species during the ORR process), showing below 10% of HO_2^- production during ORR catalyzed by B, N-graphene and 30% for h-BN/graphene. DFT calculations were used to analyze the origin of the synergistic effect. Based on the coupling interactions between pyridinic N (one of N species) and B, the mechanism could be drafted as follows: first, the C atom located between the dopants of N and B was polarized by N. The 2P orbital of the polarized C atom could then donate extra electrons to an adjacent B atom, resulting in the "activated" B atoms. With increasing numbers of activated B atoms, the electron occupancy of the 2p orbital of this activated B atom became more favorable to the adsorption of and bonding with HO_2^-. Therefore, this charge-transfer process could induce a synergistic coupling effect between N and B dopants, in which N acted as an electron-withdrawing group to indirectly activate B, resulting in the enhancement of ORR activity. The theoretical results from DFT calculations corresponded well with the experimental data, confirming the effectiveness of the two-step doping strategy in preparing novel B, N-graphene catalysts for RMABs.

Like N and P, oxygen (O) has also higher electronegativity than carbon. O has also been used in the fabrication of co-doped carbon materials with N. Despite few literature related to N, O co-doped carbons for RMABs, a novel 3D O, N co-doped carbon nanoweb (ON-CNW) material was developed by Li and Mathiram[27] as a metal-free catalyst for hybrid Li-air batteries (HLABs). In their facile synthesis, a self-synthesized polypyrrole (PPY) nanoweb was annealed at 900°C to form a N-doped carbon nanoweb (N-CNW) material. After further activation using KOH solution, the N-CNW was heated at 600°C for 0.5 hours to obtain the ON-CNW material. A reference N-doped carbon nanosphere (N-CNS) material was synthesized by annealing a self-synthesized PPY nanosphere in the same route as ON-CNW. In the characterization of structure, SEM and TEM indicated that in the activation process, the maximum mass ratio of KOH to N-CNW was fixed as 2:1 because excessive KOH can result in the collapse of the (002) planes of graphite, thus the loss of the web morphology. Based on the deconvolution of the N1s and O 1 s spectra in XPS, the content of pyridone/pyrrolic on ON-CNW could be found, as shown in Table 3.5.

From Table 3.5, it can be seen that the content of pyridone/pyrrolic on ON-CNW is about three times that of N-CNW, resulting in higher carbonyl functional groups (from pyridone). The pyridone nitrogen was reported to be able to stabilize singlet dioxygen by forming a stable adduct, weakening and breaking the bond in the oxygen molecule,[111,112] so that the activation using KOH could further improve ORR activity. When CV measurements were run in O_2- or N_2-saturated 0.1 M KOH, only a single ORR peak in O_2-saturated 0.1 M KOH displayed a lower peak potential (~0.4 V) for N-CNS than that (~0.29 V) for N-CNW, indicating the different activities induced by

TABLE 3.5

N1 s Compositional Analysis from XPS Spectra

Catalyst Sample	N Content (wt.%)	Pyridinic N (Atom %)	Pyridone N/ Pyrrolic N (Atom %)	Quaternary N (Atom %)	Pyridinic N-O (Atom %)
N-CNS	6.71	28.72	9.62	44.5	17.16
N-CNW	8.70	36.3	7.05	45.11	11.54
ON-CNW	4.72	32.16	20.14	36.87	10.83

Source: Modified from Li, L. and Manthiram, A. *Adv. Energy Mater.* **2014**, *4*, 1301795.[27]

the different morphologies and structures of these two PPY-derived carbon materials. N-CNS possessed a high particle resistance and reduced the mass transport in packed carbon spheres while the 3D interconnected N-doped carbon fibers in N-CNW benefitted electron transport, oxygen transport, and electrolyte diffusion. In a further comparison of the ORR peak potentials of ON-CNW and N-CNW, ON-CNW presented a more positive shift, probably due to the higher pyridone/pyrrolic content (~20.14%) than N-CNW (~7.05%). According to the measured Koutechky-Levich plots (see Figure 3.9a) using a RDE technique, the ORR electron transfer numbers of N-CNS, N-CNW, and ON-CNW were calculated to be 2.2, 3.3, and 3.7, respectively, suggesting that ON-CNW tended to follow a four-electron transfer mechanism. Interestingly, when the performances of N-CNW and ON-CNW samples were evaluated as an air-electrode in an assembled hybrid Li-air cell, the discharge voltage profiles of the cell revealed an apparent decrease in the difference between ON-CNW and Pt/C. The increased current densities reflected the excellent catalytic activity of ON-CNW. At current densities of 0.5, 1.0, 1.5, and 2.0 mA cm^{-2} (see Figure 3.9b), the discharge plateau of ON-CNW was lower than that of Pt/C but higher than those of N-CNW and acetylene black, suggesting that ON-CNW has higher ORR activity than N-CNW and carbon black. After a compared cycling test was carried out in the hybrid Li-air cell with a constant current density of 0.5 mA cm^{-2}, the round-trip overpotential for N-CNW was increased from 1.00 to 1.43 V while the round-trip overpotential for ON-CNW was increased from 0.92 to 1.02 V, demonstrating the

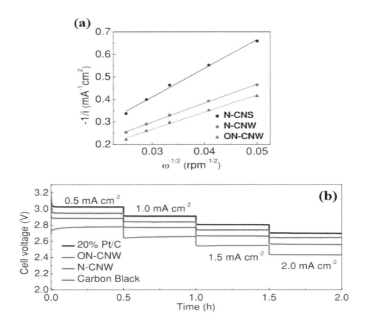

FIGURE 3.9 (a) Discharge voltage profiles of hybrid Li-air cells with Pt/C, N-CNW, ON-CNW, or carbon black as the ORR catalyst at different current densities. Cycling performance of hybrid Li-air cells with (b) N-CNW or (c) ON-CNW as the ORR catalyst. (Modified from Li, L.; Manthiram, A. *Adv. Energy Mater.* **2014**, *4*, 1301795.[27])

improved activity and stability induced by co-doping of O and N, and confirming ON-CNW as a promising ORR catalyst for hybrid Li-air cells.

As discussed above, co-doping N and other high-electronegativity elements such as B, P, or O to carbon to form co-doped catalysts can give higher ORR catalytic activities due to unique electronic structures coupled with synergistic coupling effects between heteroatom dopants than corresponding single atom doped counterparts. The co-doping of N and S[113–117] has been well documented to have effect on the redistribution of spin and charge densities, favoring the formation of more ORR active sites, regardless of the fact that S has similar electronegativity to carbon. Among N, S co-doped carbon materials, a certain amount of mesoporous distribution and pyridinic nitrogen on the edges have been considered to be two striking factors to enhance ORR activity.[5] To obtain simultaneous optimization of both porous structure and surface functionalities of N and S co-doped carbon, Wu et al.[113] designed and fabricated a polyquaternium derived heteroatom (N and S) co-doped hierarchically porous carbon (N-S-HPC) catalyst for ORR via a simple, large-scale and green synthetic route. To avoid the environmental impact related to hydrofluoric acid and ammonia, they not only combined a simple silicate templated two-step graphitization of impregnated carbon derived from nitrogen-enriched polyquaternium-2 (PQ-2) without needing any NH_3 activation, but also replaced hydrofluoric acid with sodium hydroxide to remove silica in the product. For experimental comparisons, two reference samples, N-HPC and N-S-PC, were prepared using the same synthesis routes without the S-containing agent and silica template. In the investigation of morphology and structure of N-S-HPC, a combination of SEM and TEM revealed that, with a shell thickness of about 1 nm, the N-S-HPC could exhibit a sponge-like structure with a disordered mixture of mesopores (~20 nm) and macropores (50–100 nm). This was also confirmed with XRD and BJH. Two broad diffraction peaks (23° and 43°) corresponded to the (002) and (004) planes of carbon with low graphitization, indicating an entirely amorphous structure. BJH analysis in N_2 adsorption-desorption plots revealed the mesoporous sized, centered at 10, 15, and 28 nm while some pores were around 1 nm, suggesting the presence of micropores. Correspondingly, N-S-HPC possessed a high BET surface area (~1201 m^2 g^{-1}) that was higher than that of N-S-PC (630 m^2 g^{-1}) and confirmed the effects of the silica template. This result also suggested that the hierarchical porous structure should be composed of micropores, mesopores, and macropores, in which the mesoporous structure was the dominant porous structure. In further evaluations of the co-doping of N and S using XPS, the N1 s associated with graphitic N and pyridinic N species indicated two peaks at 163.9 eV and 165.1 eV in the fitting of S 2p were assigned to S $2p_{3/2}$ and S $2p_{1/2}$, respectively, corresponding to thiophene-S due to their spin-orbit coupling.[118] A comparison between the N-HPC and N-S-HPC samples showed that the latter had a higher graphitic N content (~78%) than the former (~59.2%), suggesting that the co-doping of N and S could not only improve the porous structure but also dramatically modify the surface functionalities, favoring the enhancement of ORR activity. When electrochemical measurements were performed to examine the ORR activities of the N-S-HPC catalyst, it was found that in the LSV curves, obtained in O_2-saturated 0.1 M KOH at 1600 rpm with a scan rate of 5 mV s^{-1} and a catalyst loading of 500 µg cm^{-2}, N-S-HPC could give a higher onset potential (~0.99 V) and a higher half-wave potential (~0.86 V) than N-HPC and N-S-PC samples, demonstrating its higher ORR activity. In particular, even at a low catalyst loading of 100 µg cm^{-2}, N-S-HPC could still exhibit a higher half-wave potential (~0.83 V) than that of commercial 20 wt% Pt/C catalysts (~0.82 V), indicating that N-S-HPC had a comparable ORR performance to Pt/C in alkaline solution. Moreover, further ORR tests for N-S-HPC revealed an electron transfer number of ~3.97, indicating a complete reduction of oxygen to water, which is similar to the same four-electron oxygen reduction catalyzed by Pt/C. Interestingly, instead of 0.1 M KOH solution, an acidic medium (i.e., 0.5 M H$_2$SO$_4$ solution) was also used in the evaluation of the ORR performance of N-S-HPC. The CV curves in O$_2$-saturated 0.5 M H$_2$SO$_4$ solution demonstrated that N-S-HPC had a slightly lower ORR activity than Pt/C in acidic solution, as evidenced by the half-wave potential (~0.73 V) compared to the half-wave potential (~0.78 V) of Pt/C. However, N-S-HPC in acidic solution still gave a four-electron ORR mechanism and also showed remarkable stability evidenced by a small negative shift of about 30 mV for the half-wave potential after 5000 cycling. To demonstrate the potential of N-S-HPC catalysts in MABs, a Zn-air battery with 6 M KOH was assembled for further electrochemical measurements using N-S-HPC as the air-electrode. At 1.0 V, N-S-HPC reached a high current density of 317 mA cm^{-2}. This was much higher than that of Pt/C (~103 mA cm^{-2}). The peak power density using N-S-HPC was ~536 mW cm^{-2}, also significantly superior to Pt/C (~145 mW cm^{-2}). Both performances in the Zn-air battery demonstrated that N-S-HPC outperformed state-of-the-art Pt-based catalysts, suggesting that the co-doping of N, S could result in an optimized combination of porous structures and surface functionalities of N and S, and thus the enhancement of electrocatalytic activity.

Compared to the carbon material derived from nitrogen-enriched polyquaternium-2 (PQ-2) for preparation of the N-S-HPC air-electrode, the use of recycled bio-waste to produce non-precious metal carbon-catalysts for MABs is an innovative way to combine biomass engineering and the electrocatalytic field.[48,49,115] Electrochemical measurements of sweet potato vine-derived porous carbon co-doped with N and S by Gao et al.[115] were carried out in 0.1 M KOH solution, the result showed a superior electrocatalytic ORR activity and durability when compared to commercial Pt/C. This was due to the combined effects induced by in-situ N and S co-doping, providing porous graphitic structures along with a high surface area and excellent conductivity. These biomass-derived carbon-based electrocatalysts can not only be used in MABs and fuel cells but also in supercapacitors, providing a new avenue for the development of high-performance multi-functional carbon-based electrochemical materials for large-scale applications.

In the studies focused on carbon materials co-doped with N and another element (e.g., B, P, O, or S) that exhibited higher synergistic effect on activity than corresponding N-doped carbon materials, Zhang et al.[95] carried out an experiment in synthesizing and comparing three different co-doped carbon catalysts: N, S co-doped graphdiyne (NSGD), N, B co-doped graphdiyne (NBGD), and N, F co-doped graphdiyne (NFGD), respectively. The selection of graphdiyne (GD) as the carbon source is attributed to its special structure in which each benzene ring is connected to six adjacent benzene rings through two acetylenic bonds, resulting in a flat 2D structure.[119] After a series of electrochemical measurements in an assembled primary Zn-air battery, the ORR activity of NFGD was found to be comparable to that of commercial Pt/C (20 wt%

Pt on Vulcan XC-72) and much better than those of the other two dual-doped GDs. In particular, NFGD possessed better stability as well as higher tolerance to methanol crossover and CO poisoning effects than commercial Pt/C. Except for co-doped carbon materials with N as a dopant for MABs, the first principle study was also carried out to theoretically demonstrate the synergistically enhanced catalytic effect of B, P co-doping in B, P co-doped graphene when compared with B-doped graphene and P-doped graphene. Interestingly, a yolk-shell N, P, B ternary-doped biocarbon derived from yeast cells[120] was recently found to exhibit enhanced ORR activities due to the optimized content of N, P, B co-doping. Using metal as one of the dopants and other co-doped carbon materials for MABs will be discussed in the next subsection.

3.3   COMPOSITES OF CARBON AND METAL

3.3.1   COMPOSITES OF CARBON AND NOBLE METAL OR NOBLE METAL-ALLOY

3.3.1.1   Composites of Carbon and Pt or Pt-Alloy

As a representative noble metal, Pt is well known in the catalytic field, especially in the application of fuel cells. The composites of carbon and Pt have been commonly utilized as carbon-supported Pt fuel cell electrocatalysts towards ORR, in which different carbon materials can act as high-surface area substrates for proper dispersion of Pt nanoparticles to form carbon-supported Pt-based catalysts. These catalysts offer the highest catalytic activity of all carbon-supported pure metals.[13,121] However, it has been well documented that Pt/C alone is not an efficient bifunctional electrocatalyst due to its poor activity towards OER in MABs.[122–124] With increasing interests in MABs due to its higher energy densities, as compared to current rechargeable batteries, composites of carbon and Pt have been used to develop and explore novel bifunctional composite catalysts for MABs.

To address the primary challenges of OER, metals such as Au,[125–127] Co,[128] Zn,[129] Ir,[130–132] Pd,[132,133] and Ru[131] have been used to form carbon-supported Pt-alloy catalysts in various MABs. Because Au can have an active effect on the surface modification of Pt, Lu et al.[125] combined Au and Pt onto the surfaces of carbon nanoparticles and evaluated the ORR and OER activities of PtAu/C in Li-O_2 cells. Using $HAuCl_4$ and H_2PtCl_6 as Au and Pt sources, they first prepared PtAu nanoparticles and then loaded these nanoparticles onto Vulcan XC-72 (VC) to form a 40 wt% PtAu/C catalyst. The observed TEM images revealed a uniform distribution of PtAu nanoparticles ($\sim$6.8 $\pm$ 1.4 nm) on carbon while the XRD data (see Figure 3.10a) indicated a solid-solution composed of Pt and Au corresponding to $Pt_{0.5}Au_{0.5}$. When the CV curves (see Figure 3.10b) were tested for PtAu/C, the calculated electrochemical surface area (ESA) of Pt and Au was 38 $\pm$ 4 m^2 g$^{-1}_{PtAu}$, which was in reasonable agreement with the dispersion of PtAu observed by TEM. Under further investigations of ORR and OER activities in an assembled Li-O_2 cell, comparisons of first discharge and charge voltages, as in Figure 3.10c, show that the discharge voltages of 40 wt%PtAu/C are comparable to those of 40 wt% Au/C, while the charge voltages of 40 wt%PtAu/C are comparable to those of 40 wt%Pt/C. Moreover, 40 wt%PtAu/C as the air-electrode displays a higher round-trip efficiency than pure carbon, even though they possess a similar specific capacity in the first cycle ($\sim$1200 mA g$^{-1}_{carbon}$). This suggests the incorporation of Au into Pt surfaces can

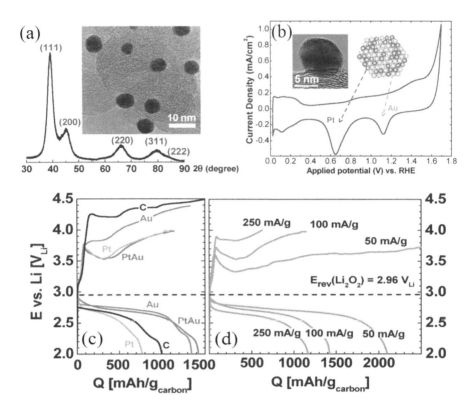

FIGURE 3.10 (a) TEM image (top right) and XRD data of 40 wt%PtAu/C, (b) cyclic voltammograms of 40 wt%PtAu/C collected in Ar-saturated 0.5 M H_2SO_4 between 0.05 and 1.7 V vs RHE (room temperature and 50 mV/s). Insets: (left) HRTEM image of 40 wt%PtAu/C and (right) schematic representation of Pt-Au, with arrows indicating the CV signatures for Pt and Au, (c) Li-O_2 cell first discharge/charge profiles of carbon at 85 mA g_{carbon}^{-1} and of 40 wt%Au/C, 40 wt%Pt/C, and 40 wt%PtAu/C at 100 mA g_{carbon}^{-1}, (d) Li-O_2 cell discharge/charge profiles (first cycle) of 40 wt%PtAu/C at 50, 100, and 250 mA g_{carbon}^{-1}. (From Lu, Y.-C. et al. *J. Am. Chem. Soc.* **2010**, *132*, 12170–12171.[125])

enhance both the ORR and OER kinetics; however, with increasing charging cycles, the charge voltages of 40 wt% PtAu/C are lower than those of 40 wt% Pt/C. When the current densities are decreased, the discharge and charge voltages are considerably reduced (See Figure 3.10d). This research demonstrates that PtAu/C is responsible for both ORR and OER enhancements after the incorporation of Au into Pt atoms on the nanoparticle surface, presenting a reasonable strategy to develop bifunctional catalysts for MABs. To gain deeper understanding and to find the effects of Pt and Au on both ORR and OER, a Pt-Au alloy nanoparticle catalyst supported on carbon black (Vulcan XC-72) was also prepared by Yin et al.[126] They demonstrated that PtAu/C could play an important role in the charge/discharge performance of rechargeable LOBs owing to their nanoscale alloying and phase properties.

Similar to Au, Co was also used to form Pt-Co alloy nanoparticles supported on Vulcan XC-72 carbon for LABs. To fabricate Pt_xCo_y/Vulcan XC-72 catalysts with a loading of 20 wt%, Su et al.[128] conducted a chemical reduction method to deposit a series of Pt_xCo_y (x: y = 4, 2, 1, and 0.5) alloy nanoparticles onto Vulcan XC-72, using $H_2PtCl_6 \cdot 6H_2O$ and $CoCl_2 \cdot 6H_2O$ as the Pt and Co sources. In structural characterizations, the XRD patterns showed that in the solid-solution of Pt-Co, all the Pt_xCo_y samples had sharp peaks, matching the previously reported face-center cubic (fcc) phase of Pt. The diffraction peaks at 40°, 47°, and 68° were assigned to Pt (111), (200), and (220), respectively, and the diffraction peak at ∼25° was associated with the appearance of amorphous Vulcan XC-72 carbon. Meanwhile, no characteristic diffraction peaks were observed for metallic Co or cobalt oxides, suggesting the complete incorporation of Co into Pt. With increasing Co content, major diffraction peaks of Pt_xCo_y were shifted to higher angles due to the incorporation of Co into the fcc crystal structure of Pt. Correspondingly, the calculated crystallite size was decreased with increasing Co content. Based on the XRD spectra, the average crystallites for all catalyst samples measured to be in the range of 5–8 nm, which is in agreement with TEM and HRTEM analysis (see Figure 3.11) in which the observed crystallinity possessed a particle size below 10 nm.

The combination of SEM and TEM showed that with a similar morphology, all Pt_xCo_y/C samples had carbon nanoparticles with sizes of 50–100 nm, in which the Pt-Co alloy nanoparticles displayed uniform distribution. When electrochemical performances were tested in LAMs using a bare Vulcan XC-72 carbon and a typical $PtCo_2$/C as air-electrode, it was found that after 5 cycles, the discharge capacity of $PtCo_2$/C was decreased from 2578 to 2074 mAh g^{-1} whereas Vulcan XC-72 had a

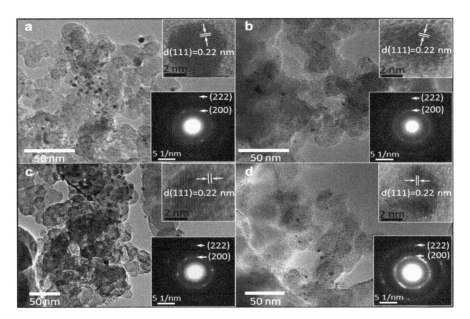

FIGURE 3.11 TEM images for Pt_xCo_y/Vulcan XC-72 carbon, (a) Pt_4Co, (b) Pt_2Co, (c) PtCo, and (d) $PtCo_2$. The insets are the corresponding select area diffraction patterns and the right upper corner inset of (d) is the lattice resolved high-resolution TEM image of a single $PtCo_2$ nanoparticle. (From Su, D. et al. *J. Power Sources* **2013**, *244*, 488–493.[128])

degraded capacity from 965 to 376 mAh g^{-1}, suggesting that PtCo$_2$/C possessed a much higher capacity retention. Among all Pt$_x$Co$_y$/C samples, PtCo$_2$/C reached a maximum capacity of 3040 mAh g^{-1} which was higher than that for the PtAu/C (1200 mAh g^{-1}) developed in a previous study.[125] This indicated that when compared to Au, Co should be better in Pt-alloys for the improvement of both ORR and OER. In a further analysis of discharge and charge curves in the second cycle, the Pt$_x$Co$_y$/C sample possessed a much lower charge voltage ($\sim$3.94 V) and over-potential ($\sim$1.21 V) than Vulcan XC-72. Moreover, the specific capacities were increased positively with increasing Co content, indicating the improved catalytic activity towards ORR and OER in LABs. It was also demonstrated that with lower electronegativity, Co atoms sitting near Pt atoms could increase the electron density of Pt and weaken the strength of the metal-oxygen bond, resulting in the improvement of OER and ORR activities. Along with the increased Co content near Pt atoms, the surface Pt electron density and segregation were also increased, leading to the enhancement of OER and ORR activities. Cycling tests in LABs also confirmed the better cyclability performance of Pt$_x$Co$_y$/C as compare to bare Vulcan XC-72. To further understand and achieve high ORR and OER activities from Pt-alloys with non-noble metals, Zhang et al.[129] synthesized PtZn/carbon aerogel alloy catalysts and compared the electrochemical performance of this catalyst with PtCo/carbon aerogel, Pt/carbon aerogel and Pt/carbon black in magnesium-air batteries. The tested CVs showed that the ORR performance of the catalysts was ordered: PtZn/carbon aerogel ($\sim$293.0 mA g^{-1}) > PtCo/carbon aerogel ($\sim$232.3 mA g^{-1}) > Pt/carbon aerogel ($\sim$191.6 mA g^{-1}) > Pt/carbon black ($\sim$85.5 mA g^{-1}). When magnesium-air batteries were assembled with the synthesized catalysts, PtZn/carbon aerogel demonstrated higher specific discharge capacity (1349 mAh g^{-1}) than PtCo/carbon aerogel ($\sim$1283 mAh g^{-1}), Pt/carbon aerogel ($\sim$1113 mAh g^{-1}), and Pt/carbon black ($\sim$997 mAh g^{-1}), suggesting that PtZn/carbon aerogel possessed better ORR and OER activities than the other three catalysts. They believed that by reducing the leaching problem in alkaline solutions, Zn could promote the formation of Pt-rich surfaces, resulting in more d-band vacancies than Co.

As a noble metal, iridium (Ir)was also used to alloy with Pt to prepare carbon-supported Pt-Ir catalysts for RMABs.[132] In the synthesized Pt-Ir/C catalyst, Pt and Ir were not in a Pt-Ir alloy form but in metallic Pt and Ir forms. Using H$_2$PtCl$_6$·H$_2$O and IrCl$_3$ as Pt and Ir sources, they experimentally prepared Pt-Ir/C, Pt/C, and Ir/C catalysts via an impregnation reduction method and compared their performances. Based on the EDS analysis, the catalyst loading was $\sim$40 wt% and the ratio of Pt to Ir was near 1. XRD confirmed the presence of Pt and Ir, with calculated crystallite sizes of 0.91, 1.67, and 1.15 nm for Ir/C, Pt/C, and Pt-Ir/C catalysts, respectively. When XPS was employed to investigate the chemical states of Pt and Ir, the spectrum of Pt 4f for Pt/C exhibited a doublet peak (i.e., 4f$_{7/2}$ and 4f$_{5/2}$) corresponding to the characteristic values for metallic Pt. A similar doublet peak in Ir 4f for Ir/C was associated with Ir 4f$_{7/2}$ and 4f$_{5/2}$, reflecting the characteristic values for metallic Ir. For Pt-Ir/C, both doublet peaks could be observed, which was in agreement with the peaks in metallic Pt and Ir, suggesting the absence or negligible electronic interactions between Pt and Ir atomic orbitals, as these interactions could result in the decreasing binding energies of Pt 4f electrons.[133] This indicated that the shift in XRD reflections observed in

Pt-Ir/C could be attributed to the peak overlapping from Pt and Ir, and not to their alloy effects. In the OER activity comparison, the current-potential curves tested in a Ar-saturated 0.1 M KOH solution revealed that both Pt/C and commercial Pt/C (60 wt%, Alfa Aesar) gave a higher onset potential ($\sim$0.66 V) than Pt-Ir/C ($\sim$0.53 V) and Ir/C ($\sim$0.50 V), indicating that Pt-Ir/C had a better OER than the other two Pt/C catalysts. When discharge-charge measurements were run in a Li-air battery, the comparison of the first cycle discharge-charge potential profiles indicated that Pt-Ir/C showed the lowest discharge overpotential ($\sim$0.15 V) while at the end of the charging process, Pt-Ir/C had the lowest potential difference ($\sim$0.727 V) between the discharge and charge potentials. Therefore, with the combination of Ir and Pt, Pt-Ir/C clearly exhibited higher ORR and OER activities that that of Pt/C. Other noble metals such as Pd and Ru were also used to alloy with Pt, and the detailed comparisons between these carbon-supported Pt-alloy catalysts (e.g., PtPd/C and PtRu/C) and carbon-supported non-Pt containing alloy catalyst (i.e., PdIr/C) in the application of lithium-air batteries[131] will be discussed in the next section.

3.3.1.2 Composites of Carbon and Other Noble Metal or Alloy

Other than Pt and Pt-alloys, other noble metals (Ir, Ru, Pd, Au, and Ag) or alloys have also been used to prepare noble metal-based electrocatalysts supported on carbon materials for RMABs. It is known that carbon as an air-electrode material can not only conduct electrons but also accommodate the deposition of Li_2O_2 to ensure the diffusion of oxygen and electrolyte in rechargeable Li-O$_2$ batteries (RLOBs), the unavoidable carbon corrosion during charging and discharge, as well as the generation of by-products (e.g., Li_2CO_3) on the interface of the electrode/electrolyte, have been serious problems restricting the performance of a battery.[134,135] Moreover, pure carbon is insufficient in promoting the decomposition of Li_2O_2, resulting in high overpotentials during charging and therefore limiting the reversibility and rate capability of the electrode.[136]

To address these issues in RMOBs, nano-sized iridium (Ir) catalysts were introduced into a three-dimensional porous graphene for RLOBs with a non-aqueous electrolyte,[137] in which Ir acted as a catalytic noble metal that could functionalize the catalytic surface of carbon. In a vacuum-promoted exfoliation method followed by a deoxygenation process, as shown in Figure 3.12a, iridium-functionalized deoxygenated hierarchical graphene (DHG) is synthesized and labeled as Ir@DHG. In the absence of Ir, DHG as a reference sample was prepared via the same method. Ir@DHG was tested to have a high BET surface area ($\sim$372.5 m^2 g^{-1}) and its nitrogen adsorption-desorption curves showed pore sizes ranging from 2 to 200 nm, indicating the presence of fertile mesoporous and macroporous architectures favoring the transport of oxygen, electrolytes, and electrons to and from catalytic sites during discharge and charge. When XPS was used to investigate the surface chemistry, the ratio of O/C showed that DHG possessed a largely reduced form of graphene due to the removal of oxygen functional groups. Ir@DHG however, displayed smaller peaks for C1 s possibly because Ir closely covered the graphene and decreased the exposure of the carbon surface. Importantly, the O1 s intensity of Ir@DHG was slightly smaller than that of DHG, suggesting the graphene in Ir@DHG could maintain a highly deoxygenated surface and enhance the stability of the electrode/electrolyte

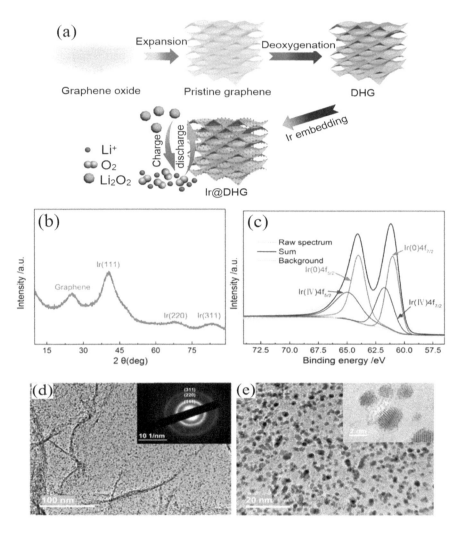

FIGURE 3.12 (a) Illustration of the synthesis process of Ir@DHG composites, (b) XRD pattern of Ir@DHG, (c) XPS of Ir 4f spectrum of Ir@DHG, (d) TEM image and (d) the corresponding SEAD pattern of Ir@DHG, and (e) high resolution TEM images of Ir@DHG. (Modified from Zhou, W. et al. *J. Mater. Chem. A* **2015**, *3*, 14556–14561.[137])

interface.[134,138] In further investigations of Ir nanoparticles using a combination of XRD, XPS, and TEM (see Figure 3.12b–e), the XRD diffraction peaks at 40.4, 68.5, and 83.4 can be attributed to Ir(111), (220), and (311), respectively, which matches the face-centered cubic crystal structure of iridium, whose high resolution XPS spectrum of Ir 4f shows the dominating Ir(0) $4f_{5/2}$ and Ir(0) $4f_{7/2}$ with factional Ir(IV) $4f_{5/2}$ and Ir(IV) $4f_{7/2}$; this suggests a certain degree of surface oxidation. In particular, TEM images (Figure 3.12) show that Ir nanoparticles (∼2.08 nm) are uniformly anchored onto the entire plane without any aggregation, resulting in the homogeneous modification of graphene. To evaluate ORR and OER activities for

Ir@DHG, CV tests observed in RLOB at 1 mV s^{-1} showed that Ir@DHG possessed a higher peak, indicating better ORR activity than DHG, while two strong oxidation peaks at 3.25 and 4.37 V indicated the much higher anodic currents of Ir@DHG. This illustrated that Ir@DHG possessed superior OER activity compared to DHG. After an evaluation of electrochemical performance using galvanostatic charge-discharge, it was shown that Ir@DHG exhibited a narrower discharge-charge potential gap than both DHG and conventional Super P carbon, demonstrating the outstanding rate capability and excellent reversibility of Ir@DHG. Interestingly, cycling performances at a current density of 2000 mA g^{-1} with a limited capacity of 1000 mAh g^{-1} revealed that Ir@DHG could run 150 cycles with a terminal voltage higher than 2.5 V, directly confirming its excellent cycling stability. It was also demonstrated, by the charge-discharge processes investigated through a combination of SEM, TEM and electrochemical impedance spectroscopy (EIS), that discharge products for Ir@DHG were almost completely decomposed during the charge process, confirming that Ir@DHG had a higher reversibility and cycling performance than DHG. In addition, Ir nanoparticles-anchored reduced graphene oxide (Ir-RGO) catalysts were prepared for air-electrode catalysts in RLOBs.[137] A discharge capacity to 9529 mAh g^{-1} (11.36 mAh cm^{-2}) at a current density of 0.5 mA cm^{-2} (393 mA g^{-1}) could be obtained, but the stability seemed to be insufficient.

Instead of Ir, ruthenium (Ru)[139,140] and palladium (Pd)[141–143] have also been used to functionalize carbon to reduce overpotential and capacity decay during round-trip cycling in RMABs. Using a kind of carbonized bacterial cellulose (CBC) as a carbon support, Tong et al.[139] prepared a 5 wt% Ru/CBC composite as a binder-free air-electrode in non-aqueous RLOBs in which Ru nanoparticles could offer active sites for both ORR and OER, and CBC could offer transport pathways for both electrons and oxygen. This Ru-functionalized carbon was expected to provide abundant space for Li$_2$O$_2$ deposition during discharge, reduce the volume effect, and obtain good stability. To evaluate CBC as a viable carbon support, BET and TEM measurements showed a high surface area of 397.6 m^2 g^{-1} and a porous network structure, favoring the distribution of active sites and mass transport. After 5 wt% Ru was introduced into the CBC, SEM and TEM images showed a uniform distribution of Ru nanoparticles on the CBC surface. The obtained galvanostatic discharge-recharge curves at a current density of 200 mA g^{-1} displayed a narrower discharge-charge potential gap for Ru/CBC than CBC, suggesting that 5 wt% Ru/CBC possessed better ORR and OER activities. Meanwhile, the SEM images for 5 wt% Ru/CBC after charging showed a uniform deposition of the discharge product (i.e., Li$_2$O$_2$) on the CBC surface, demonstrating a strong interaction between Li$_2$O$_2$ and CBC, which was favorable to the stability of 5 wt% Ru/CBC in the tested RLOB. The discharge-recharge cycle curve was almost stabilized in the 25th cycle, showing the acceptable stability of 5 wt% Ru/CBC. Similar to Ru-functionalized CBC catalysts, Pd-catalyzed carbon nanofibers (Pd/CNF) catalysts were also prepared by Alegre et al.[141] for alkaline RMABs. Using Na$_6$Pd(SO$_3$)$_4$ as the Pd source, they performed a chemically colloidal method to fabricate the Pd/CNFs catalyst where CNFs were synthesized at 700°C. Commercial carbon black (i.e., Vulcan) was employed as a reference support material for the Pd nanoparticles. In comparison to carbon support properties, CNFs displayed higher graphitic characteristics and conductivity than

Vulcan, even though its surface area ($\sim$97 m^2 g^{-1}) was lower than that of Vulcan ($\sim$224 m^2 g^{-1}). Further comparisons of structure and morphology carried out using XRD, TEM, and SEM found that Pd/CNFs and Pd/Vulcan catalysts possessed the typical fcc structure characteristics of Pd while the calculated crystallite sizes of Pd/CNFs and Pd/Vulcan were 6.1 and 6.5 nm, respectively. TEM and SEM images revealed the uniform dispersion of Pd nanoparticles on CNFs which had a few micrometers in length and hollow cavitied carbon filaments with average diameters of 55 $\pm$ 15 nm. To evaluate the electrocatalytic activity towards ORR and OER, both Pd/CNFs and Pd/Vulcan catalysts were tested in a half-cell configuration. Two full polarization curves obtained from 1.2 to 0.3 V under ORR and from 1.2 to 2 V under OER showed that Pd/CNFs was slightly more active for both ORR and OER than Pd/Vulcan, as shown by the lower overpotential of Pd/CNF when compared to that of Pd/Vulcan. This was possibly attributed to the higher electrical conductivity and/or smaller Pd size of Pd/CNFs, which could favor charge transfers and thus the improvement in ORR and OER kinetics. For the evaluation of carbon corrosion, a stress test based on chronoamperometric measurements at a high potential (1.8 V vs. RHE) for 8 hours revealed that Pd/CNFs possessed a lower OER/ORR onset potential gap ($\sim$700 mV) than that of Pd/Vulcan, showing higher carbon corrosion in Pd/Vulcan. Further charge/discharge cycle tests showed that Pd/CNFs could run 20 repeated cycles compared to 5 cycles of Pd/Vulcan, indicating the higher stability of Pd/CNFs.

Au[144] and Ag[145] were also used to functionalize carbon materials in the application of RMABs, and the performances are similar to Ir, Ru, and Pd. Their composites with carbon, such as carbon black (Ketjenblack) and CNTs, have shown enhanced activity and improved battery performance. Although the use of noble metals in carbon-based composites was shown to improve ORR and OER activities, a recent study[140] of Ru- and Pd-catalyzed carbon nanotube fabrics revealed that Li-O$_2$ cells with Ru- and Pd-CNT cathodes, in particular the latter, seemed not to work according to the desired Li-O$_2$ electrochemistry. Based on the combined analysis using differential electrochemical mass spectrometry and Fourier transform infrared spectroscopy, it was observed that the presence of noble metal catalysts could impair the reversibility of the cells due to a decrease in O$_2$ recovery efficiency (the ratio of the amount of O$_2$ evolved during recharge that was consumed in the preceding discharge) coupled with an increase in CO$_2$ evolution during charging.

Because carbon supported Pt-based binary metal catalysts can give ORR and OER catalytic activities, carbon supported non-noble metal-based binary metal catalysts such as Pd$_3$Pb/C[146] and PdIr/C[147] have also been studied in the exploration of advanced carbon-based bifunctional catalysts in RMABs. To replace expensive Pt-based alloy catalysts, a structurally ordered Pd$_3$Pb/C catalyst via a modified impregnation-reduction method was synthesized;[146] the experimental conditions (e.g., temperature, time, reduction agent, metal precursor, etc.) to obtain a structurally ordered intermetallic phase were strictly controlled, providing uniform active sites on the same surface plane. Two reference samples, Pd/C and Pt/C, were also fabricated by a similar procedure and annealed at 600°C for 12 hours. As monitored by XRD and TEM, 600°C was the suitable annealing temperature for the preparation of ordered Pd$_3$Pb/C because lower temperatures (below 600°C) could result in disordered phases while higher temperatures (above 600°C) could cause the agglomeration of particles. XRD confirmed the formation of ordered Pd_3Pb/C. TEM images showed that the uniform distribution of ordered Pd_3Pb nanoparticles possessed an average particle size of 7.2 ± 0.5 nm while Pd/C and Pt/C had particle sizes of 6.3 ± 0.4 and 5.1 ± 0.3 nm, respectively. Polarization curves in O_2 saturated-0.1 M KOH solution at 1600 rpm revealed that the ordered Pd_3Pb/C was more active than both Pd/C and Pt/C, as evidenced by its more positive onset potential of the ORR than those of Pd/C and Pt/C. Moreover, the ordered Pd_3Pb/C exhibited a higher half-wave potential ($\sim$0.92 V) than Pd/C ($\sim$0.88 V) and Pt/C ($\sim$0.88 V), indicating a better ORR activity for the ordered Pd_3Pb/C. After an accelerated durability test with 5000 continuous cycles, the ordered Pd_3Pb/C displayed a degradation of only 10 mV in its half-wave potential. This was much lower than Pd/C (30 mV) or Pt/C (40 mV), illustrating the superior durability of Pd_3Pb/C. Importantly, when the discharge and charge voltage profiles were measured in an assembled Zn-air battery with the ordered Pd_3Pb/C and Pt/C, the initial round-trip overpotentials for the ordered Pd_3Pb/C was increased from 0.72 V at the first cycle to 0.86 V at the 135th cycle, while the initial round-trip overpotential for Pt/C increased from 0.77 V at the first cycle to 1.34 V at the 135th cycle. This suggested that after the incorporation of Pd, the ordered Pd_3Pb/C gained a higher catalytic activity towards both ORR and OER, and also better durability than Pt/C. Aside from the Pd_3Pb/C catalyst, PdIr/C was also studied in Ko et al.[131] Using a typical impregnation-reduction method, they synthesized a PdIr/C alloy catalyst, as well as PtPd/C and PtRu/C alloy catalysts with the ratio of the two metals and the content of metal loading at 1:1 and 44%, respectively. In the comparison of their ORR and OER activities based on the initial charge-discharge behaviors, PtRu/C possessed a lower overpotential for both charging and discharging than the other two catalysts, while the capacity followed an order of: PtRu/C ($\sim$346 mAh g^{-1}) > PtPd/C ($\sim$153 mAh g^{-1}) > PdIr/C ($\sim$135 mAh g^{-1}). This indicated that when compared to PtPd/C and PdIr/C, PtRu/C possessed higher ORR and OER performances. Cycling performance tests showed that PtRu/C possessed a stable cycling performance with higher cyclability for over 40 cycles.

3.3.2 COMPOSITES OF CARBON AND NON-NOBLE METAL

Non-noble metal-composited carbon materials have also been explored for enhancing both ORR and OER and reduce carbon corrosion to improve the reversibility and rate capability of RMABs. They are also inexpensive, easily fabricated, and possess good interactions between non-noble metal(s) and carbon. For example, Yu et al.[1] explored Tantalum (Ta) as a promoter to reduce battery polarization and overpotential by developing vertically aligned CNTs (VACNTs) with ultra-lengths on permeable Ta foils as the air-electrode via a thermal chemical vapor deposition (TCVD) method for non-aqueous rechargeable Li-O_2 batteries. A CNT powder (CNT-P, 20–40 μm in length and 15 nm in outer diameter) and a commercial VACNT on a stainless steel (SS) mesh substrate (VACNTs-SS) were used as the two reference samples in the characterization of structure and property. The combination of SEM and TEM showed that in the VACNTs-Ta, all the CNTs were perpendicular to the Ta foil substrate and all the CNTs were highly aligned and uniform at the macroscopic level with a superficial height of approximately 240 μm. The tubes of the VACNTs-Ta were also found to be well-crystallized, uniform, and multi-walled with an outer diameter of ~12 nm. Compared to those of VACNTs-SS and entangled CNT-P,[1] the lower ratio of D band and G band in Raman spectroscopy indicated that VACNTs-Ta had less surface defects and lower reactivity with O_2 reduction species than the other two samples. Interestingly, the BET surface area of VACNTs-Ta (206 m^2 g^{-1}) was higher than that of VACNTs-SS (80 m^2 g^{-1}) due to the hierarchical structure of VACNTs-Ta with tortuosity. BJH analysis of the pores revealed the co-existence of micropores, mesopores, and macropores in the VACNTs-Ta, with mesopores and macropores being the majority. In the evaluation and comparison of electrochemical properties, the first discharge and charge behaviors (see Figure 3.13a) of an assembled Li-O_2 cell show that VACNTs-Ta air-electrode can deliver a larger gravimetric specific capacity (~4300 mAh g^{-1}) at 200 mA g^{-1} than both VACNTs-SS (~3200 mAh g^{-1}) and CNT-P (~700 mAh g^{-1}), indicating that VACNTs-Ta may have both higher ORR and OER activities than the other two samples. This may be because there are three advantages over VACNTs-SS: (1) single orientated CNTs can provide favorable pathways for oxygen and Li ion transport, (2) large surface area benefit more active sites for ORR,

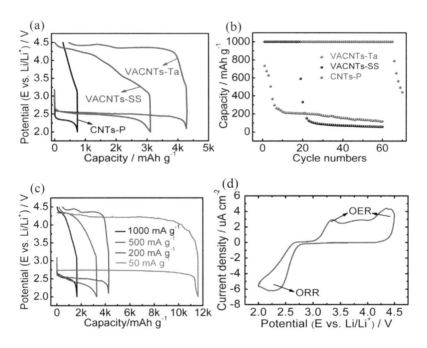

FIGURE 3.13 Electrochemical performance of VACNTs-Ta in the electrolyte of 1 M bis(trifluoromethane) sulfonamide lithium salt (LiTFSI) in tetraethylene glycol dimethyl (TEGDME): (a) the first discharge-charge profiles of VACNTs-Ta, VACNTs-SS, and CNT-P, (b) cycling performance of VACNTs-Ta at a current density of 200 mAh g^{-1}, (c) rate performance of VACNTs-Ta under current densities of 50, 200, 500, and 1000 mA g^{-1}, (d) CV of VACNTs-Ta cathode tested in Ar/O_2 (80:20 v/v) atmosphere between 2.0 and 4.5 V vs. Li/Li+ at a scan rate of 20 mV s^{-1}. (From Yu, R. et al. *J. Power Sources* **2016**, *306*, 402–407.[1])

and (3) less defects and high electrical conductivity and stability resist the attack of oxidative species in the electrolyte. In two further tests of cycling (see Figure 3.13b) and rate capability (see Figure 3.13c), VACNTs-Ta exhibits higher cycling performance (65 cycles at 200 mA g^{-1} as well as a curtailed specific capacity of 1000 mAh g^{-1}) and rate capability (10,000 mAh g^{-1} at 50 mA g^{-1}), suggesting that VACNT-Ta possesses more favorable architecture and stability than those of VACNT-SS or CNT-P. In a CV curve (see Figure 3.13d) evaluating the ORR and OER of VACNT-Ta, it can be found that the anodic area (corresponding to the OER) is larger than the cathodic area (corresponding to the ORR), indicating more electrons stayed in the OER than in the ORR due to the discharge and side products formed during discharge. The easier decomposition of discharge and side products can result in the extended cycle life for VACNT-Ta. It has been demonstrated that the use of Ta not only favors the fabrication of VACNT but also creates good interactions between Ta and VACNT, benefiting co-transportation/reaction and thus the reduction of discharge product (e.g., Li_2O_2) aggregation. To develop high-performance non-noble metal-functionalized carbon air-electrode materials, Su et al.[148] studied and compared composites of carbon with Ni, Co, and Cu quantum dots (QDs) with a MnO/carbon composite. Based on their research results of capacitance and rate performance, they proposed that Ni, Co, and Cu QDs composited with carbon should be a new and important path in developing advanced carbon-based air-electrode materials in Li-air batteries.

Based on studies on single non-noble metal-composited materials for novel carbon-based composite catalysts, two non-noble metals (e.g., transition metals such as Co, Ni, Fe, Cu, and Mn) composited materials have also been explored for BMABs due to the stronger synergistic effect between the two metals and carbon. Based on the assumption that graphene, as a supporter, may reduce the aggregation of active metal sites, that metal nanoparticles may hinder the restacking of graphene layers, and that increasing the space between layers[147] and providing macropores favorable to the transport of oxygen, lithium ions, and electrolyte, Chen et al.[149] designed and synthesized Co-Cu bimetallic yolk-shelled nanoparticles supported on graphene (CoCu/graphene) as the air-electrode for LOBs. In the synthesis, graphene was prepared through a modified Hummers method.[150] CuCl$_2$·2H$_2$O and CoCl$_2$·6H$_2$O were used as the Cu and Co sources to obtain CoCu/graphene through a typical hydrothermal route followed by calcination. Co/graphene and Cu/graphene samples were also prepared for references. After XRD confirmed the coexistence of Co and Cu along with graphene, TEM images showed highly distributed CoCu nanoparticles on graphene sheets without aggregation, and the yolk-shell structured CoCu nanoparticles had an average diameter of 10–20 nm. As estimated by ICP-MS, the molar ratio of Co to Cu was 43:12. When electrochemical measurements were run to investigate the electrocatalytic activity of the CoCu/graphene, it was found that the CoCu/graphene showed the larger discharge capacity (14,821 mAh g^{-1}) and coulombic efficiency (92%) when compared with Co/graphene and Cu/graphene cathodes, suggesting that CoCu/graphene possessed better electrocatalytic activity. Moreover, Co/graphene presented much higher coulombic efficiencies than Cu/graphene, demonstrating that Co could efficiently accelerate OER while Cu might have a positive influence on ORR, even though Co/graphene displayed a lower discharge capacity than Cu/graphene. In a comparison of the tested CV curves of

CoCu/graphene, it displayed the most positive ORR onset potential and the highest ORR/OER peak current densities, confirming the higher catalytic activity of CoCu/graphene than those of Co/graphene and Cu/graphene. A further comparison of the rate performance at three different current densities of 200, 400, and 800 mA g^{-1} indicated the higher rate capability and round-trip efficiency of CoCu/graphene should be resulted from the synergistic effect of the bimetallic CoCu nanoparticles for improving both ORR and OER. When cycling performances were measured in a RLOB with a current density of 200 mA g^{-1} and a cutoff capacity of 1000 mAh g^{-1}, the discharge voltage of Co/graphene and Cu/graphene degraded to lower than 2.5 V after only 71 and 37 cycles, respectively, while the discharge voltage of CoCu/graphene remained above 2.5 V for 122 cycles, demonstrating the superior stability of ORR activity of CoCu/graphene catalyst. Particularly, during the initial 20 cycles, CoCu/graphene could maintain a charge platform lower than 4.1 V, which was superior to both Co/graphene and Cu/graphene, demonstrating its excellent OER activity. When the cycling test was extended at a higher current density of 500 mAh g^{-1} with a cutoff capacity of 1000 mAh g^{-1}, CoCu/graphene could stably run 204 cycles with a discharge terminal voltage of 2.0 V. This was better than both Co/graphene ($\sim$144 cycles) and Cu/graphene ($\sim$101 cycles). The cycling performance of CoCu/graphene was shown to be comparable to those of the best cycling performances in LOBs.[151,152] The improved electrochemical performance was attributed to the synergistic effects of the bimetallic nanoparticles, as well as its combination with graphene. Therefore, graphene plays a key role in the construction of highly conductive grids while Co and Cu tremendously enhance the OER and ORR kinetics in RLOBs, demonstrating a promising strategy to use non-noble metals-composited carbon as air-electrode. A similar work in which CNTs composited with Fe and Co (FeCo-CNTs) was used as an air-electrode catalyst for LOBs[153] was also reported. The bimetallic Fe and Co, coupled with a small amount of oxidized states, were confirmed in the FeCo-CNTs catalyst, as evidenced by the characterization using SEM, TEM, XRD, EDS, and SAED patterns (See Figure 3.14). It can be seen that the FeCo-CNT air-electrode in an assembled LOB exhibits higher capacity ($\sim$3600 mAh g^{-1}) than that of pristine CNTs ($\sim$1276 mAh g^{-1}) and higher round-trip efficiency (72.15% vs. 62.5%). This result indicates that FeCo-CNTs possesses superior ORR and OER activities to that of pristine CNTs.

Aside from CoCu/graphene and FeCo-CNTs composites, electrospun graphitic carbon nanofibers with in-situ encapsulated Co-Ni nanoparticles were also developed,[154] in which the encapsulation of metal catalysts into nanocarbon could suppress aggregation, presenting more active sites for both ORR and OER.[155,156] In a typical electrospinning synthesis route, Cobalt(II) acetate tetrahydrate (Co(Ac)$_2$·4H$_2$O) and Nickel(II) acetate tetrahydrate (Ni(Ac)$_2$·4H$_2$O) were used as Co and Ni sources to form Co-Ni/CNFs catalyst samples. In the characterization of structure and morphology, SEM and TEM showed that for Co-Ni/CNFs, the Co-Ni metal nanoparticles (25–70 nm) were evenly distributed along the length of the nanofibers with diameters of 80–200 nm. Moreover, due to the catalytic effect of metallic Co-Ni nanoparticles, a highly graphitized carbon structure was formed and appeared as onion-like graphene layers with a d-spacing of 0.33 nm, corresponding to the (002) plane of graphite. With both mesopores and macropores, these graphene

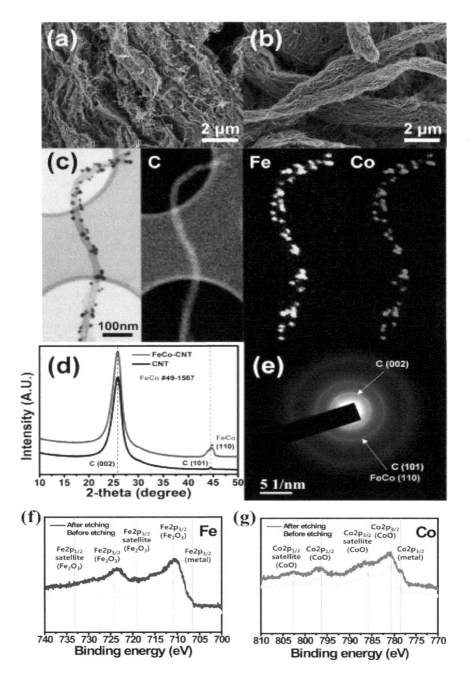

FIGURE 3.14 Morphology and structure of the synthesized FeCo-CNTs composite. SEM images of (a) FeCo-CNTs and (b) pristine CNTs; (c) TEM image of FeCo-CNTs with EDS mapping images; (d) XRD patterns of FeCo-CNTs and pristine CNTs; (e) SAED patterns of FeCo-CNTs. XPS spectra of the FeCo-CNTs after exposure to air for a week: (f) Fe2p, and (g) Co2p. (Modified from Kwak, W.-J. et al. *J. Mater. Chem. A* **2016**, *4*, 7020–7026.[153])

layers could prevent the agglomeration of Co-Ni nanoparticles during long-term electrochemical circumstances.[155] XPS measurements confirmed the presence of CoNi alloys, along with a small amount of oxides. Further information, such as conductivity, surface area, and pore volume, are provided in Table 3.6. It can be seen that compared to CNFs, Co-Ni nanoparticles have higher electrical conductivity, surface area, and pore volume due to the incorporation of Co and Ni into CNFs. To evaluate the electrocatalytic activity, electrochemical tests were run in an assembled non-aqueous LOB without any binders or additives. A cyclic test with an upper-limit capacity of 1000 mAh g^{-1} at a current density of 200 mA g^{-1} showed that the Co-Ni/CNFs cathode running for 60 cycles displayed initial overpotentials of only 0.22 and 0.70 V for ORR and OER, respectively. Particularly, in the first discharge/charge curves obtained at a current density of 200 mA g^{-1}, the Co-Ni/CNFs air-electrode displayed a first discharge capacity of 8635 mAh g^{-1}. This was higher than that of CNFs electrode, suggesting that the in-situ formation of Co-Ni could improve the electrochemical performance and enhance both the ORR and OER. Moreover, in the investigation of the reversible formation and decomposition of Li_2O_2 during the electrochemical process, the combination of ex-situ XRD and TEM revealed that the Co-Ni/CNFs air-electrode could effectively remove the deposited Li_2O_2 layer and then retain the original structure with the encapsulation of Co and Ni so that the agglomeration of the metal catalysts was completely suppressed in the structure. It was also demonstrated that the inherently interconnected, conductive network of Co-Ni/CNFs should be sufficient for LOBs without any binders and additives. Other than Co-Ni catalyzed CNFs, which could result in high graphite carbon and interactive CNFs, the as-prepared CuFe catalyzed carbon black (Ketjenblack) materials were found to drastically increase the catalytic site density, resulting in improvement of ORR kinetics in lithium-air batteries.

3.4 COMPOSITES OF CARBON AND OXIDES

In recent years, metal oxides have been reported to exhibit promising activity for ORR and OER reactions. However, their self-passivity and low electrical conductivity can decrease active sites and hinder charge transports, leading to low catalytic activity.[157,158] The composites of conductive carbon and oxides have become an effective strategy

TABLE 3.6
Electrical Conductivities, Surface Areas and Pore Volumes of Co-Ni/CNFs and CNFs

Samples	Conductivity (S cm^{-1})	Surface Area (m^2 g^{-1})	Pore Volume (cm^3 g^{-1})	I_D/I_G
Co-Ni/CNF	1.58	206	0.43	1.17
CNF	1.10×10^{-2}	23	0.07	1.46

Source: Huang, J. et al. Carbon 2016, 100, 329–336.[154]

to enhance catalytic activity because these composites can overcome the drawbacks of low conductivity and corrosion.[4,159,160]

3.4.1 COMPOSITES OF CARBON AND PEROVSKITE OXIDE

Perovskite oxides, with a general formula of ABO_3[161] (Figure 3.15, the A-site is a rare alkaline earth metal cation and the B-site is a 3d transition metal cation), have attracted increasing interest from researchers due to their defective structures, low cost, excellent oxygen mobility, and outstanding activities towards both ORR and OER.[162–165]

Although carbon-based electrocatalysts have demonstrated enhanced ORR activities, they still suffer from poor stability during the OER process.[27] In general, the OER activity determines the reversibility of the catalyst and also the cycling performance of MABs. To improve overall energy efficiency and retain stability, perovskite oxides have been proposed because the composites of carbon and perovskite oxide can improve ORR and OER activities in alkaline environments when compared to pure perovskite and carbon-based on three factors: the active oxide, the conductive carbon support, and the synergistic effects between them.[166]

To improve the electrical conductivity of perovskite oxides and investigate the effects of perovskite oxides on both ORR and OER of a carbon-perovskite oxide composite, Xu et al.[167] designed and investigated $BaMnO_3$-carbon composites and their bifunctional electrocatalytic capabilities for both ORR and OER. In a typical experiment, carbon-coated $BaMnO_3$ nanorod ($BaMnO_3$@5%C) samples were synthesized using a coating method in which $BaCl_2$ and MnO_2 were used as Ba and Mn sources with a ratio of Ba to Mn being 1:1. SEM, TEM, HRTEM, transmission electron microscope (FEI), and EDS were combined to characterize the $BaMnO_3$@5%C samples. EDS tests confirmed the atomic ratio of Ba:Mn to be 1:1, while SEM images showed that $BaMnO_3$ nanorods had diameters of 100–200 nm

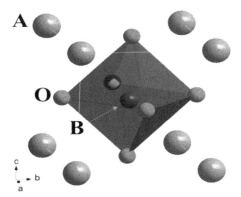

A: Rare earth/alkali metals
B: Transition metals

FIGURE 3.15 A unit cell in perovskite showing relative positions of different ions. (Modified from Gupta, S. et al. *Chem. Asian J.* **2016**, *11*, 10–21.[161])

and lengths of 1–4 μm. TEM revealed that, with a thickness of 10 nm, the carbon was compactly and uniformly coated along the $BaMnO_3$ nanorods. Moreover, the comparison of $BaMnO_3$ and $BaMnO_3$@5%C samples in XRD indicated the presence of perovskite $BaMnO_3$ structures with a small amount of MnO_2 resulting from the residual raw reactant. In the investigation of ORR and OER, the obtained CV curves in O_2-saturated 0.1 M KOH solution (see Figure 3.16a) showed that the peak current density (∼2.5 mA cm^{-2}) of $BaMnO_3$@5%C was much higher than that of $BaMnO_3$ (∼0.94 mA cm^{-2}), demonstrating the superior electrocatalytic activity of $BaMnO_3$@5%C. This was further confirmed when $BaMnO_3$@5%C displayed obvious positive shifting from −0.19 to −0.12 V. At the same time, the tested LSV curves (see Figure 3.16b) reveal that, when compared to $BaMnO_3$, $BaMnO_3$@5%C has a higher positive half-wave potential, again confirming the higher activity of $BaMnO_3$@5%C. Interestingly, $BaMnO_3$@5%C possesses an electron transfer number (see Figure 3.16c) of ∼3.8, which is higher than that of $BaMnO_3$ (3.4–3.7) and close to that of Pt/C (∼3.9). Particularly, when anodic linear scanning voltammograms (see Figure 3.16d) in N_2-saturated 0.1 M KOH solution at a rotation speed of 1600 rpm is used to evaluate the OER, it is found that $BaMnO_3$@5%C presents a higher current density and a more negative onset potential than $BaMnO_3$ and Pt/C, demonstrating the better OER activity of $BaMnO_3$@5%C. Further durability tests (see Figure 3.16e,f) are run for ORR and OER using a chronoamperometric method for 12 hours in O_2-and N_2-saturated 0.1 M KOH at 1600 rpm. These tests show that at −0.35 V (vs. Ag/AgCl), the current density of $BaMnO_3$@5%C only decays less than 1%, which is less than those of Pt/C (∼50%) and $BaMnO_3$ (∼6%), demonstrating the higher ORR stability of $BaMnO_3$@5%C when compared to the other two catalysts. Meantime, at 0.7 V (vs. Ag/AgCl), the current densities are decreased by 26.2%, 60.1%, and 70.3% for $BaMnO_3$@5%C, $BaMnO_3$, and Pt/C catalysts, respectively, indicating that $BaMnO_3$@5%C possesses the best OER durability. Therefore, with the addition of the carbon coating, the fabricated composite of $BaMnO_3$ and carbon can improve both ORR and OER activity and durability, presenting promising potentials in RMABs.

However, not all carbon materials can enhance both ORR and OER after they are composited with perovskite oxides. When $LaNiO_3$ and $LaMnO_3$ perovskite oxides were mixed with carbon (C-NERGY Super C65®) to form $LaNiO_3$-C and $LaMnO_3$-C composites,[168] the electrochemical measurements showed that, although their ORR activities were enhanced, $LaMnO_3$ exhibited a poor OER activity with the addition of carbon, while $LaNiO_3$ had an improved OER with the addition of carbon. This suggests that the activity for carbon-perovskite oxide composites strongly is dependent on the selection of carbon and perovskite oxides, as well as their fabrication methods.

With regard to perovskite oxides with a general formula of ABO_3, the substitution of the A-site and/or B-site metal cations and the generation of oxygen deficiency/vacancy have large effects on their electronic structure and coordination chemistry, which can determine their ORR and OER activities.[161] As two typical representatives of perovskite oxides with a partial substitution of the A-site, $La_{0.6}Ca_{0.4}CoO_3$ (LCC) and $Sr_{0.5}Sm_{0.5}CoO_{3-\delta}$ (SSC) have been combined with carbon black to form different carbon-supported perovskite oxides for RMABs.[169–172] To compare the bifunctionality of their composites, Velraj and Zhu[169,170] conducted a glycine nitrate process to prepare C-LCC composites using graphitized Vulcan XC-72R carbon powder

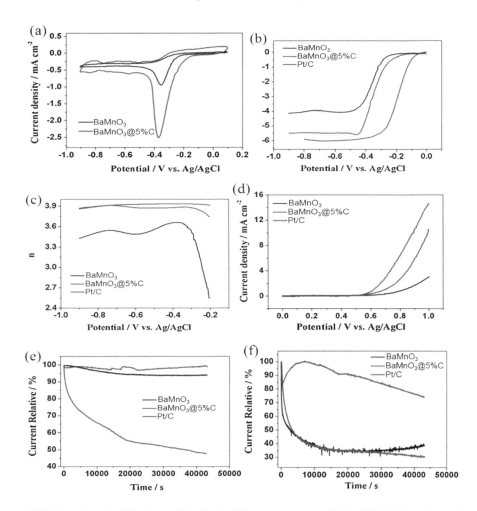

FIGURE 3.16 (a) CV plots of $BaMnO_3$ (black curve) and $BaMnO_3$@5%C (red curve) nanorods in O_2 saturated 0.1 M KOH solution; (b) Linear sweeping voltammograms (LSVs) on the rotating ring-disk electrode for bare $BaMnO_3$, $BaMnO_3$@5%C and commercial Pt/C in O_2 saturated 0.1 M KOH solution at 1600 rpm; (c) The corresponding calculated electron transfer number (n); (d) The LSVs for bare $BaMnO_3$, $BaMnO_3$@5%C and commercial Pt/C in N_2 saturated 0.1 M KOH solution at 1600 rpm; (e) Current-time (i-t) chronoamperometric responses in O_2-saturated 0.1 M KOH at -0.35 V (vs. Ag/AgCl) for ORR at a rotating speed of 1600 rpm; (f) Current-time (i-t) chronoamperometric responses in N_2-saturated 0.1 M KOH at 0.7 V (vs. Ag/AgCl) for OER at a rotating speed of 1600 rpm. (Modified from Xu, Y. et al. *Electrochim. Acta* **2015**, *174*, 551–556.[167])

as a support material. SSC nano-powder was deposited on the same support as a reference sample (C-SSC). After the single perovskite phase was confirmed by XRD for LCC, electrochemical measurements were run in a three-electrode arrangement with an 8.5 M KOH solution. The obtained cathodic polarization curves revealed that the C-SSC air-electrode possessed a better ORR activity than C-LCC while both exhibited higher ORR activities than the graphitized Vulcan XC-72R electrode. However, no significant difference in OER activity was found between C-SSC and C-LCC. Both C-SSC and C-LCC air-electrode displayed higher OER activity than the graphitized Vulcan XC-72R electrode. Interestingly, when the cycling performance was tested under a current density of 53 mA cm^{-2}, C-SSC could run 106 cycles at -0.3 V. This was much longer than that of C-LCC ($\sim$68 cycles at -0.3 V). The graphitized Vulcan XC-72R air-electrode exhibited a very poor cycling performance because of its poor OER capabilities. These results show that with the incorporation of carbon, C-LCC can not only enhance both the ORR and OER activities but also improve the durability with C-SSC possessing similar catalytic activity.

In addition to the partial substitution of A-sites, the partial substitution of B-sites was successfully conducted to form new perovskite oxides in the application of carbon-based composites as RMAB air-cathode catalysts. For example, Yuasa et al.[173] prepared carbon-supported $LaMn_{0.6}Fe_{0.4}O_3$ (C-$LaMn_{0.6}Fe_{0.4}O_3$) electrocatalysts via a reverse homogeneous precipitation (RHP) method, and investigated the effect of $LaMn_{0.6}Fe_{0.4}O_3$ on discharge and charge properties in a LAB. In a typical experimental route, $La(NO_3)_3$, $Mn(NO_3)_2$, and $Fe(NO_3)_3$ were used as La, Mn, and Fe sources. In the structural characterization, XRD revealed the perovskite-phase of the orthorhombic system without any other impurity phases in the C-$LaMn_{0.6}Fe_{0.4}O_3$ with a calculated crystalline size of 17.4 nm. TEM images exhibited a uniform distribution of $LaMn_{0.6}Fe_{0.4}O_3$ on the carbon support. The observed $LaMn_{0.6}Fe_{0.4}O_3$ had a size range of 15–20 nm in diameter, matching with the XRD results. According to BET and BJH analysis, the surface area ($\sim$566 m^2 g^{-1}) and pore volume ($\sim$0.8470 cm^3 g^{-1}) of the carbon supported $LaMn_{0.6}Fe_{0.4}O_3$ was lower than those of pure carbon ($\sim$910 m^2 g^{-1}, $\sim$1.3157 cm^3 g^{-1}). When an assembled Li-air cell (see Figure 3.17) is used to test the discharge and charge performance, the obtained four charge/discharge curves at a current density of 0.5 mA cm^{-2} show that when compared to C-$LaMn_{0.6}Fe_{0.4}O_3$, carbon presents a unstable charge voltage due to the oxidation corrosion by anodic polarization.[174] This carbon corrosion resulted in a decrease in active sites and a degradation of the carbon electrode with discharge/charge cycles. With the addition of $LaMn_{0.6}Fe_{0.4}O_3$, C-$LaMn_{0.6}Fe_{0.4}O_3$ exhibited stable discharge/charge curves because

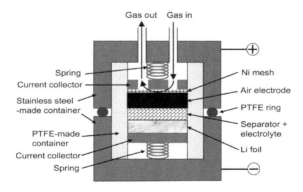

FIGURE 3.17 Schematic cell configuration of Li-air batteries. (From Yuasa, M. et al. *J. Power Sources* **2013**, *242*, 216–221.[173])

the higher OER activity of $LaMn_{0.6}Fe_{0.4}O_3$ could effectively prevent oxidation corrosion from anodic polarization. The discharge voltage of $C-LaMn_{0.6}Fe_{0.4}O_3$ was higher than that of carbon, demonstrating that $LaMn_{0.6}Fe_{0.4}O_3$ could also act as an ORR catalyst in the non-aqueous electrolyte as well as in the alkaline aqueous solution due to Mn^{3+} or Mn^{4+} ions acting as active sites for ORR.[175,176] This research demonstrated that $C-LaMn_{0.6}Fe_{0.4}O_3$ could enhance both ORR and OER, resulting in improved discharge and charge properties of Li-air cells. In addition, carbon composited with perovskite oxides derived from the substitution of both A- and B-sites and their electrochemical properties as air-electrodes were also investigated.[177,178] All these researches demonstrate that carbon-perovskite oxides are promising candidates for alternative bifunctional non-noble metal-based catalysts.

3.4.2 COMPOSITES OF CARBON AND SPINEL OXIDE

As substitutes for expensive noble metal catalysts, spinel oxides[67] (denoted as $A_xB_{3-x}O_4$ (A, B = Co, Zn, Ni, Fe, Cu, Mn, etc.)) have gained much attention due to their advantages including low-cost, considerable activity, high abundance (stability), and environmental friendliness.[179–182] However, to achieve better catalytic activities towards both ORR and OER, net spinel oxides are usually attached onto conducting carbon substrates with the aim at assuring fast electron transport and good interactions between oxides and carbons.[179]

Using ordered mesoporous carbon as a substrate base on its high specific surface, excellent electrical conductivity, environmental friendliness, low-cost, and adjustable mesopore sizes,[8] Li et al.[181] performed an in-situ growth of spinel $CoFe_2O_4$ nanoparticles on rod-like ordered mesoporous carbon (CFO/RC) via a hydrothermal treatment process with an annealing procedure. In their typical experiment, RC was first prepared via a conventional nanocasting strategy using rod-like ordered mesoporous silica as a hard template. Then, $Co(NO_3)_2 \cdot 6H_2O$ and $Fe(NO_3) \cdot 9H_2O$ were used as Co and Fe sources in a hydrothermal treatment to obtain the CFO/RC composite. At different annealing temperatures of 300°C, 400°C, 500°C, and 600°C, the final composites were obtained and labeled as CFO/RC-300, CFO/RC-400, CFO/RC-500, and CFO/RC-600, respectively. Pristine CFO without RC was prepared using the same method as reference. Structure characterizations carried out by XRD, TEM, TG, BET, BJH, and XPS, and the comparison of various CFO/RC composites with pure CFO in the XRD pattern confirmed the presence of the cubic spinel phase without other impurity phases, matching with the crystallite structure in the analysis of HRTEM images. TEM images showed that after the addition of RC, the nanoparticle size of CFO was decreased from 30 to 20 nm, possibly because the RC support could provide a large surface and functional groups (e.g., oxygen-containing) to prevent the agglomeration and growth of CFO nanoparticles. Moreover, the RC support in the CFO/RC composite still maintained its regularly ordered nanowire arrays with mesoporous channels, which could promote the fast diffusion of reactants, ions, electrons, and products, thus possibly improving the electrocatalytic activity.[8] The BJH analysis based on N_2 adsorption-desorption isotherm of the CFO/RC-400 composite revealed two size centers at 3.38 and 19.1 nm, respectively, confirming two sets of uniform mesopore systems. The CFO/RC-400 composite also had a much lower surface area of 150.3 m^2 g^{-1} than RC ($\sim$1103 m^2 g^{-1}) because of the very low content of RC in the CFO/RC-400 composite (5.7 wt% RC according to TGA measurement). For the examination of surface chemical composition and cation oxidation states using XPS, no shift was found for the Co 2p peaks for pure CFO and CFO/RC-400 catalysts, indicating that the Co cations in both catalysts possessed the same chemical surroundings, whereas for the Fe 2p$_{3/2}$ and Fe 2p1/2 of the Fe 2P spectra, CFO/RC-400 showed two small shifts to higher binding energy compared to pure CFO, suggesting a strong coupling between CFO and RC. In a further analysis, the O1 s peak of CFO/RC-400 was seen to shift toward higher binding energies compared with pure CFO, confirming a strong coupling between CFO and RC from the lattice oxygen in the Co/Fe-O framework during the hydrothermal treatment. To assess electrocatalytic properties, electrochemical measurements were run using a three-electrode electrochemical system. The LSV curves in O$_2$-saturated 0.1 M KOH solution at a rotation rate of 1600 rpm showed that all CFO/RC composites exhibited superior onset potentials (-0.10 to -0.13 V) when compared to both CFO (-0.23 V) and RC (-0.29 V) due to the synergistic effect of the two components as well as a large amount of defected CFO nanoparticles providing more active sites. For the Tafel plots in the low overpotential region, the calculated Tafel slope (99 mV per dec) of the CFO/RC-400 composite was lower than those of CFO (129 mV per dec) and RC (125 mV per dec). Both results indicated that CFO/RC had enhanced ORR activity after the incorporation of CFO into RC. Meanwhile, in the comparison of the various CFO/RC composites, CFO/RC-400 displayed a higher onset potential ($\sim$$-0.10$ V) than CFO/RC-600 ($\sim$$-0.125$ V), CFO/RC-500 ($\sim$$-0.110$ V), and CFO/RC-300 ($\sim$$-0.130$ V). Further, LSV measurements at different rotation rates showed that CFO/RC could give a higher ORR electron transfer number ($\sim$4.00) than CFO/RC-300 ($\sim$3.51), CFO/RC-500 ($\sim$3.78), and CFO/RC-600 ($\sim$2.98), suggesting that the interaction of CFO and RC with a high specific surface area and uniform mesopore size could not only significantly promote the electrocatalytic activity of CFO/RC-400 for ORR but also modify the ORR catalytic pathway.[183] For the examination of OER, the anodic LSV curves were recorded in N$_2$-saturated 0.1 M KOH solution at a rotation rate of 1600 rpm. Both the onset potential and Tafel slope (in the low overpotential region) of CFO/RC-400 ($\sim$0.41 V, 92 mV per decade) were smaller than those of CFO ($\sim$0.43 V, 112 mV per decade), RC ($\sim$0.75 V, 308 mV per decade), and commercial 20 wt% Pt/C ($\sim$0.54 V, 105 mV per decade), while at 1.0 V, CFO-RC-400 displayed a much higher current density of 39.6 mA cm^{-2} than CFO ($\sim$34.5 mA cm^{-2}), RC ($\sim$3.37 mA cm^{-2}), and commercial Pt/C ($\sim$27.8 mA cm^{-2}) catalysts, demonstrating a much higher OER activity for CFO/RC-400 as compared to CFO, RC, and Pt/C catalysts. This was mainly due to the hierarchical mesoporous structure of the RC matrix and the strong coupling and synergistic effects between CFO and RC. The stabilities of CFO/RC-400 and Pt/C catalysts were also examined and compared in current-time chronoamperometric response curves in O$_2$- and N$_2$-saturated 0.1 M KOH solution with a rotation rate of 1600 rpm, the decay of ORR current density for CFO/RC-400 was $\sim$9.5% at -0.6 V over 20,000 seconds of continuous operation; this was lower than that of Pt/C ($\sim$29.9%). Compared with a decrease of 34.5% for Pt/C, CFO/RC-400 had a lower decrease of OER current density ($\sim$6.8%) as well. Both durability tests showed that CFO/RC-400 possessed a more stable durability for both ORR and OER than Pt/C. Based on the obtained ORR and OER activities, such as the onset potential, CFO/RC-400 composite was found to outperform other carbon-spinel oxide composite catalysts such as NiCo$_2$O$_4$/graphene,[184] CoFe$_2$O$_4$/graphene,[185] and CoFe$_2$O$_4$/biocarbon,[186] suggesting that ordered mesoporous carbon rods were more suitable for loading spinel oxide nanoparticles to form highly active electrocatalysts for both ORR and OER than other carbon matrixes, such as CNTs, graphene, and biocarbon materials.

Similar to CoFe$_2$O$_4$ and NiCo$_2$O$_4$ ternary spinel oxides, Co$_3$O$_4$,[187] a binary spinel oxide, has also been investigated in the development of carbon-spinel oxide composite bifunctional catalysts. For example, using a shape-controlled spinel Co$_3$O$_4$, Liu et al.[187] fabricated a cubic Co$_3$O$_4$ and multi-walled carbon nanotube (cCo$_3$O$_4$/MWCNT) composite via a facile hydrothermal route in which MWCNTs were acid-functionalized as structure directing/oxidizing agents. In a typical experiment, Co(CH$_3$COO)$_2$·4H$_2$O was used as the Co source and dissolved into ultrapure water. After the pH was adjusted to 8 with ammonia, the acid-treated MWCNTs were added and a heat-treatment at 150°C for 5 hours was carried out. Post-procedures such as washing and drying were carried out and resulted in a black product of cCo$_3$O$_4$/MWCNT. For comparison, pure cCo$_3$O$_4$, acid-treated MWCNT, and a physical mixture of cCo$_3$O$_4$ + MWCNT acted as the three reference samples. In structural characterizations, the comparison of XRD patterns of cCo$_3$O$_4$ and cCo$_3$O$_4$/MWCNT showed that the peaks of cCo$_3$O$_4$ were shifted slightly to a larger 2θ angle after the addition of MWCNT, suggesting a slight lattice contraction possibly due to the variations in crystal sizes[188,189] or the interactions between cCo$_3$O$_4$ and MWCNT. SEM and TEM images not only showed the cubic morphology of cCo$_3$O$_4$ but also exhibited the attachment of MWCNT to cCo$_3$O$_4$ surface without free cubic particles. This observation suggested an effective tethering between MWCNT and cCo$_3$O$_4$ where the acid-functionalized MWCNT served as an oxidizing/structure directing agent to oxidize cobalt ions into spinel cobalt oxides and regulated the formation of cubic cobalt oxide. In the cCo$_3$O$_4$/MWCNT composite, cCo$_3$O$_4$ had a content of ~54% according to TGA data. This was the same as the physical mixture of cCo$_3$O$_4$ + MWCNT. To investigate the bifunctional activities toward both ORR and OER, electrochemical measurements were run in a three-electrode system with 0.1 M KOH solution. The LSV curves in O$_2$-saturated 0.1 M KOH solution showed that among all the tested catalysts, cCo$_3$O$_4$/MWCNT had the most positive onset potential and the highest current density (~−2.91 mA cm^{-2} at −0.4 V), indicating that cCo$_3$O$_4$/MWCNT possessed superior ORR activity to cCo$_3$O$_4$, MWCNT, and cCo$_3$O$_4$ + MWCNT. cCo$_3$O$_4$ + MWCNT did, however, show some improved activity compared to cCo$_3$O$_4$ and MWCNT catalysts. The LSV curves in N$_2$-saturated 0.1 KOH solution also showed the highest OER activity for cCo$_3$O$_4$/MWCNT, evidenced by the highest current density (~16.0 mA cm^{-2} at 0.7 V) in comparison with the other catalysts. A further cycling test with 500 continuous CV cycles revealed that the residual ORR current density of cCo$_3$O$_4$/MWCNT could give about 4, 34, and 3 times improvements, respectively, in comparison with MWCNT, cCo$_3$O$_4$, and cCo$_3$O$_4$-MWCNT catalysts, while the OER current density of cCo$_3$O$_4$/MWCNT was 49% higher than that of cCo$_3$O$_4$ + MWCNT and the individual components. This demonstrated that cCo$_3$O$_4$/MWCNT could produce a stronger interaction between cCo$_3$O$_4$ and MWCNT, resulting in a possible coupling effect and thus higher ORR and OER activities.

3.4.3 Composites of Carbon and Other Oxides

Apart from perovskite oxides and spinel oxides, other oxides have also been used to prepare effective carbon-based composite bifunctional catalysts to tackle the sluggish kinetics of ORR and OER in RMABs. As well known, manganese oxides are of interest due to their favorable catalytic activity, low cost, abundance, and eco-friendliness.[190] Due to its excellent electrocatalytic activity towards O_2 reduction in alkaline media, MnO_2 has been widely used as electrocatalysts in MABs.[191,192] A combination of the unique structural and electronic features of carbon materials and the interesting electrocatalytic properties of MnO_2 can impart significant synergetic effects on the bifunctional activities of ORR and OER, and then MAB performances with MnO_2-Carbon composites as air-electrodes.[193–197]

To improve the catalytic activities for both ORR and OER and enhance the round-trip efficiency in non-aqueous lithium-oxygen batteries, Cao et al.[193] developed a carbon-embedded α-MnO_2@graphene nanosheet (α-MnO_2@GN) composite in which 8 wt% graphene was covered by α-MnO_2 to prevent side reactions on the surface of the graphene, and reduce the formation of Li_2CO_3. In a typical experimental route, $KMnO_4$ was used as the Mn source to react with graphene in the preparation of α-MnO_2@GN composite while the graphene was synthesized from a thermal reduction of home-made graphene oxide. For comparison, α-MnO_2 acted as the reference sample. In the characterization of structure and morphology, the combination of XRD and SEM revealed the tetragonal α-MnO_2 phase (space group I4/m, JCPDS01-072-1982) in an urchin-like α-MnO_2@ GN composite with $\sim$500 nm in diameter, as well as uniform nanorods of $\sim$100 nm in length and 10 nm in width for the pure α-MnO_2. The MnO_2 nanorods were connected to each other and no graphene could be observed. TGA showed a carbon content of $\sim$8 wt% in the α-MnO_2@GN composite, confirmed by an elemental analysis using EDS (C: $\sim$5.35 wt%). Specifically, the high-angle annular dark-field (HAADF) images and the corresponding EDS linear scanning results showed that the graphene, which was embedded in the α-MnO_2@GN composite, was fully covered by α-MnO_2 nanorods due to the uniform growth of α-MnO_2 nanorods on the surface of the graphene. The α-MnO_2@GN composite was found to have a pore size of $\sim$20 nm, a BET surface of $\sim$124.4 m^2 g^{-1} and a pore volume of 0.732 cm^3 g^{-1}. Both the BET surface area and the pore volume were two times that of MnO_2 nanorods, indicating that the conducting graphene not only enlarged the surface area but also improved pore size, benefiting the distribution of active sites and the transport of ions, electrons, and oxygen. When the electrochemical performance of the α-MnO_2@GN composite was evaluated using a Swagelok type designed electrochemical cell with three different cathodes between 2.2 and 4.2 V (vs. Li/Li$^+$) at a current density of 50 mA g^{-1}, it was observed that α-MnO_2@GN delivered a higher capacity of $\sim$2413 mAh g^{-1} than those of graphene ($\sim$2142 mAh g^{-1}) and α-MnO_2 (1560 mAh g^{-1}). For the ORR process, α-MnO_2@GN gave an average voltage of $\sim$2.92 V, which was only $\sim$40 mV lower than the thermodynamic equilibrium potential based on the reaction $2Li^+ + O_2 + 2\ e^- \rightarrow Li_2O_2$ ($E^0 = 2.96$ V). This indicated a lower ORR overpotential of α-MnO_2@GN than those of graphene ($\sim$210 mV) and α-MnO_2 ($\sim$400 mV) air-electrodes, demonstrating that α-MnO_2@GN possessed an enhanced

ORR activity due to the incorporation of graphene. Meanwhile, α-MnO$_2$@GN also exhibited improved OER activity, evidenced by its lower average charging potential of 3.72 V, which was $\sim$0.16 and $\sim$0.33 V lower than those of graphene and α-MnO$_2$, respectively. Therefore, α-MnO$_2$@GN had the lowest overpotentials in the discharge and charge processes, resulting in the enhanced catalytic activities for both ORR and OER. To investigate the rate performance of α-MnO$_2$@GN, current densities were varied from 50 to 300 mA g^{-1} in the test. It was found that at 50, 100, 200, and 300 mA g^{-1}, all coulombic efficiencies for α-MnO$_2$@GN could reach to $\sim$100%, indicating excellent stability. For cycling performance measurements at 100 mA g^{-1} with a cut-off capacity of 1000 mAh g^{-1}, α-MnO$_2$@GN ran stably for $\sim$47 cycles, without any capacity decay and with coulombic efficiencies of $\sim$100%, again demonstrating the strong stability of the α-MnO$_2$@GN composite. Interestingly, to check whether or not there are the side reactions, the comparison of coulombic efficiency at a cut-off voltage of 4.2 V revealed that α-MnO$_2$@GN displayed a similar coulombic efficiency of $\sim$99.8% and charge capacity of $\sim$2409 mAh g^{-1} to α-MnO$_2$, indicating no obvious side reactions on the MnO$_2$ surface, while the coulombic efficiency of graphene was only $\sim$71.1%, with a charge capacity of $\sim$1500 mAh g^{-1}, suggesting the irreversible side reactions occurred mainly on the carbon surface. In further XPS analysis for the products on the electrode during discharging/charging, Li 1 s spectra of α-MnO$_2$@GN, graphene, and α-MnO$_2$ air-electrodes only showed one peak at 55.4 eV, which was assigned to the Li-O bond of Li$_2$O$_2$ during discharging (down to 2.2 V).[198] This indicated that Li$_2$O$_2$ was the main discharge product on either MnO$_2$ or carbon, without side reaction during discharging in the DMSO-based electrolyte. However, for the charging process (back to 4.2 V), the Li 2 s peaks of α-MnO$_2$@GN and α-MnO$_2$ disappeared, confirming the full decomposition of Li$_2$O$_2$ on their surface, while a new peak at $\sim$54.5 eV was seen in the Li 2 s spectrum of graphene, which could be associated with the Li-O bond of Li$_2$CO$_3$.[198] These analysis and investigation of Li 1 s of XPS spectra for the air-electrodes were in agreement with results from galvanostatic discharge/charge tests. Therefore, for the reduction of side reactions, an important strategy should be to prevent the direct contact of carbon with the electrolyte and Li$_2$O$_2$.

Similar to α-MnO$_2$@GN composites, other MnO$_2$ carbon-based composite air-electrode materials such as MnO$_2$-CNT,[194] MnO$_2$-carbon paper,[195] and MnO$_2$-graphene[196] were also reported to have an active effect on the improvement of battery performance for rechargeable LOBs, LABs, and ZABs.

Aside from MnO$_2$, CoO[10,199,200] and RuO$_2$[4,201,202] have also been composited with carbon to form carbon-oxide composites as air-electrode materials for MABs to create synergetic effects and enhance the electrocatalytic activities for both ORR and OER. For example, Gao et al.[199] proposed a novel strategy to improve the catalytic performance of CoO through the integration of dotted carbon species and oxygen vacancies (see Figure 3.18a,b) by synthesizing a carbon-doted effective CoO with oxygen vacancies (CoO/C) for LOB air-electrodes through a simple calcination of the formed pink precipitate of ethanol-mediated Co(Ac)$_2$·4H$_2$O. Their structure characterizations identified the dotted species, oxygen vacancies, and low composition of CoO$_{0.89}$ by the Rietveld refinement in XRD spectra, the shift of

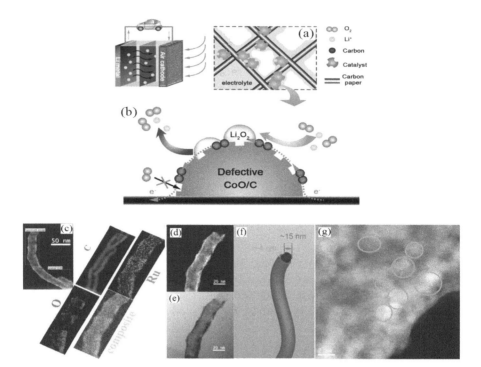

FIGURE 3.18   Schematic illustration of the synergetic effect of the dotted carbon and oxygen vacancies in CoO/C on ORR and OER: (a) schematic of the CoO/C-based air cathode; (b) synergetic mechanism of the dotted carbon and oxygen vacancies in CoO/C on ORR and OER. STEM images and EDS maps for the CNT@RuO$_2$ composite; (c) HAADF-STEM image and EDX maps (of the C, Ru, and O elements and of all three elements in the composite) of the CNT@RuO$_2$ sample: the green box highlights the spectrum image while the yellow box highlights the spatial drift; (d) HAADF-STEM; (e) BF-STEM images of a single CNT@RuO$_2$ structure (space bars: 20 nm); (f) schematic diagram of a single CNT@RuO$_2$ structure; (g) enlarged HAADF-STEM image (scale bar: 2 nm). (Modified from Jian, Z. et al. *Angew. Chem., Int. Ed.* **2014**, *53*, 442–446.[4]; Gao, R. et al. *ACS Catal.* **2016**, *6*, 400–406.[199])

Co-O vibration peak in Raman spectra, and both the shift of the edge for Co2p 1/2 peak in Co 2p spectra and the presence of O4 peak in XPS. Their results revealed the existence of oxygen deficiency in the as-prepared CoO/C after the addition of C species. In the evaluation of battery performance and electrocatalyst activities for both ORR and OER, CoO/C delivered higher capacities of ~7011 mAh g^{-1} and ~4074 mAh g^{-1} at 100 and 400 mA g^{-1}, respectively, than those of CoO (~5189 mAh g^{-1}, ~2059 mAh g^{-1}). The cyclability test in a rechargeable LOB with a current density of ~200 mA g^{-1} and a high voltage cutoff of 4.5 V revealed that CoO/C could run a higher cycle number of 50 than CoO. These results indicated that with the addition of carbon, the formed CoO/C could have a higher electrocatalytic activity towards both ORR and OER than CoO. Further investigations of the morphologies and phase compositions of Co/C and CoO electrodes under different charge/discharge states using SEM and XRD were conducted, and the results revealed that the main discharge product of Li_2O_2 could be completely decomposed by CoO/C rather than by CoO, resulting in the lower charge-discharge overpotential and higher cycling properties of CoO/C. These enhanced electrochemical performances could be attributed to the combination of carbon and CoO to induce the carbon dotting and oxygen vacancies into CoO and the synergetic effects of the two components. Parallel to this CoO/C composites for LOBs, core-shell structured CNT@RuO$_2$ composites were also studied as air-electrode catalysts for rechargeable LOBs.[4,203] The dark-field (HAADF) and bright-field (BF) STEM images (see Figure 3.18c–g) combined with EDS maps clearly confirm the presence of a uniform coating layer of RuO$_2$ around CNTs. The tested discharge/charge curves at a current 385 mA g_{carbon}^{-1} showed that as compared to CNT, CNT@RuO$_2$ had lower discharge and charge overpotentials of 0.21 and 0.51 V, respectively, resulting in a higher round-trip efficiency of ~79% and a higher specific capacity of ~4350 mAh g^{-1}, confirming its higher ORR and OER activities.

The promising strategy of compositing carbon with oxides to enhance positive effects on MAB performance has led more and more oxides such as La$_2$O$_3$,[204] zirconium doped ceria,[205] and cobalt-manganese mixed oxide (Co$_x$Mn$_{1-x}$O)[206] to be explored and created novel carbon-oxide composites with improved ORR and OER activities.

3.5 COMPOSITES OF CARBON AND NITRIDES

With increasing academic and technological interests in the development of novel carbon-based composite bifunctional catalysts for RMABs, the strategy of combining carbon and nitrides (e.g., TiN or CN) to make bifunctional catalysts has become an effective option to improve the rechargeability and round-trip efficiency of RMABs. It is well known that with a high electronic conductivity and a good electrochemical activity, titanium nitride (TiN), a typical transition metal nitride, easily enables itself to be widely applied in electrochemistry.[207] In particular, TiN particles are reported to work well as an air-electrode in the weak acidic electrolyte based hybrid Li-air batteries.[208] Micro- and nano-sized TiN particles are also comparatively studied for ORR and exhibits serial and parallel 4e- oxygen reduction processes in alkaline aqueous electrolytes.[209]

Using a template method (see Figure 3.19a),[210] Li et al.[211] prepared a nano-sized TiN supported on Vulcan XC-72 (n-TiN/VC) as a bifunctional catalyst for non-aqueous LOBs. A commercial micrometer TiN was used to prepare a referenced m-TiN/VC sample. XRD patterns revealed that the TiN in n-TiN/VC possessed a crystallite size of 4.3 nm according to the Scherrer Equation, which agreed with the observations in TEM images. The BET measurements showed the surface areas were 172, 144, and 233 m^2 g^{-1} for n-TiN/VC, m-TiN/VC, and VC samples, respectively. To investigate electrochemical properties, a LOB using n-TiN/VC, m-TiN/VC, and VC as three separate air-electrodes was assembled. The tested CVs (see Figure 3.19b) under O$_2$ from 2.0 to 4.0 V at a scan rate of 0.05 mV s^{-1} revealed that for the cathodic scan, n-TiN/VC showed higher ORR current than m-TiN/VC and VC, demonstrating a better ORR activity, while for the anodic scan, n-TiN/VC presented three oxidation peaks rather than two, as seen in the CVs of m-TiN/VC and VC electrodes. This

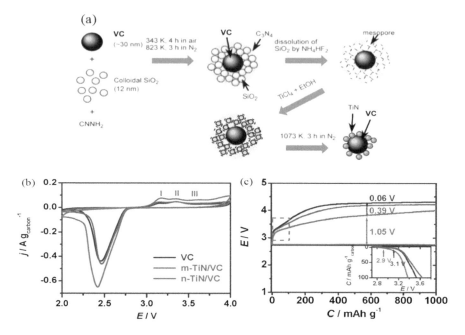

FIGURE 3.19   (a) Preparation process of n-TiN/VC; (b) CV curves of Li-O$_2$ batteries with VC, m-TiN/VC, and n-TiN/VC as cathode catalysts under an O$_2$ atmosphere from 2.0 to 4.0 V at 0.05 mV s^{-1}; (c) Discharge-recharge curves of VC, m-TiN/VC, n-TiN/VC as cathode catalysts of Li-O$_2$ batteries, and an enlarged section highlighted (inset) at 50 mA g$_{carbon}^{-1}$. (Modified from Li, F. et al. *Chem. Commun.* **2013**, 49, 1175–1177.[211])

indicated three varied OER pathways (i.e., I: Equation 3.1, II: Equation 3.2, and III: Equation 3.3) according to the elucidated decomposition mechanism:[212,213]

$$LiO_2 \rightarrow Li^+ + O_2 + e^- \qquad (3.1)$$

$$Li_2O_2 \rightarrow 2Li^+ + O_2 + 2e^- \qquad (3.2)$$

$$2LiO_2 \rightarrow 4Li^+ + 2O_2 + 4e^- \qquad (3.3)$$

In further discussions of the OER using recharge curves (see Figure 3.19c) at a current density 50 mA g$_{carbon}^{-1}$, it can be found that among the three electrodes, n-TiN/VC gives the lowest recharge voltage and, accordingly, the lowest voltage gap ($\sim$1.05 V), showing that n-TiN/VC can effectively promote OER during charge. Moreover, the initial section of the charge in the insert of Figure 3.19c shows a lower onset potential ($\sim$2.9 V) for n-TiN/VC than those for m-TiN/VC ($\sim$3.1 V) and VC ($\sim$3.1 V), matching the strong oxidation peak of LiO$_2$ for n-TiN/VC shown in Figure 3.19b and demonstrating the better OER activity of n-TiN/VC due to the interaction between nano-sized TiN and VC. Additionally, the XRD in the discharge and charge indicated the existence of both LiO$_2$ and Li$_2$O when VC electrode was used, but both m-TiN/VC and n-TiN/VC did not, suggesting VC had lower ORR and OER activities than m-TiN/VC and n-TiN/VC. When five discharge-charge cycles were run, n-TiN/VC exhibited a higher charge voltage at the 5th cycle than m-TiN/VC and VC. This was possibly attributed to the strong interaction between the remaining Li_2O_2 deposits and the TiN nanoparticles on the carbon surface, suggesting that, among the three air-electrdes, n-TiN/VC could give the highest stability. Specifically, according to the results of XRD and FTIR, TiN could be oxidized to TiO_2 and N_2 in a weak acid electrolyte when the potential went up to 4.3 V during charging, suggesting that the LOB based on n-TiN/VC air-electrode should have a limited cell voltage of 4.3 V even though it had enhanced catalytic activities for both ORR and OER. At the same time, Park et al.[214] studied bimodal mesoporous titanium nitride/carbon microfibers composites (bmp-TiN/C), with a few hundred nanometers in size, as an air-electrode for LOB. In the bmp-TiN/C composite, the porous structure could efficiently benefit mass transfer of the reactant species and create a synergistic effect between TiN and carbon, resulting in superior catalytic activities for both ORR and OER.

Aside from TiN, graphitic carbon nitride (g-CN) is a metal-free material attracting much interest due to its abundance and negligible metal pollution.[215] It has been reported that nitrogen atoms in g-CN can increase the electropositivity of adjacent carbon atoms, thus resulting in low energy barriers of oxygen adsorption and activation to produce ORR activity.[216] However, due to its poor electrical conductivity, the ORR activity of g-CN is far from satisfactory;[217] to enhance conductivity, Cheng et al.[218] combined a conducting carbon material with porous graphitic g-CN using a template-free synthesis route to obtain g-CN/C composites as catalysts for ORR in MABs. In their experiments, cheap and green D-glucose was used as the carbon source while melamine and cyanuric acid was used as the g-CN precursors. Based on the amount of D-glucose (i.e., 15.84, 39.60, and 79.20 mmol) used, three g-CN/C samples named as g-CN/C-1, g-CN/C-2, and g-CN/C-3 were obtained, respectively. For comparison, a physically mixed g-CN and carbon sample was labeled as g-CN + C. When the structure and morphology was characterized, both SEM and TEM images showed that the typical gCN/C-2 sample contained fluffy microspheres (2–3 μm in diameters) with plenty of voids and pores. It was composed of 3D interconnected nanosheets with a hazy interface between the g-CN and the carbon due to the g-CN and carbon not being well crystallized. The addition of carbon improved the BET surface area from 200 to 450 $m^2 g^{-1}$, as well as the porous structure including micropores and macropores. Further, XPS confirmed the presence of C, O, and N. Table 3.7 presents the ratio of CN in the g-CN based on the atomic ratio of carbon and nitrogen and the relative composition ratios (%) of the four carbon components derived from decomposed XPS spectra. A comparison of ORR activities was conducted for the three different composite samples (i.e., g-CN/C-1, g-CN/C-2, and g-CN/C-3) and revealed that g-CN/C-2 with 33.75 wt% of N content could deliver the most positive potential than the other two composite samples, g-CN/C-1 (~50.35 wt% N) and g-CN/C-1 (~26.09 wt% N), demonstrating that g-CN/C-2 had the best ORR activity. When the ORR performance of g-CN/C-2 was evaluated in 0.1 M KOH solution using a combination of RDE and RRDE, no redox responses were observed in the CVs for g-CN/C-2 in Ar-saturated 0.1 M KOH solution, whereas a significant ORR reduction peak at 0.73 V in the CVs in O_2-saturated 0.1 M KOH solution could be observed, indicating the good ORR catalytic activity of this catalyst. Specifically,

TABLE 3.7
The Ratio of C/N in g-CN Based on XPS Results

Samples	C1 s (atm.%)[a]	N1 s (atm.%)	C/N[b] in g-CN
g-CN	50.74	43.11	0.67
g-CN/C-1	74.60	15.61	0.53
g-CN/C-2	74.70	19.88	0.69
g-CN/C-3	76.40	16.57	0.90

Source: Modified from Fu, X. et al. *Chem. Commun.* **2016**, *52*, 1725–1728.[218]

[a] Atomic ratios of carbon and nitrogen in composites are obtained from XPS results.

[b] The ratio of C/N in g-CN is calculated via carbon present in C-N bonds divided by nitrogen, according to the equation C/N = P4 * C1 s/N1 s.

g-CN/C-2 presented a higher ORR peak current than pure carbon and the physically mixed g-CN/C, again demonstrating higher ORR activity of g-CN/C-2 than other two. Moreover, the LSV curves of g-CN/C-2 obtained in O_2-saturated 0.1 M KOH solution at 1600 rpm exhibited a significantly more positive onset potential (~0.90 V) and a larger disk current (limiting current density, 4.10 mA cm^{-2}) than those of g-CN and g-CN + C samples, again confirming the higher ORR activity of g-CN/C-2. When durability tests during ORR was carried out in O_2-saturated 0.1 M KOH solution using current-time chronoamperometric responses, it was found that after 20,000 seconds, g-CN/C-2 had lower decay (~20%), displaying better stability than g-CN + C. All the results above demonstrated the high ORR activity and stability of g-CN/C-2. Recently, doping strategies (e.g., P-doping) have also been used for g-CN to make carbon-based composite ORR catalysts for MABs. For example, Ma et al.[219] conducted an in-situ growth of P-g-C3N4 on carbon-fiber paper (PCN-CFP) as a flexible oxygen electrode for ZABs. They demonstrated the outstanding activity, stability, and reversibility of PCN-CFP for both ORR and OER in a tested ZAB featuring low overpotential and long lifetime. Their work presented a positive way to use nitrides in the exploration and creation of novel high-performance carbon-based composite cathodes for MABs.

3.6   COMPOSITES OF CARBON AND CARBIDES

Although there are few publications on the composite of carbon with carbides as air-electrode materials in various MABs, researchers have recently attempted to utilize carbides such as tungsten carbide (WC)[220] and boron carbide (B$_4$C)[221] as carbon-based composites to improve capacity, rechargeability, and round-trip efficiency of MABs. Not only do carbides possess good ORR activity but C-WC composites also exhibit high ORR performance in alkaline mediums.[222]

Because WC possesses comparable catalytic activities to noble metal catalysts,[13,223] Koo et al.[220] coated a uniform WC layer onto a carbon (Ketjenblack EC600-JD) electrode using physical vapor deposition (PVD) (see Figure 3.20a). TEM images

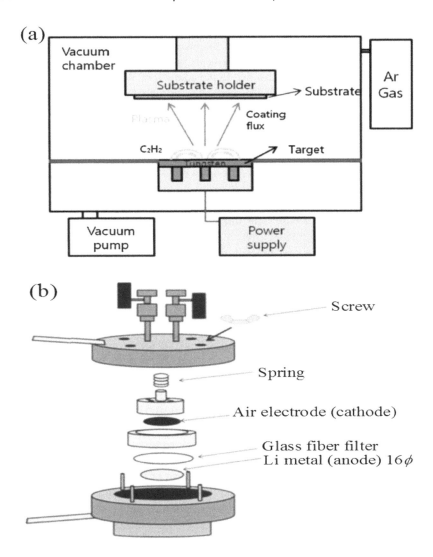

FIGURE 3.20   (a) (Color online) Illustration of PVD method for WC coating; (b) Design and configuration of the Li-O$_2$ cell. (Modified from Koo, B. S. et al. *Japan. J. Appl. Phys.* **2015**, *54*, 047101.[220])

revealed a 20-nm thick uniform WC-coating layer. Electrochemical measurements were performed in an assembled non-aqueous Li-O$_2$ cell (see Figure 3.20b). The discharge curves (down to 2.0 V) showed that the WC-coated electrode could deliver a capacity of ~7000 mAh g$_{carbon}^{-1}$. This was two-folds higher than that of carbon electrodes. Additionally, it was found that in the comparison of the 1st with 10th discharge-charge curves at a current density 100 mA g$_{carbon}^{-1}$, the WC-coated electrode had lower overpotential gaps of 700 mV and 1200 mV as compared to those of carbon ones. These results suggested that WC coating could result in an enhanced activity of both ORR and OER. For the WC-coated electrode, the increased cycle stability and round-trip efficiency clearly confirmed the improvements in reaction rates and the decrease in discharge and charge voltage gaps due to the addition of WC.[224,225] Moreover, it was shown that during the charge and discharge cycling processes, the addition of WC could result in the rapid creation and decomposition of reaction products (i.e., Li_2O_2) due to the enhanced catalytic effects. As the current density was increased from 100 to 200 mA g_{carbon}^{-1}, the WC-coated electrode also showed a decreased overpotential than the carbon one, further confirming the benefits of WC coating.

Like the WC-coated carbon air-electrode above, a B_4C nanowire-carbon nanotube (BC) composited electrode[221] also exhibited the improved rechargeability and round-trip efficiency for LOBs due to the synergistic effects of B_4C and carbon for enhancing the catalytic activities for both ORR and OER. In a typical experiment, CNTs were used as a template to react with boron nanopowders while nickel acted as the reaction catalyst. After a series of post-procedures including drying and annealing, the final BC composite was obtained. Using the same annealing process, pure CNTs mixed with additional nickel (CN) was prepared as a reference sample for comparison. When characterizing structure and morphology, the peak at 26° observed by XRD was assigned to CNTs while the two peaks at 34.9° and 37.6° were assigned to the (2131) and (1121) crystalline planes of B_4C with a Norbide phase (JCSPDF: 35-0798). In the Raman spectra of BC composite, the detected bands below 1200 cm⁻¹ matched the characteristics of B_4C,[226,227] confirming the presence of B_4C. Field emission scanning electron microscopy (FESEM) showed a large amount of B4C nanowires growing from the CNT aggregation, as confirmed by both STEM and EDS. These CNT aggregations possessed diameters in the range of 40–100 nm, as well as lengths above 2 μm. In the electrochemical measurements using RDE techniques in O_2-saturated 0.1 M KOH at a scan rate of 10 mV s⁻¹, the LSV curves at 900 rpm showed that the BC composite could give a more positive onset potential and a larger current density, corresponding to a higher ORR activity than those of Pt/C and the CN composite. Based on the Koutecky-Levich plots, the BC composite was found to have a nearly four-electron ORR process, similar to that of commercial 20 wt% Pt/C (n = 4.0), but different from that of the CN composite (3.2–3.5). Meantime, the comparison of OER curves revealed that the BC composite had a smaller Tafel slope of ~70 mV per decade than Pt/C (~123 mV per decade), indicating that BC possessed a superior OER activity to 20 wt% Pt/C. Interestingly, as CV tests were run in O_2-saturated $LiCF_3SO_3$ in TEGDME (molar ratio = 1:4) electrolyte with a rotation speed of 900 rpm and a scan rate of 10 mV s⁻¹, the BC composite presented smaller overpotential than those of CN and Pt/C. When electrochemical properties were tested in an assembled battery with BC and CN composite air-electrodes, it was displayed that when compared to CN, BC not only showed higher ORR potential and lower OER potential, but also exhibited better cycling performance with a higher round-trip efficiency. This was probably because both B_4C and carbon provide efficient synergistic effects for electrocatalytic reactions. In the discharge/charge stages, XRD coupled with FESEM measurements further demonstrated that the BC composite could effectively deal with the formation and decomposition of Li_2O_2 nanorods, resulting in excellent rechargeability performance.

3.7 OTHER CARBON-BASED COMPOSITES

Aside from the carbon-based composites discussed above, other carbon-based composites can simply be classified into carbon-based binary composites and ternary composites. Carbon-based binary composites contain carbon as one of the components while ternary composites include one or two carbon components. These carbon-based composites have been found to provide strong synergistic effects for enhancing catalytic activities for both ORR and OER in the application of RMABs. Carbon-based ternary composites can also provide stronger synergistic effects for electrocatalytic performances due to the interactions between its three components.

3.7.1 OTHER CARBON-BASED BINARY COMPOSITES

To find high-performance bifunctional catalysts and solve the carbon corrosion issue in RMABs, other functional materials such as sulfides,[228,229] β-FeOOH,[230,231] soil,[232] polyimides,[233] and Fe phthalocyanine (FePc),[234] have been used to form carbon-based binary composites as air-electrode materials for MABs.

Among all non-noble metal chalcogenides, low-cost cobalt sulfides have demonstrated the highest ORR activity in aqueous solutions.[235] Lyu et al.[228] developed a novel CoS_2 nanoparticles-reduced graphene oxide (CoS_2/RGO) composite for aprotic Li-O_2 batteries. In their experiments, a hydrothermal route was employed to prepare the CoS_2/RGO composite in which cobalt acetate ($Co(Ac)_2$) and thioacetamide were used as Co and sulfur sources. In a composition test, TGA showed 78 wt% CoS_2 while XRD found that the patterns of Cobalt sulfide on the graphene sheet were a typical cubic CoS_2 phase (JCPDS 00-41-1471). In the investigation of electrochemical performance in an aprotic Li-O_2 cell using $LiClO_4$-DMSO as the electrolyte and bare RGO, commercial Vulcan XC-72 carbon, and CoS_2/RGO as the Air-electrodes, it was found that among the three electrodes, CoS_2/RGO could deliver a higher ORR onset potential and a notably higher OER current peak (at ~3.75 V), corresponding to higher catalytic activities for both ORR and OER. The first full discharge-charge curves in the range of 2.3–4.3 V at 100 mA g^{-1} showed that, compared to bare RGO and commercial Vulcan XC-72 carbon, CoS_2/RGO exhibited higher discharge potentials and lower charge potentials, representing a lower overpotential. XRD, XPS, and SEM were used to investigate the formation and decomposition of Li_2O_2 and the behavior of Li_2CO_3 as a side product during discharging/charging when CoS_2/RGO was used as the catalyst. In the measurements of rate capability and cyclability of the Li-O_2 cell, it was found that although rate capabilities were increased with rates from 50 to 500 mA g^{-1}, CoS_2/RGO was found to have a limited capacity of 500 mAh g^{-1}, indicating a possible incomplete decomposition of side products, (e.g., Li_2CO_3). Instead of CoS_2, $NiCo_2S_4$ was combined with rGO to synthesize $NiCo_2S_4$-rGO composites by Wu et al.[229] Their research demonstrated that due to the synergistic effect between $NiCo_2S_4$ and rGO, the $NiCo_2S_4$-rGO composite exhibited superior ORR performance to commercial 20 wt% Pt/C, providing a potential opportunity to be a bifunctional catalyst in the application of MABs.

Similar to α-MnO_2, β-FeOOH possess a special 2 × 2 tunnel structure (containing [$Fe^{3+}O_6$] octahedral),[236] which was reported to favor the accommodation of Li^+

and O^{2-} ions within its inner tunnels and thus enabled the reversible formation/decomposition of Li_2O_2 in LOBs.[175,237,238] Moreover, β-FeOOH was also reported to have not only high ORR activity in fuel cells, but also good OER performance for water splitting. To develop novel low-cost, easy-fabrication carbon-composite air-electrode materials, Chen et al.[230] and Li et al.[231] designed and studied β-FeOOH-carbon aerogels (FCAs) and β-FeOOH-MWCNTs (β-FeOOH/MWCNTs) composite electrode materials for RLOBs. β-FeOOH-carbon aerogels (FCAs) were synthesized using an in-situ synthesis route in which β-FeOOH was grown on the surface of carbon aerogels (CAs) with porous structure and surface area, while β-FeOOH-MWCNTs (β-FeOOH/MWCNTs) were synthesized by coating β-FeOOH on MWCNTs via a wet chemical method. After the well-crystallized tetragonal β-FeOOH phase was confirmed using XRD and TEM, the electrochemical performances were conducted using an assembled battery with 1 M LITFSI in TEGDME electrolyte. Based on the obtained CVs, Chen et al.[230] found that due to the strong synergistic effect between high active β-FeOOH and high porosity CAs, the FCAs exhibited higher ORR and OER activities than the referenced β-FeOOH-Super-P (FSP) sample. Meantime, Li et al.[231] demonstrated that compared to their referenced Ketjenblack (KB) sample, the β-FeOOH/MWCNTs composite possessed higher catalytic activity for both ORR and OER due to the addition of β-FeOOH. Specifically, at a current density of 0.1 mA cm^{-2}, FCAs could give the highest specific capacity of ~10,230 mAh g^{-1}, while at a current density of 80 mA g^{-1} it delivered a specific capacity of ~6182 mAh g^{-1}. It seemed that compared to MWCNTs, carbon aerogels had better synergistical effects with β-FeOOH.

To find new effective second materials for preparing high-performance carbon-based binary composite bifunctional catalysts for MABs, Hu et al.[232] used a simple mixing method to fabricate a soil/Vulcan XC-72 carbon (soil/C) composite catalyst in which carbon species were separated by soil particles (see Figure 3.21a,b). They believed that the inactive soil particles, whose main content is SiO_2, could act as avoid volume expanders to suppress surficial passivation of the carbon electrode and thus expose more area and catalytic sites for enlarging capacity and accelerating battery reactions. Several fixed mass-ratios of soil to carbon tested were 0:10, 0.2:10, 1:10, 5:10, and 10:10, respectively. SEM and element mapping revealed the uniform distribution of soil in Vulcan XC-72 carbon. Electrochemical measurements were conducted to evaluate the electrochemical performances of the soil/C composites. It was found that in the soil/carbon composites, the optimal content of carbon for ORR/OER activity was dependent on the conductivity, distribution, and active sites. Moreover, the soil/C composite ran for 50 stable cycles in 1 M LITFSI in TEGDME, indicating a stability in an organic system.[239–241] Importantly, when the soil/C composite with a ratio of 10:1 was measured as air-electrode in RLOBs, large discharge capacity (~7640 mAh g^{-1}), robust cyclability (~100 cycles at 0.2 mA cm^{-2}), and superior rate capability were obtained. This demonstrates that solid/C composites are a potential low-cost and metal-free electrode material for application in RLOBs. Other than the use of soil in carbon-based binary composite electrode materials, polyimide (PI)[233] and Fe phthalocyanine (FePc)[234] were also combined with carbon to form carbon-PI and carbon-FePc composites for RLABs. It was reported that carbon-PI composite could improve cycling performances due to the reduced side reactions resulting from

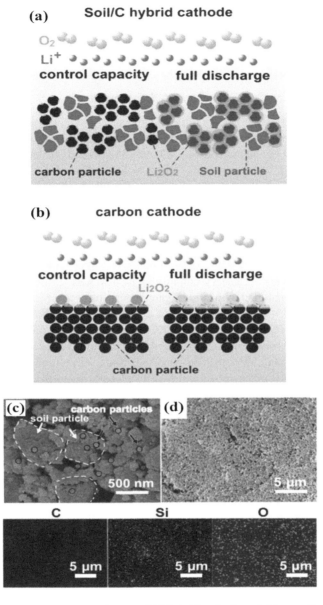

FIGURE 3.21 Scheme pictures for (a) soil/C hybrid cathode and (b) pure C cathode. The void volume of the C cathode can be expanded by soil particles, which is inactive for Li-O$_2$ battery chemistry and hardly deposited or blocked by the discharge product Li$_2$O$_2$. With controlled capacity, the discharge product Li$_2$O$_2$ is scattered and in small dimensions; even fully discharged, a much larger amount of discharge product can also deposit on the deeper inside of the soil/C hybrid cathode. On the contrary, the discharge product Li$_2$O$_2$ will quickly block the surface of pure carbon cathodes and even be piled up on the cathode surface. Part (c) SEM image of the soil/C hybrid; (d) SEM of the soil/C hybrid electrode and element mapping for C, Si, and O. (Modified from Hu, X. et al. *Inorg. Chem. Front.* **2015**, *2*, 1006–1010.[232])

the addition of PI, and that the carbon-FePc composite exhibited comparable ORR catalytic activity to 20 wt% Pt/CB with a 4-electron ORR pathway. These results demonstrated a new way to prepare high-performance carbon-composite electrode materials for MABs.

3.7.2   OTHER CARBON-BASED TERNARY COMPOSITES

Based on the synergistic effects of three components, as well as their inter-correlation, some researchers have recently focused on developing carbon-based ternary composites as advanced air-electrode materials for RMABs. Porous carbon, with excellent porous structures and high surface areas, is an important component in ternary composites, while the selection of the other two components are often dependent on their catalytic performance (i.e., ORR and OER) and physiochemical properties (e.g., active sites, thermal ability, conductivity, and chemical/physical stability). For example, it was experimentally demonstrated that the composite catalysts, combined with different types of electrocatalysts, could offer improved catalytic activity towards both ORR and OER that was much better than the individual components.[242–248]

Among carbon-based ternary composites for RMABs, carbon coupled with an oxide is usually combined with a third component such as a noble metal,[242,243] a non-noble metal,[244–246] a polyelectrolyte,[247] or a second carbon,[248] to give its enhanced catalytic activities towards ORR and/or OER in the applications of RMABs. For example, to reduce the loading of Pt and improve the OER activity of Pt/C, Su et al.[243] designed and developed Pt/C-LiCoO$_2$ composite bifunctional catalysts. In their ultrasonic mixture route, three Pt/C-LiCoO$_2$ catalysts were obtained according to different mass ratios of Pt:LiCoO$_2$: 1:9, 1:19, and 1:49, respectively. After the phase structure of LiCoO$_2$ was confirmed for the three composites by XRD patterns, the catalytic activity and stability were evaluated using a three-electrode system with 0.1 M KOH solution. In the CVs obtained in O$_2$-saturated 0.1 M KOH solution, it was found that with decreasing content of LiCoO$_2$, the ORR peak for the Pt/C-LiCoO$_2$ composite shifted greatly to positive potentials. At Pt/LiCoO$_2$ ratios of 1:19 (i.e., 2 wt% Pt in Pt-LiCoO$_2$), the ORR peak position of the Pt/C-LiCoO$_2$ composite was near the ORR peak position ($\sim$0.81 V) of Pt/C, suggesting that the Pt/C-LiCoO$_2$ composite had a good catalytic activity towards ORR, which was also evidenced by all the LSV results obtained at a rotating electrode rate of 1600 rpm. This also indicated that all three Pt/C-LiCoO$_2$ composites had higher ORR activities than pure LiCoO$_2$. Using RRDE techniques, the calculated electron transfer number for the typical PtC-LiCoO$_2$ (2 wt% Pt in Pt-LiCoO$_2$) was 3.92, indicating a four-electron transfer in the ORR mechanism. For OER, the LSV curves at 1600 rpm showed that all Pt/C-LiCoO$_2$ composites gave higher OER currents than Pt/C or pure LiCoO$_2$, demonstrating their enhanced catalytic activity towards OER, confirmed by the calculated Tafel slopes for pristine LiCoO$_2$, Pt/C-LiCoO$_2$ (1:9), Pt/C-LiCoO$_2$ (1:19), and Pt/C-LiCoO$_2$ (1:49), which were 89, 87, 82, and 90 mV dec^{-1}, respectively. In their evaluation of bifunctional activity of the catalysts, the potential difference between ORR (potential at -3 mA cm^{-2}) and OER (potential at 10 mA cm^{-2}) was used as a common indicator,[249] defined as $\Delta E = E_{OER}@10$ mA cm$^{-2} - E_{ORR}@$-3 mA cm^{-2}. The calculated potential difference for Pt/C, pure LiCoO$_2$, Pt/C-LiCoO$_2$ (1:9),

Pt/C-LiCoO$_2$ (1:19), and Pt/C-LiCoO$_2$ (1:49) were 0.99, 1.15, 0.86, 0.87, and 0.91 V, respectively, indicating that Pt/C-LiCoO$_2$ composites had better bifunctional activities. This research clearly showed that Pt/C- LiCoO$_2$ composites had great potential in the application of RMABs. In another study, Wu et al.[244] prepared and studied Ni modified MnO$_x$/C (Ni-MnO$_x$/C) composites towards ORR and their application in RZABs with Ni as a typical non-noble metal. They demonstrated that at an optimal Ni/Mn atomic ratio of 1:2, the Ni-MnO$_x$/C electrode in a tested RZAB could give a high-power density of $\sim$122 mW cm^{-2}. This was better than MnO$_x$/C ($\sim$89 mW cm^{-2}) and slightly higher than the referenced Pd/C ($\sim$121 mW cm^{-2}) and Pt/C ($\sim$120.5 mW cm^{-2}) air-electrodes.

Because poly (diallyl dimethylammonium chloride [PDDA]), a positively charged polyelectrolyte, can modify carbon supports by providing available functional groups favorable to the formation of more active sites, Zhai et al.[247] prepared PDDA functionalized CNTs and then combined it with spinel CoMn$_2$O$_4$ to form a ternary CoMn$_2$O$_4$/PDDA-CNTs composite as an ORR electrocatalyst. Their results showed that with increasing the loading of CoMn$_2$O$_4$ from 36 wt% to 83 wt%, the CoMn$_2$O$_4$/PDDA-CNTs composites exhibited increased ORR current densities in both alkaline and neutral conditions with a 4e reduction pathway, which even outperformed Pt/C catalysts at medium overpotential due to the non-covalent coupling effects between CoMn$_2$O$_4$ and PDDA-CNTs. Moreover, when the optimal content of CoMn$_2$O$_4$ was 36 wt% in the CoMn$_2$O$_4$/PDDA-CNTs composite, the potential difference between the ORR (Potential at -3 mA cm^{-2}) and OER (Potential at 10 mA cm^{-2}) was $\sim$0.849 V, showing a potential bifunctional activity. To find a suitable third component, Wang and Ma[248] selected two different oxides to fabricate α-MnO$_2$-LaNiO$_3$/CNTs composites in which the combination of α-MnO$_2$ and LaNiO$_3$ was believed to play a synergistic effect on bifunctional activity. Normally, α-MnO$_2$ can act as a good ORR catalyst while LaNiO$_3$ should be a good OER catalyst.[163,170,250,251] The assembled RZAB using MnO$_2$-LaNiO$_3$/CNTs composite could achieve a maximum power density of 55.1 mW cm^{-2} at a current density of 81 mA cm^{-1}, and the voltage polarization exhibiting a 1.4% decrease in discharge and 4.8% increase in charge after 75 charge-discharge cycles due to its good bifunctional activity and durability.

Other carbon-based ternary composites such as Fe/Fe$_3$C-CNFs[252] and Fe-doped NiOOH grown on graphene-encapsulated FeNi$_3$ nanodots (FeNi$_3$@GR@Fe-NiOOH) composites[253] were also studied.[253] The design of the Fe/Fe$_3$C-CNFs composite focused on the incorporation of iron-based catalysts into microporous carbon with high electronic/ionic conductivity and catalytic activity.[254,255] The FeNi$_3$@GR@Fe-NiOOH composite, composed of a core of bimetallic Fe-Ni, a middle shell of graphene, and an outer shell of Fe-NiOOH, was prepared for enhancing catalytical activity of OER based on electron injection by tuning the Fermi surface of the catalyst to match the potential of oxygen evolution. Electrochemical measurements found the synergistic effects of Fe/Fe$_3$C-CNFs, giving a high specific capacity ($\sim$6920 mAh g^{-1} at 80 mA g^{-1}). This composite, on the other hand, delivered superior OER activity in 0.1 M KOH solution to that of RuO$_2$, with an optimal onset overpotential of 220 mV (vs. RHE), an overpotential of 290 mV at 10 mA cm^{-2}, and a Tafel slope of 65 mV dec^{-1}. The above results suggested the two carbon-based ternary composites could provide new design strategies in developing high-performance electrocatalyst materials for MABs.

Interestingly, a carbon-based four-component bifunctional catalyst[42] that included a NiFe layered double hydroxide (NiFe LDH), cobalt, amorphous carbon, and acetylene black, was recently reported as an air-electrode material in a RZAB to exhibit an unexpectedly high cycling stability of 200 cycle periods with stable charge/discharge voltages of 2.05 V and 1.05 V when the battery test was performed at a current density of 40 mA cm^{-2} with 2 hours per cycle.

3.8 CHAPTER SUMMARY

In this chapter, carbon-composited bifunctional catalysts for RMABs are comprehensively reviewed in terms of their material selection, synthesis method, structural characterization, and electrochemical performance. Compared with single material catalysts, composited catalysts possess synergistic effects on structural and electrical properties resulting from different components, leading to enhanced ORR and OER performances in MAB air-electrode catalysts. In the design of advanced carbon-based composites as bifunctional catalysts, the proper selection of carbon materials is crucial because the properties of the carbon material (e.g., porous structure, surface area, and electronic conductivity) can affect the transport of ionic/electronic/gas, electron conductivity, and the distribution of catalytic sites. As concluded, the strong interaction between the carbon material with other component(s) can determine the bifunctional activities of ORR and OER and their stability/durability.

As a special type of carbon material, graphene tends to hold active sites at the edge rather than the mesoporous structure. However, the restacking of graphene layers is an issue that needs to be resolved in the development of advanced graphene-composited bifunctional catalysts for MABs. It is should also be emphasized that not all non-carbon material(s) are able to act as useful component(s) in carbon-composited bifunctional catalysts in terms of producing synergistic interactions and overcoming carbon corrosion/oxides in MABs. Other material components such as oxides and metals can have more effective influences on the electrocatalytic activity and cycling performance of carbon-composite catalyst; in particular, perovskite oxides exhibit bifunctionality for catalytic ORR and OER in alkaline solutions. To solve issues of carbon corrosion and oxidation in MABs, novel strategies such as carbon-composited ternary bifunctional catalysts have shown promising effects on the improvement of catalytic performances. Compared with individual components, these ternary and/or trinary composite catalysts offer improved catalytic performances and strong resistances to carbon corrosion in MABs.

REFERENCES

1. Yu, R.; Fan, W.; Guo, X.; Dong, S. Highly ordered and ultra-long carbon nanotube arrays as air cathodes for high-energy-efficiency Li-oxygen batteries. *J. Power Sources* **2016**, *306*, 402–407.
2. Meng, W.; Liu, S.; Wen, L.; Qin, X. Carbon microspheres air electrode for rechargeable Li-O$_2$ batteries. *RSC Adv.* **2015**, *5*, 52206–52209.

3. Wang, F.; Xu, Y.-H.; Luo, Z.-K.; Pang, Y.; Wu, Q.-X.; Liang, C.-S.; Chen, J.; Liu, D.; Zhang, X.-H. A dual pore carbon aerogel-based air cathode for a highly rechargeable lithium-air battery. *J. Power Sources* **2014**, *272*, 1061–1071.

4. Jian, Z.; Liu, P.; Li, F.; He, P.; Guo, X.; Chen, M.; Zhou, H. Core-shell-structured CNT@RuO$_2$ composite as a high-performance cathode catalyst for rechargeable Li-O$_2$ batteries. *Angew. Chem., Int. Ed.* **2014**, *53*, 442–446.

5. Zheng, Y.; Jiao, Y.; Ge, L.; Jaroniec, M.; Qiao, S. Z. Two-step boron and nitrogen doping in graphene for enhanced synergistic catalysis. *Angew. Chem. Int. Ed.* **2013**, *52*, 3110–3116.

6. Li, B.; Geng, D.; Lee, X. S.; Ge, X.; Chai, J.; Wang, Z.; Zhang, J.; Liu, Z.; Andy Hor, T. S.; Zong, Y. Eggplant-derived microporous carbon sheets: Towards mass production of effcient bifunctional oxygen electrocatalysts at low cost for rechargeable Zn–air batteries. *Chem. Commun.* **2015**, *51*, 8841–8844.

7. Puig, A.; Perez-Munuera, I.; Carcel, J. A.; Hernando, I.; Garcia-Perez, J. V. Moisture loss kinetics and microstructural changes in eggplant (Solanum melongena L.) during conventional and ultrasonically assisted convective drying. *Food Bioprod. Process.* **2012**, *90*, 624–632.

8. Qie, L.; Chen, W.; Xu, H.; Xiong, X.; Jiang, Y.; Zou, F.; Hu, X.; Xin, Y.; Zhang, Z.; Huang, Y. Synthesis of functionalized 3D hierarchical porous carbon for high-performance supercapacitors. *Energy Environ. Sci.* **2013**, *6*, 2497–2504.

9. Zhu, Y.; Murali, S.; Stoller, M. D.; Ganesh, K. J.; Cai, W.; Ferreira, P. J.; Pirkle, A. et al. Carbon-based supercapacitors produced by activation of graphene. *Science* **2011**, *332*, 1537–1541.

10. Li, Y.; Gong, M.; Liang, Y.; Feng, J.; Kim, J. E.; Wang, H.; Hong, G.; Zhang, B.; Dai, H. Advanced zinc-air batteries based on high-performance hybrid electrocatalysts. *Nat. Commun.* **2013**, *4*, 1805.

11. Prabu, M.; Ketpang, K.; Shanmugam, S. Hierarchical nanostructured NiCo$_2$O$_4$ as an efficient bifunctional non-precious metal catalyst for rechargeable zinc-air batteries. *Nanoscale* **2014**, *6*, 3173–3181.

12. Pekala, R. W.; Calif, P. H. *U.S. Patent* 4,873,218 **1989**.

13. Wang, Y.-J.; Fang, B.; Li, H.; Bi, X. T.; Wang, H. Progress in modified carbon support materials for Pt and Pt-alloy cathode catalysts in polymer electrolyte membrane fuel cells. *Prog. Mater. Sci.* **2016**, *82*, 445–498.

14. Liu, Y.; Wu, P. Graphene quantum dot hybrids as efficient metal-free electrocatalyst for the oxygen reduction reaction. *ACS Appl. Mater. Interfaces* **2013**, *5*, 3362–3369.

15. Li, Y.; Zhao, Y.; Cheng, H.; Hu, Y.; Shi, G.; Dai, L.; Qu, L. Nitrogen-doped graphene quantum dots with oxygen-rich functional groups. *J. Am. Chem. Soc.* **2012**, *134*, 15–18.

16. Hu, C.; Yu, C.; Li, M.; Wang, X.; Dong, Q.; Wang, G.; Qiu, J. Nitrogen-doped carbon dots decorated on graphene: A novel all-carbon hybrid electrocatalyst for enhanced oxygen reduction reaction. *Chem. Commun.* **2015**, *51*, 3419–3422.

17. Zhou, X.; Tian, Z.; Li, J.; Ruan, H.; Ma, Y.; Yang, Z.; Qu, Y. Synergistically enhanced activity of graphene quantum dot/multi-walled carbon nanotube composites as metal-free catalysts for oxygen reduction reaction. *Nanoscale* **2014**, *6*, 2603–2607.

18. Wang, M.; Fang, Z.; Zhang, K.; Fang, J.; Qin, F.; Zhang, Z.; Li, J.; Liu, Y.; Lai, Y. Synergistically enhanced activity of graphene quantum dots/graphene hydrogel composites: A novel all-carbon hybrid electrocatalyst for metal/air batteries. *Nanoscale* **2016**, *8*, 11398–11402.

19. Peng, J.; Gao, W.; Gupta, B. K.; Liu, Z.; Romero-Aburto, R.; Ge, L.; Song, L. et al. Graphene quantum dots derived from carbon fibers. *Nano Lett.* **2012**, *12*, 844–849.

20. Yu, D.; Zhang, Q.; Dai, L. Highly efficient metal-free growth of nitrogen-doped single-walled carbon nanotubes on plasma-etched substrates for oxygen reduction. *J. Am. Chem. Soc.* **2010**, *132*, 15127–15129.

21. Luo, G.; Huang, S.-T.; Zhao, N.; Cui, Z.-H.; Guo, X.-X. A super high discharge capacity induced by a synergetic effect between high-surface-area carbons and a carbon paper current collector in a lithium-oxygen battery. *Chin. Phys. B* **2015**, *24*, 088102.

22. Zhang, Y.; Zhang, H.; Li, J.; Wang, M.; Nie, H.; Zhang, F. The use of mixed carbon materials with improved oxygen transport in a lithium-air battery. *J. Power Sources* **2013**, *240*, 390–396.

23. Cheon, J. Y.; Kim, K.; Sa, Y. J.; Sahgong, S. H.; Hong, Y.; Woo, J.; Yim, S.-D.; Jeong, H. Y.; Kim, Y.; Joo, S. H. Graphitic nanoshell/mesoporous carbon nanohybrids as highly efficient and stable bifunctional oxygen electrocatalysts for rechargeable aqueous Na-air batteries. *Adv. Energy Mater.* **2016**, *6*, 1501794.

24. Song, M. J.; Shin, M. W. Fabrication and characterization of carbon nanofiber@mesoporous carbon core-shell composite for the Li-air battery. *Appl. Surf. Sci.* **2014**, *320*, 435–440.

25. Zhao, Z.; Li, M.; Zhang, L.; Dai, L.; Xia, Z. Design principles for heteroatom-doped carbon nanomaterials as highly efficient catalysts for fuel cells and metal-air batteries. *Adv. Mater.* **2015**, *27*, 6834–6840.

26. Ren, X.; Wang, B.; Zhu, J.; Liu, J.; Zhang, W.; Wen, Z. The doping effect on the catalytic activity of graphene for oxygen evolution reaction in a lithium-air battery: A first-principles study. *Phys. Chem. Chem. Phys.* **2015**, *17*, 14605–14612.

27. Li, L.; Manthiram, A. O- and N-doped carbon nanowebs as metal-free catalysts for hybrid Li-air batteries. *Adv. Energy Mater.* **2014**, *4*, 1301795.

28. Lin, X.; Lu, X.; Huang, T.; Liu, Z.; Yu, A. Binder-free nitrogen-doped carbon nanotubes electrodes for lithium-oxygen batteries. *J. Power Sources* **2013**, *242*, 855–859.

29. Wang, Z.; Xiong, X.; Qie, L.; Huang, Y. High-performance lithium storage in nitrogen-enriched carbon nanofiber webs derived from polypyrrole. *Electrochim. Acta.* **2013**, *106*, 320–326.

30. Li, H.; Kang, W.; Wang, L.; Yue, Q.; Xu, S.; Wang, H.; Liu, J. Synthesis of three-dimensional flowerlike nitrogen-doped carbons by a co-pyrolysis route and the effect of nitrogen species on the electrocatalytic activity in oxygen reduction reaction. *Carbon* **2013**, *54*, 249–257.

31. Mi, R.; Liu, H.; Wang, H.; Wong, K.-W.; Mei, J.; Chen, Y. Lau, W.-M., Yan, H. Effects of nitrogen-doped carbon nanotubes on the discharge performance of Li-air batteries. *Carbon* **2014**, *67*, 744–752.

32. Shin, W. H.; Jeong, H. M.; Kim, B. G.; Kang, J. K.; Choi, J. W. Nitrogen-doped multiwall carbon nanotubes for lithium storage with extremely high capacity. *Nano Lett.* **2012**, *12*, 2283–2288.

33. Antonietti, M.; Fechler, N.; Fellinger, T.-P. Carbon aerogels and monoliths: Control of porosity and nanoarchitecture via sol-gel routes. *Chem. Mater.* **2014**, *26*, 196–210.

34. Ma, J.-L.; Zhang, X.-B. Optimized nitrogen-doped carbon with a hierarchically porous structure as a highly efficient cathode for Na-O$_2$ batteries. *J. Mater. Chem. A* **2016**, *4*, 10008–10013.

35. Sakaushi, K.; Fellinger, T.-P.; Antonietti, M. Bifunctional metal-free catalysis of mesoporous noble carbons for oxygen reduction and evolution reactions. *ChemSusChem* **2015**, *8*, 1156–1160.

36. Men, Y.; Siebenbürger, M.; Qiu, X.; Antonietti, M.; Yuan, J. Low fractions of ionic liquid or poly(ionic liquid) can activate polysaccharide biomass into shaped, flexible and fire-retardant porous carbons. *J. Mater. Chem. A* **2013**, *1*, 11887–11893.

37. Balach, J.; Wu, H.; Polzer, F.; Kirmse, H.; Zhao, Q.; Wei, Z.; Yuan, J. Poly(ionic liquid)-derived nitrogen-doped hollow carbon spheres: Synthesis and loading with Fe$_2$O$_3$ for high-performance lithium ion batteries. *RSC Adv.* **2013**, *3*, 7979–7986.

38. Kichambare, P.; Kumar, J.; Rodrigues, S.; Kumar, B. Electrochemical performance of highly mesoporous nitrogen doped carbon cathode in lithium-oxygen batteries. *J. Power Sources* **2011**, *196*, 3310–3316.

39. Liu, Z.; Zhang, G.; Lu, Z.; Jin, X.; Chang, Z.; Sun, X. One-step scalable preparation of N-doped nanoporous carbon as a high-performance electrocatalyst for the oxygen reduction reaction. *Nano Research* **2013**, *6*, 293–301.

40. Hadidi, L.; Davari, E.; Lqbal, M.; Purkait, T. K.; Lvey, D. G.; Veinot, J. G. C. Spherical nitrogen-doped hollow mesoporous carbon as an efficient bifunctional electrocatalyst for Zn-air batteries. *Nanoscale* **2015**, *7*, 20547–20556.

41. Cong, K.; Radtke, M.; Stumpf, S.; Schröter, B.; McMillan, D. G. G.; Rettenmayr, M.; Ignaszak, A. Electrochemical stability of the polymer-derived nitrogen-doped carbon: An elusive goal? *Mater. Renew Sustain Energy* **2015**, *4*, 5.

42. Wang, Q.; Zhou, D.; Yu, H.; Zhang, Z.; Bao, X.; Zhang, F.; Zhou, M. NiFe layered double-hydroxide and cobalt-carbon composite as a high-performance electrocatalyst for bifunctional oxygen electrode. *J. Electrochem. Soc.* **2015**, *162*, A2362–A2366.

43. Long, G.; Wan, K.; Liu, M.; Li, X.; Liang, Z.; Piao, J. Effect of pyrolysis conditions on nitrogen-doped ordered mesoporous carbon electrocatalysts. *Chin. J. Catal.* **2015**, *36*, 1197–1204.

44. Eisenberg, D.; Stroek, W.; Geels, N. J.; Sandu, C. S.; Heller, A.; Yan, N.; Rothenberg, G. A simple synthesis of an N-doped carbon ORR catalyst: Hierarchical micro/meso/macro porosity and graphitic shells. *Chem. Eur. J.* **2016**, *22*, 501–505.

45. Jing, Y.; Zhou, Z. Computational insights into oxygen reduction reaction and initial Li_2O_2 nucleation on pristine and N-doped graphene in Li-O_2 batteries. *ACS Catal.* **2015**, *5*, 4309–4317.

46. Trogadas, P.; Ramani, V.; Strasser, P. Fuller, T. F.; Coppens, M. O. Hierarchically structured nanomaterials for electrochemical energy conversion. *Angew. Chem. Int. Ed.* **2016**, *55*, 122–148.

47. Zhao, B.; Collinson, M. M. Well-defined hierarchical templates for multimodal porous material fabrication. *Chem. Mater.* **2010**, *22*, 4312–4319.

48. Wang, M.; Lai, Y.; Fang, J.; Li, J.; Qin, F.; Zhang, K.; Lu, H. N-doped porous carbon derived from biomass as an advanced electrocatalyst for aqueous aluminum/air battery. *Inter. J. Hydrog. Energy* **2015**, *40*, 16230–16237.

49. Zeng, X.; Leng, L.; Liu, F.; Wang, G.; Dong, Y.; Du, L.; Liu, L.; Liao, S. Enhanced Li-O_2 battery performance, using graphene-like nori-derived carbon as the cathode and adding LiI in the electrolyte as a promoter. *Electrochim. Acta* **2016**, *200*, 231–238.

50. Yang, H. B.; Miao, J.; Hung, S.-F.; Chen, J.; Tao, H. B.; Wang, X.; Zhang, L. et al. Identification of catalytic sites for oxygen reduction and oxygen evolution in N-doped graphene materials: Development of highly efficient metal-free bifunctional electrocatalyst. *Sci. Adv.* **2016**, *2*, e1501122.

51. Wang, L.; Yin, F.; Yao, C. N-doped graphene as a bifunctional electrocatalyst for oxygen reduction and oxygen evolution reactions in an alkaline electrolyte. *Inter. J. Hydrog. Energy* **2014**, *39*, 15913–15919.

52. Zhao, C.; Yu, C.; Liu, S.; Yang, J.; Fan, X.; Huang, H.; Qiu, J. 3D porous N-doped graphene frameworks made of interconnected nanocages for ultrahigh-rate and long-life Li-O_2 batteries. *Adv. Funct. Mater.* **2015**, *25*, 6913–6920.

53. He, M.; Zhang, P.; Liu, L.; Liu, B.; Xu, S. Hierarchical porous nitrogen doped three-dimensional graphene as a free-standing cathode for rechargeable lithium-oxygen batteries. *Electrochim. Acta* **2016**, *191*, 90–97.

54. Liu, Y.; Li, J.; Li, W.; Li, Y.; Zhan, F.; Tang, H.; Chen, Q. Exploring the nitrogen species of nitrogen doped graphene as electrocatalysts for oxygen reduction reaction in Al-air batteries. *Inter. J. Hydrog. Energy* **2016**, *41*, 10354–10365.

55. Zhang, Z.; Peng, B.; Chen, W.; Lai, Y.; Li, J. Nitrogen-doped carbon nanotubes with hydrazine treatment as cathode materials for lithium-oxygen batteries. *J. Solid State Electrochem.* **2015**, *19*, 195–200.
56. Yadav, R. M.; Wu, J.; Kochandra, R.; Ma, L.; Tiwary, C. S.; Ge, L.; Ye, G.; Vajtai, R.; Lou, J.; Ajayan, P. M. Carbon nitrogen nanotubes as efficient bifunctional electrocatalysts for oxygen reduction and evolution reactions. *ACS Appl. Mater. Interfaces* **2015**, *7*, 11991–12000.
57. Zhu, S.; Chen, Z.; Li, B.; Higgins, D.; Wang, H.; Li, H.; Chen, Z. Nitrogen-doped carbon nanotubes as air cathode catalysts in zinc-air battery. *Electrochim. Acta* **2011**, *56*, 5080–5984.
58. Li, Y.; Wang, J.; Li, X.; Liu, J.; Geng, D.; Yang, J.; Li, R.; Sun, X. Nitrogen-doped carbon nanotubes as cathode for lithium-air batteries. *Electrochem. Commun.* **2011**, *13*, 668–672.
59. Liu, Q.; Wang, Y.; Dai, L.; Yao, J. Scalable fabrication of nanoporous carbon fiber films as bifunctional catalytic electrodes for flexible Zn-air batteries. *Adv. Mater.* **2016**, *28*, 3000–3006.
60. Liang, H.-W.; Wu, Z.-Y.; Chen, L.-F.; Li, C.; Yu, S.-H. Bacterial cellulose derived nitrogen-doped carbon nanofiber aerogel: An efficient metal-free oxygen reduction electrocatalyst for zinc-air battery. *Nano Energy* **2015**, *11*, 366–376.
61. Shui, J.; Du, F.; Xue, C.; Li, Q.; Dai, L. Vertically aligned N-doped coral-like carbon fiber arrays as efficient air electrodes for high-performance non-aqueous Li-O$_2$ batteries. *ACS Nano* **2014**, *8*, 3015–3022.
62. Park, G. S.; Lee, J.-S.; Kim, S. T.; Park, S.; Cho, J. Porous nitrogen doped carbon fiber with churros morphology derived from electrospun bicomponent polymer as highly efficient electrocatalyst for Zn-air batteries. *J. Power Sources* **2013**, *243*, 267–273.
63. Liu, J.; Wang, Z.; Zhu, J. Binder-free nitrogen-doped carbon paper electrodes derived from polypyrrole/cellulose composite for Li-O$_2$ batteries. *J. Power Sources* **2016**, *306*, 559–566.
64. Su, Y.; Zhang, Y.; Zhuang, X.; Li, S.; Wu, D. Zhang, F.; Feng, X. Low-temperature synthesis of nitrogen/sulfur co-doped three-dimensional graphene frameworks as efficient metal-free electrocatalyst for oxygen reduction reaction. *Carbon* **2013**, *62*, 296–301.
65. Yan, H. J.; Xu, B.; Shi, S. Q.; Ouyang, C. Y. First-principles study of the oxygen adsorption and dissociation on graphene and nitrogen doped graphene for Li-air batteries. *J. Appl. Phys.* **2012**, *112*, 104316.
66. Zhang, Z.; Bao, J.; He, C.; Chen, Y. N.; Wei, J. P.; Zhou, Z. Hierarchical carbon-nitrogen architectures with both mesopores and macrochannels as excellent cathodes for rechargeable Li-O$_2$ batteries. *Adv. Funct. Mater.* **2014**, *24*, 6826–6833.
67. Zhou, W.; Zhang, H.; Nie, H.; Ma, Y.; Zhang, Y.; Zhang, H. Hierarchical micron-sized mesoporous/macroporous graphene with well-tuned surface oxygen chemistry for high capacity and cycling stability Li-O$_2$ battery. *ACS Appl. Mater. Interfaces* **2015**, *7*, 3389–3397.
68. Yang, Y.; Sun, Q.; Li, Y.-S.; Li, H.; Fu, Z.-W. A CoO$_x$/carbon double-layer thin film air electrode for non-aqueous Li-air batteries. *J. Power Sources* **2013**, *223*, 312–318.
69. Zhang, K.; Zhang, L.; Chen, X.; He, X.; Wang, X.; Dong, S.; Han, P. et al. Mesoporous cobalt molybdenum nitride: A highly active bifunctional electrocatalyst and its application in lithium-O$_2$ batteries. *J. Phys. Chem. C* **2013**, *117*, 858–865.
70. Lu, H.-J.; Li, Y.; Zhang, L.-Q.; Li, H.-N.; Zhou, Z.-X.; Liu, A.-R.; Zhang, Y.-J.; Liu, S.-Q. Synthesis of B-doped hollow carbon spheres as efficient non-metal catalyst for oxygen reduction reaction. *RSC Adv.* **2015**, *5*, 52126–52131.
71. Zheng, Y.; Jiao, Y.; Jaronic, M.; Jin, Y. G.; Qiao, S. Z. Nanostructured metal-free electrochemical catalysts for highly efficient oxygen reduction. *Small* **2012**, *8*, 3550–3566.

72. Yang, L. J.; Jiang, S. J.; Zhao, Y.; Zhu, L.; Chen, S.; Wang, X. Z.; Wu, Q.; Ma, J.; Ma, Y. W.; Hu, Z. Boron-doped carbon nanotubes as metal-free electrocatalysts for the oxygen reduction reaction. *Angew. Chem. Int. Ed.* **2011**, *50*, 7132–7135.

73. Shu, C.; Lin, Y.; Zhang, B.; Hamid, S. B. A.; Su, D. Mesoporous boron-doped onion-like carbon as long-life oxygen electrode for sodium-oxygen batteries. *J. Mater. Chem. A* **2016**, *4*, 6610–6619.

74. Su, J.; Cao, X.; Wu, J.; Jin, C.; Tian, J.-H.; Yang, R. One-pot synthesis of boron-doped ordered mesoporous carbon as efficient electrocatalysts for the oxygen reduction reaction. *RSC Adv.* **2016**, *6*, 24728–24737.

75. Lin, Y. M.; Su, D. S. Fabrication of nitrogen-modified annealed nanodiamond with improved catalytic activity. *ACS Nano* **2014**, *8*, 7823–7833.

76. Wang, R.; Sun, X.; Zhang, B.; Sun, X.; Su, D. Hybrid nanocarbon as a catalyst for direct dehydrogenation of propane: Formation of an active and selective core-shell sp^2/sp^3 nanocomposite structure. *Chem. –Eur. J.* **2014**, *20*, 6324–6331.

77. Lin, Y.; Pan, X.; Qi, W.; Zhang, B.; Su, D. S. Nitrogen-doped onion-like carbon: A novel and efficient metal-free catalyst for epoxidation reaction. *J. Mater. Chem. A* **2014**, *2*, 12475–12483.

78. Hsu, W. K.; Firth, S.; Redlich, P.; Terrones, M.; Terrones, H.; Zhu, Y. Q.; Grobert, N. et al. Boron-doping effect in carbon nanotubes. *J. Mater. Chem.* **2000**, *10*, 1425–1429.

79. Liu, X. M.; Romero, H. E.; Gutierrez, H. R.; Adu, K.; Eklund, P. C. Transparent boron-doped carbon nanotube films. *Nano Lett.* **2008**, *8*, 2613–2619.

80. Cermignani, W.; Paulson, T. E.; Onneby, C.; Pantano, C. G. Synthesis and characterization of boron-doped carbons. *Carbon* **1995**, *33*, 367–374.

81. Lin, Y.; Zhu, Y.; Zhang, b.; Kim, Y. A.; Endo, M.; Su, D. S. Boron-doped onion-like carbon with enriched substitutional boron: The relationship between electronic properties and catalytic performance. *J. Mater. Chem. A* **2015**, *3*, 21805–21814.

82. Lu, J.; Lau, K. C.; Sun, Y.-K.; Curtiss, L. A.; Amine, K. Review-understanding and mitigating some of the key factors that limit non-aqueous lithium-air battery performance batteries and energy storage. *J. Electrochem. Soc.* **2015**, *162*, A2439–A2446.

83. Cruz-Silva, E.; Urípez-Urías, F.; Muñoz-Sandoval, E.; Sumpter, B. G.; Terrones, H.; Charlier, J.-C.; Meunier, V.; Terronest, M. Electronic transport and mechanical properties of phosphorus- and phosphorus-nitrogen-doped carbon nanotubes. *ACS Nano* **2009**, *3*, 1913–1921.

84. Cruz-Silva, E.; Urípez-Urías, F.; Muñoz-Sandoval, E.; Sumpter, B. G.; Terrones, H.; Charlier, J.-C.; Meunier, V.; Terronest, M. Phosphorus and phosphorus-nitrogen doped carbon nanotubes for ultrasensitive and selective molecular detection. *Nanoscale* **2011**, *3*, 1008–1013.

85. Wu, J.; Yang, Z.; Sun, Q.; Li, X.; Strasser, P.; Yang, R. Synthesis and electrocatalytic activity of phosphorus-doped carbon xerogel for oxygen reduction. *Electrochim. Acta.* **2014**, *127*, 53–60.

86. Zhu, Y.-P.; Liu, Y.; Liu, Y.-P.; Ren, T.-Z.; Chen, T.; Yuan, Z.-Y. Carbon materials for efficient electrocatalytic oxygen reduction. *ChemCatChem* **2015**, *7*, 2903–2909.

87. Wu, J.; Yang, Z.; Li, X.; Sun, Q.; Jin, C.; Strasser, P.; Yang, R. Phosphorus-doped porous carbons as efficient electrocatalysts for oxygen reduction. *J. Mater. Chem. A* **2013**, *1*, 9889–9896.

88. Dake, L. S.; Baer, D. R.; Friedrich, D. M. Auger parameter measurements of phosphorus compounds for characterization of phosphazenes. *J. Vac. Sci. Technol. A.* **1989**, *7*, 1634.

89. Paraknowitsch, J. P.; Zhang, Y. J.; Wienert, B.; Thomas, A. Nitrogen- and phosphorus-co-doped carbons with tunable enhanced surface areas promoted by the doping additives. *Chem. Commun.* **2013**, *49*, 1208–1210.

90. Li, R.; Wei, Z.; Gou, X.; Xu, W. Phosphorus-doped graphene nanosheets as efficient metal-free oxygen reduction electrocatalysts. *RSC Adv.* **2013**, *3*, 9978–9984.

91. Molinasabio, M.; Rodriguezreinoso, F.; Caturla, F.; Sellés, M. J. Porosity in granular carbons activated with phosphoric acid. *Carbon* **1995**, *33*, 1105–1113.
92. Puziy, A. M.; Poddubnaya, O. I.; Socha, R. P.; Gurgul, J.; Wisniewski, M. XPS and NMR studies of phosphoric acid activated carbons. *Carbon* **2008**, *46*, 2113–2123.
93. Zhang, M.; Dai, L. Carbon nanomaterials as metal-free catalysts in next generation fuel cells. *Nano Energy* **2012**, *1*, 514–517.
94. Liu, Z.-W.; Peng, F.; Wang, H.-J.; Yu, H.; Zheng, W.-X.; Yang, J. Phosphorus-doped graphite layers with high electrocatalytic activity for the O_2 reduction in an alkaline medium. *Angew. Chem., Int. Ed.* **2011**, *50*, 3257–3261.
95. Zhang, S.; Cai, Y.; He, H.; Zhang, Y.; Liu, R.; Cao, H.; Wang, M. et al. Heteroatom doped graphdiyne as efficient metal-free electrocatalyst for oxygen reduction reaction in alkaline medium. *J. Mater. Chem. A.* **2016**, *4*, 4738–4744.
96. Zhao, Y.; Yang, L.; Chen, S.; Wang, X.; Ma, Y.; Wu, Q.; Jiang, Y.; Qian, W.; Hu, Z. Can boron and nitrogen co-doping improve oxygen reduction reaction activity of carbon nanotubes? *J. Am. Chem. Soc.* **2013**, *135*, 1201–1204.
97. Zheng, Y.; Jiao, Y.; Ge, L.; Jaroniec, M.; Qiao, S. Z. Two-step boron and nitrogen doping in graphene for enhanced synergistic catalysis. *Angew. Chem.* **2013**, *125*, 3192–3198.
98. Choi, C. H.; Park, S. H.; Woo, S. I. Binary and ternary doping of nitrogen, boron, and phosphorus into carbon for enhancing electrochemical oxygen reduction activity. *ACS Nano* **2012**, *6*, 7084–7091.
99. Wang, S.; Lyyamperumal, E.; Roy, A.; Xue, Y.; Yu, D.; Dai, L. Vertically aligned BCN nanotubes as efficient metal-free electrocatalysts for the oxygen reduction reaction: A synergetic effect by Co-doping with boron and nitrogen. *Angew. Chem. Int. Ed.* **2011**, *50*, 11756–11760.
100. Liang, J.; Jiao, Y.; Jaroniec, M.; Qiao, S. Z. Sulfur and nitrogen dual-doped mesoporous graphene electrocatalyst for oxygen reduction with synergistically enhanced performance. *Angew. Chem. Int. Ed.* **2012**, *51*, 11496–11500.
101. Xue, Y.; Yu, D.; Dai, L.; Wang, R.; Li, D.; Roy, A.; Lu, F.; Chen, H.; Liu, Y.; Qu, J. Three-dimensional B, N-doped graphene foam as a metal-free catalyst for oxygen reduction reaction. *Phys. Chem. Chem. Phys.* **2013**, *15*, 12220–12226.
102. Jiao, Y.; Zheng, Y.; Jaroniec, M.; Qiao, S. Z. Origin of the electrocatalytic oxygen reduction activity of graphene-based catalysts: A roadmap to achieve the best performance. *J. Am. Chem. Soc.* **2014**, *136*, 4394–4403.
103. Zhang, J.; Zhao, Z.; Xia, Z.; Dai, L. A metal-free bifunctional electrocatalyst for oxygen reduction and oxygen evolution reactions. *Nat. Nanotechn.* **2015**, *10*, 444–452.
104. Jiang, H.; Zhu, Y.; Feng, Q.; Su, Y.; Yang, X.; Li, C. Nitrogen and phosphorus dual-doped hierarchical porous carbon foams as efficient metal-free electrocatalysts for oxygen reduction reactions. *Chem. Eur. J.* **2014**, *20*, 3106–3112.
105. Gong, X.; Liu, S.; Ouyang, C.; Strasser, P.; Yang, R. Nitrogen- and phosphorus-doped biocarbon with enhanced electrocatalytic activity for oxygen reduction. *ACS Catal.* **2015**, *5*, 920–927.
106. Gorham, J.; Torres, J.; Wolfe, G.; d'Agostino, A.; Fairbrother, D. H. Surface reactions of molecular and atomic oxygen with carbon phosphide films. *J. Phys. Chem. B* **2005**, *109*, 20379–20386.
107. Bao, X.; Nie, X.; von Deak, D.; Biddinger, E. J.; Luo, W.; Asthagiri, A.; Ozkan, U. S.; Hadad, C. M. A first-principles study of the role of quaternary-N doping on the oxygen reduction reaction activity and selectivity of graphene edge sites. *Top. Catal.* **2013**, *56*, 1623–1633.
108. Song, L.; Liu, Z.; Reddy, A. L. M.; Narayanan, N. T.; Taha-Tijerina, J.; Peng, J.; Gao, G.; Lou, J.; Vajtai, R.; Ajayan, P. M. Binary and ternary atomic layers built from carbon, boron, and nitrogen. *Adv. Mater.* **2012**, *24*, 4878–4895.

109. Wang, S. Iyyamperumal, E.; Roy, A.; Xue, Y.; Yu, D.; Dai, L. Vertically aligned BCN nanotubes as efficient metal-free electrocatalysts for the oxygen reduction reaction: A synergetic effect by co-doping with boron and nitrogen. *Angew. Chem.* **2011**, *123*, 11960–11964.

110. Wang, S. Zhang, L.; Xia, Z.; Roy, A.; Chang, D. W.; Baek, J. B.; Dai, L. BCN graphene as efficient metal-free electrocatalyst for the oxygen reduction reaction. *Angew. Chem. Int. Ed.* **2012**, *51*, 4209–4212.

111. Pels, J. R.; Kapteijin, F.; Moulijn, J. A.; Zhu, Q.; Thomas, K. M. Evolution of nitrogen functionalities in carbonaceous materials during pyrolysis. *Carbon* **1995**, *33*, 1641–1653.

112. Matsumoto, M.; Yamada, M.; Watanabe, N. Reversible 1,4-cycloaddition of singlet oxygen to N-substituted 2-pyridones: 1,4-endoperoxide as a versatile chemical source of singlet oxygen. *Chem. Commun.* **2005**, 483–485.

113. Wu, M.; Qiao, J.; Li, K.; Zhou, X.; Liu, Y.; Zhang, J. A large-scale synthesis of heteroatom (N and S) co-doped hierarchically porous carbon (HPC) derived from polyquaternium for superior oxygen reduction reactivity. *Green Chem.* **2016**, *18*, 2699–2709.

114. Yazdi, A. Z.; Roberts, E. P. L.; Sundararaj, U. Nitrogen/sulfur co-doped helical graphene nanoribbons for efficient oxygen reduction in alkaline and acidic electrolytes. *Carbon* **2016**, *100*, 99–108.

115. Gao, S.; Li, L.; Geng, K.; Wei, X.; Zhang, S. Recycling the bio-waste to produce nitrogen and sulfur self-doped porous carbon as an efficient catalyst for oxygen reduction reaction. *Nano Energy* **2015**, *16*, 408–418.

116. Qu, K.; Zheng, Y.; Dai, S.; Qiao, S. Z. Graphene oxide-polydopamine derived N, S-codoped carbon nanosheets as superior bifunctional electrocatalyst for oxygen reduction and evolution. *Nano Energy* **2016**, *19*, 373–381.

117. Paraknowitsch, J. P.; Thomas, A. Doping carbons beyond nitrogen: An overview of advanced heteroatom doped carbons with boron, Sulphur and phosphorus for energy applications. *Energy Environ. Sci.* **2013**, *6*, 2839–2855.

118. Liu, Z.; Nie, H.; Yang, Z.; Zhang, J.; Jin, Z.; Lu, Y.; Xiao, Z.; Huang, S. Sulfur-nitrogen co-doped three-dimensional carbon foams with hierarchical pore structures as effcient metal-free electrocatalysts for oxygen reduction reactions. *Nanoscale* **2013**, *5*, 3283–3288.

119. Li, G.; Li, Y.; Liu, H.; Guo, Y.; Li, Y.; Zhu, D. Architecture of graphdiyne nanoscale film. *Chem. Comun.* **2010**, *46*, 3256–3258.

120. Zheng, X.; Cao, X.; Wu, J.; Tian, J.; Jin, C.; Yang, R. Yolk-shell N/P/B ternary-doped biocarbon derived from yeast cells for enhanced oxygen reduction reaction. *Carbon* **2016**, *107*, 907–916.

121. Wang, Y.-J.; Zhao, N.; Fang, B.; Li, H.; Bi, X.; Wang, H. Carbon-supported Pt-based alloy electrocatalysts for the oxygen reduction reaction in polymer electrolyte membrane fuel cells: Particle size, shape, and composition manipulation and their impact to activity. *Chem. Rev.* **2015**, *115*, 3433–3467.

122. Pettersson, J.; Ramsey, B.; Harrison, D. A review of the latest developments in electrodes for unitized regenerative polymer electrolyte fuel cells. *J. Power Sources* **2006**, *157*, 28–34.

123. Ikezawa, A.; Miyazaki, K.; Fukutsuka, T.; Abe, T. Investigation of electrochemically active regions in bifunctional air electrodes using partially immersed platinum electrodes. *J. Electrochem. Soc.* **2015**, *162*, A1646–A1653.

124. Wang, L.; Ara, M.; Wadumesthrige, K.; Salley, S.; Simon Ng, K. Y. Graphene nanosheet supported bifunctional catalyst for high cycle life Li-air batteries. *J. Power Sources* **2013**, *234*, 8–15.

125. Lu, Y.-C.; Xu, Z.; Gasteiger, H. A.; Chen, S.; Hamad-Schifferili, K.; Shao-Horn, Y. Platinum-Gold Nanoparticles: A highly active bifunctional electrocatalyst for rechargeable lithium-air batteries. *J. Am. Chem. Soc.* **2010**, *132*, 12170–12171.

126. Yin, J.; Fang, B.; Luo, J.; Wanjala, B.; Mott, D.; Loukrakpam, R.; Shan Ng, M. et al. Nanoscale alloying effect of gold-platinum nanoparticles as cathode catalysts on the performance of a rechargeable lithium-oxygen battery. *Nanotechnology* **2012**, *23*, 305404.

127. Terashima, C.; Iwai, Y.; Cho, S.-P.; Ueno, T.; Zettsu, N.; Saito, N.; Takai, O. Solution plasma sputtering processes for the synthesis of PtAu/C catalysts for Li-air batteries. *Int. J. Electrochem. Sci.* **2013**, *8*, 5407–5420.

128. Su, D.; Kim, H.-S.; Kim, W.-S.; Wang, G. A study of Pt_xCo_y alloy nanoparticles as cathode catalysts for lithium-air batteries with improved catalytic activity. *J. Power Sources* **2013**, *244*, 488–493.

129. Zhang, Y.; Wu, X.; Fu, Y.; Shen, W.; Zeng, X.; Ding, W. Carbon aerogel supported Pt-Zn catalyst and its oxygen reduction catalytic performance in magnesium-air batteries. *J. Mater. Res.* **2014**, *29*, 2863–2870.

130. Fujiwara, N.; Yao, M.; Siroma, Z.; Senoh, H.; Ioroi, T.; Yasuda, K. Reversible air electrodes integrated with an anion-exchange membrane for secondary air batteries. *J. Power Sources* **2011**, *196*, 808–813.

131. Ko, B. K.; Kim, M. K.; Kim, S. H.; Lee, M. A.; Shim, S. E.; Baeck, S.-H. Synthesis and electrocatalytic properties of various metals supported on carbon for lithium-air battery. *J. Mol. Catal. A-Chem.* **2013**, *379*, 9–14.

132. Ke, F.-S.; Solomon, B. C.; Ma, S.-G.; Zhou, X.-D. Metal-carbon nanocomposites as the oxygen electrode for rechargeable lithium-air batteries. *Electrochim. Acta.* **2012**, *85*, 444–449.

133. Chen, W.; Chen, S. Iridium-platinum alloy nanoparticles: Composition-dependent electrocatalytic activity for formic acid oxidation. *J. Mater. Chem.* **2011**, *21*, 9169–9178.

134. Ottakam Thotiyl, M. M.; Freunberger, S. A.; Peng, Z. Q.; Bruce, P. G. The carbon electrode in non-aqueous $Li-O_2$ cells. *J. Am. Chem. Soc.* **2013**, *135*, 494–500.

135. McCloskey, B. D.; Speidel, A.; Scheffler, R.; Miller, D. C.; Viswanathan, V.; Hummelshoj, J. S.; Norskov, J. K.; Luntz, A. C. Twin problems of interfacial carbonate formation in non-aqueous $Li-O_2$ batteries. *J. Phys. Chem. Lett.* **2012**, *3*, 997–10001.

136. Sun, B.; Huang, X.; Chen, S.; Munroe, P.; Wang, G. Porous graphene nanoarchitectures: An efficient catalyst for low charge-overpotential long life, and high capacity lithium-oxygen batteries. *Nano Lett.* **2014**, *14*, 3145–3152.

137. Zhou, W.; Cheng, Y.; Yang, X.; Wu, B.; Nie, H.; Zhang, H.; Zhang, H. Iridium incorporated into deoxygenated hierarchical graphene as a high-performance cathode for rechargeable Li-O2 batteries. *J. Mater. Chem. A.* **2015**, *3*, 14556–14561.

138. Itkis, D. M.; Semenenko, D. A.; Kataev, E. Y.; Belova, A. I.; Neudachina, V. S.; Sirotina, A. P.; Hävecker, M. et al. Reactivity of carbon in lithium-oxygen battery positive electrodes. *Nano Lett.* **2013**, *13*, 4697–4701.

139. Tong, S.; Zheng, M.; Lu, Y.; Lin, Z.; Zhang, X.; He, P.; Zhou, H. Binder-free carbonized bacterial cellulose-supported ruthenium nanoparticles for $Li-O_2$ batteries. *Chem. Commun.* **2015**, *51*, 7302–7304.

140. Ma, S.; Wu, Y.; Wang, J.; Zhang, Y.; Zhang, Y.; Yan, X.; Wei, Y. et al. Reversibility of noble metal-catalyzed aprotic $Li-O_2$ batteries. *Nano Lett.* **2015**, *15*, 8084–8090.

141. Alegre, C.; Modica, E.; Lo Vecchio, C.; Sebastán, Lázaro, J.; Aricò, A. S.; Baglio, V. Carbon nanofibers as advanced Pd catalyst supports for the air electrode of alkaline metal-air batteries. *ChemPlusChem* **2015**, *80*, 1384–1388.

142. McKerracher, R. D.; Alegre, C.; Baglio, V.; Aricò, A. S.; Ponce de León, C.; Mornaghini, F.; Rodlert, M.; Walsh, F. C. A nanostructured bifunctional Pt/C gas-diffusion electrode for metal-air batteries. *Electrochim. Acta.* **2015**, *174*, 508–515.

143. Cheng, H.; Scott, K. Selection of oxygen reduction catalyst for rechargeable lithium-air batteries-metal or oxide? *Appl. Catal. B Environ.* **2011**, *108–109*, 140–151.

144. Marinaro, M.; Riek, U.; Eswara Moorthy, S. K.; Bernhard, J.; Kaiser, U.; Wohlfahrt-Mehrens, M.; Jörissen, L. Au-coated carbon cathodes for improved reduction and evolution kinetics in aprotic Li-O_2 batteries. *Electrochem. Commun.* **2013**, *37*, 53–56.

145. Wang, T.; Kaempgen, M.; Nopphawan, P.; Wee, G.; Mhaisalkar, S.; Srinivasan, M. Silver nanoparticle-decorated carbon nanotubes as bifunctional gas-diffusion electrodes for zinc-air batteries. *J. Power Souces* **2010**, *195*, 4350–4355.

146. Cui, Z.; Chen, H.; Zhao, M.; DiSalvo, F. High-performance Pd_3Pb intermetallic catalyst for electrochemical oxygen reduction. *Nano Lett.* **2016**, *16*, 2560–2566.

147. Tien, H.-W., Huang, Y.-L., Yang, S.-Y., Wang, J.-Y., Ma, C.-C. M. The production of graphene nanosheets decorated with silver nanoparticles for use in transparent, conductive films. *Carbon* **2011**, *49*, 1550–1560.

148. Su, L.; Hei, J.; Wu, X.; Wang, L.; Wang, Y. Highly-dispersed Ni-QDs/mesoporous carbon nanoplates: A universal and commercially applicable approach based on corn straw piths and high capacitive performances. *ChemElectroChem* **2015**, *2*, 1897–1902.

149. Chen, Y.; Zhang, Q.; Zhang, Z.; Zhou, X.; Zhong, Y.; Yang, M.; Xie, Z.; Wei, J.; Zhou, Z. Two better than one: Cobalt-copper bimetallic yolk-shell nanoparticles supported on graphene as excellent cathode catalyst for Li-O_2 batteries. *J. Mater. Chem. A* **2015**, *3*, 17874–17879.

150. Stankovich, S.; Dikin, D. A.; Piner, R. D.; Kohlhaas, K. A.; Kleinhammes, A.; Jia, Y.; Wu, Y.; Nguyen, S. T.; Ruoff, R. S. Synthesis of graphene-based nanosheets via chemical reduction of exfoliated graphite oxide. *Carbon* **2007**, *45*, 1558–1565.

151. Xu, J.-J.; Wang, Z.-L.; Xu, D.; Zhang, L.-L.; Zhang, X.-B. Tailoring deposition and morphology of discharge products towards high-rate and long-life lithium-oxygen batteries. *Nat. Commun.* **2013**, *4*, 2438.

152. Xu, J.-J.; Wang, Z.-L.; Xu, D.; Men, F.-Z.; Zhang, X.-B. 3D ordered macroporous $LaFeO_3$ as efficient electrocatalyst for Li-O_2 batteries with enhanced rate capability and cyclic performance. *Energy Environ. Sci.* **2014**, *7*, 2213–2219.

153. Kwak, W.-J.; Kang, T.-G.; Sun, Y.-K.; Lee, Y. J. Iron-cobalt bimetal decorated carbon nanotubes as cost-effective cathode catalysts for Li-O2 batteries. *J. Mater. Chem. A* **2016**, *4*, 7020–7026.

154. Huang, J.; Zhang, B.; Xie, Y. Y.; Lye, W. W. K.; Xu, Z.-L.; Abouali, S.; Garakani, M. A. et al. Electrospun graphitic carbon nanofibers with in-situ encapsulated Co-Ni nanoparticles as freestanding electrodes for Li–O2 batteries. *Carbon* **2016**, *100*, 329–336.

155. Guo, X.; Liu, P.; Han, J.; Ito, Y.; Hirata, A.; Fujita, T.; Chen, M. 3D nanoporous nitrogen-doped graphene with encapsulated RuO_2 nanoparticles for Li–O_2 batteries. *Adv. Mater.* **2015**, *27*, 6137–6143.

156. Ren, X.; Zhang, S. S.; Tran, D. T.; Read, J. Oxygen reduction reaction catalyst on lithium/air discharge performance. *J. Mater. Chem.* **2011**, *21*, 10118–10125.

157. Esswein, A. J.; McMurdo, M. J.; Ross, P. N.; Bell, A. T.; Tilley, T. D. Size-dependent activity of Co_3O_4 nanoparticle anodes for alkaline water electrolysis. *J. Phys. Chem. C.* **2009**, *113*, 15068–15072.

158. Koza, J. A.; He, Z.; Miller, A. S.; Switzer, J. A. Electrodeposition of crystalline Co3O4-A catalyst for the oxygen evolution reaction. *Chem. Mater.* **2012**, *24*, 3567–3573.

159. Li, Q.; Cao, R.; Cho, J.; Wu, G. Nanostructured carbon-based cathode for non-aqueous lithium-oxygen batteries. *Phys. Chem. Chem. Phys.* **2014**, *16*, 13568–13582.

160. Li, Q.; Cao, R.; Cho, J.; Wu, G. Nanocarbon electrocatalysts for oxygen reduction in alkaline media for advanced energy conversion and storage. *Adv. Energy Mater.* **2014**, *4*, 1301415.

161. Gupta, S.; Kellogg, W.; Xu, H.; Liu, X.; Cho, J.; Wu, G. Bifunctional perovskite oxide catalysts for oxygen reduction and evolution in alkaline media. *Chem. Asian J.* **2016**, *11*, 10–21.

162. Suntivich, J.; Gasteiger, H. A.; Yabuuchi, N.; Nakanishi, H.; Goodenough, J. B.; Shao-Horn, Y. Design principles for oxygen-reduction activity on perovskite oxide catalysts for fuel cells and metal–air batteries. *Nat. Chem.* **2011**, *3*, 647.

163. Suntivich, J.; May, K. J.; Gasteiger, A.; Goodenough, J. B.; Shao-Horn, Y. A perovskite oxide optimized for oxygen evolution catalysis from molecular orbital principles. *Science* **2011**, *334*, 1383–1385.

164. Hardi, W. G.; Mefford, J. T.; Slanac, D. A.; Patel, B. B.; Wang, X. Q.; Dai, S.; Zhao, X.; Ruoff, R. S.; Johnston, K. P.; Stevenson, K. J. Tuning the electrocatalytic activity of perovskites through active site variation and support interactions. *Chem. Mater.* **2014**, *26*, 3368–3376.

165. Jung, J.-I.; Jeong, H. Y.; Lee, J.-S.; Kim, M. G.; Cho, J. A bifunctional perovskite catalyst for oxygen reduction and evolution. *Angew. Chem.* **2014**, *126*, 4670–4674.

166. Liu, J.; Liu, H.; Wang, F.; Song, Y. Composition-controlled synthesis of $Li_xCo_{3-x}O_4$ solid solution nanocrystals on carbon and their impact on electrocatalytic activity toward oxygen reduction reaction. *RSC Adv.* **2015**, *5*, 90785–90796.

167. Xu, Y.; Tsou, A.; Fu, Y.; Wang, J.; Tian, J.-H.; Yang, R. Carbon-coated perovskite $BaMnO_3$ porous nanorods with enhanced electrocatalytic properties for oxygen reduction and oxygen evolution. *Electrochim. Acta* **2015**, *174*, 551–556.

168. Nishio, K.; Molla, S.; Okugaki, T.; Nakanishi, S.; Nitta, I.; Kotani, Y. Effects of carbon on oxygen reduction and evolution reactions of gas-diffusion air electrodes based on perovskite-type oxides. *J. Power Sources* **2015**, *298*, 236–240.

169. Velraj, S.; Zhu, J. H. Cycle life limit of carbon-based electrodes for rechargeable metal-air battery application. *J. Electroanal. Chem.* **2015**, *736*, 76–82.

170. Velraj, S.; Zhu, J. H. $Sr_{0.5}Sm_{0.5}CoO_{3-\delta}$—A new bi-functional catalyst for rechargeable metal-air battery applications. *J. Power Sources* **2013**, *227*, 48–52.

171. Li, X. X.; Qu, W.; Zhang, J.; Wang, H. Electrocatalytic activities of $La_{0.6}Ca_{0.4}CoO_3$ and $La_{0.6}Ca_{0.4}CoO_3$-carbon composites toward the oxygen reduction reaction in concentrated alkaline electrolytes. *J. Electrochem. Soc.* **2011**, *158*, A597–A604.

172. Arai, H.; Müller, S.; Haas, O. AC impedance analysis of bifunctional air electrodes for metal-air batteries. *J. Electrochem. Soc.* **2000**, *147*, 3584–3591.

173. Yuasa, M.; Matsuyoshi, T.; Kida, T.; Shimanoe, K. Discharge/charge characteristic of Li-air cells using carbon-supported $LaMn_{0.6}Fe_{0.4}O_3$ as an electrocatalyst. *J. Power Sources* **2013**, *242*, 216–221.

174. Ross, P. N.; Sokol, H. The corrosion of carbon black anodes in alkaline electrolyte. I. acetylene black and the effect of cobalt catalyzation. *J. Electrochem. Soc.* **1984**, *131*, 1742–1750.

175. Débart, A.; Paterson, A. J.; Bao, J.; Bruce, P. G. α-MnO_2 nanowires: A catalyst for the O_2 electrode in rechargeable lithium batteries. *Angew. Chem. Int. Ed.* **2008**, *47*, 4521–4524.

176. Chung, K. B.; Shin, J. K.; Jang, T. Y.; Noh, D. K.; Tak, Y.; Baeck, S.-H. Preparation and analyses of MnO_2/carbon composites for rechargeable lithium-air battery. *Rev. Adv. Mater. Sci.* **2011**, *28*, 54–58.

177. Yuasa, M.; Tachibana, N.; Shimanoe, K. Oxygen reduction activity of carbon-supported $La_{1-x}Ca_xMn_{1-y}Fe_yO_3$ nanoparticles. *Chem. Mater.* **2013**, *25*, 3072–3079.

178. Yuasa, M.; Shimanoe, K.; Teraoka, Y.; Yamazoe, N. High-performance oxygen reduction catalyst using carbon-supported La-Mn-based perovskite-type oxide. *Electrochem. Solid-State Lett.* **2011**, *14*, A67–A69.

179. Liang, Y.; Li, Y.; Wang, H.; Zhou, J.; Wang, J.; Regier, T.; Dai, H. Co_3O_4 nanocrystals on graphene as a synergistic catalyst for oxygen reduction reaction. *Nat. Mater.* **2011**, *10*, 780–786.

180. Cheng, F.; Shen, J.; Peng, B.; Pan, Y.; Tao, Z.; Chen, J. Rapid room-temperature synthesis of nanocrystalline spinels as oxygen reduction and evolution electrocatalysts. *Nature Chem.* **2011**, *3*, 79–84.

181. Li, P.; Ma, R.; Zhou, Y.; Chen, Y.; Zhou, Z.; Liu, G.; Liu, Q.; Peng, G.; Liang, Z.; Wang, J. In situ growth of spinel $CoFe_2O_4$ nanoparticles on rod-like ordered mesoporous carbon for bifunctional electrocatalysis of both oxygen reduction and oxygen evolution. *J. Mater. Chem. A* **2015**, *3*, 15598–15606.

182. Yan, W.; Bian, W.; Jin, C.; Tian, J.-H.; Yang, R. An efficient bi-functional electrocatalyst based on strongly coupled $CoFe_2O_4$/carbon nanotubes hybrid for oxygen reduction and oxygen evolution. *Electrochim. Acta* **2015**, *177*, 65–72.

183. Lee, J. S.; Lee, T.; Song, H. K.; Cho, J.; Kim, B.-S. Ionic liquid modified graphene nanosheets anchoring manganese oxide nanoparticles as efficient electrocatalysts for Zn-air batteries. *Energy Environ. Sci.* **2011**, *4*, 4148–4154.

184. Lee, D. U.; Kim, B. J.; Chen, Z. One-pot synthesis of a mesoporous $NiCo_2O_4$ nanoplatelet and graphene hybrid and its oxygen reduction and evolution activities as an efficient bi-functional electrocatalyst. *J. Mater. Chem. A* **2013**, *1*, 4754–4762.

185. Bian, W.; Yang, Z.; Strasser, P.; Yang, R. A $CoFe_2O_4$/graphene nanohybrid as an efficient bi-functional electrocatalyst for oxygen reduction and oxygen evolution. *J. Power Sources* **2014**, *250*, 196–203.

186. Liu, S.; Bian, W.; Yang, Z.; Tian, J.; Jin, C.; Shen, M.; Zhou, Z.; Yang, R. A facile synthesis of $CoFe_2O_4$/biocarbon nanocomposites as efficient bi-functional electrocatalysts for the oxygen reduction and oxygen evolution reaction. *J. Mater. Chem. A.* **2014**, *2*, 18012–18017.

187. Liu, Y.; Higgins, D. C.; Wu, J.; Fowler, M.; Chen, Z. Cubic spinel cobalt oxide/multi-walled carbon nanotube composites as an efficient bifunctional electrocatalyst for oxygen reaction. *Electrochem. Commun.* **2013**, *34*, 125–129.

188. Jiang, Q.; Liang, L. H.; Zhao, D. S. Lattice contraction and surface stress of fcc nanocrystals. *J. Phys. Chem. B.* **2001**, *105*, 6275–6277.

189. Lopes, I.; El Hassan, N.; Guerba, H.; Wallez, G.; Davidson, A. Size-induced structural modifications affecting Co_3O_4 nanoparticles patterned in SBA-15 silicas. *Chem. Mater.* **2006**, *18*, 5826–5828.

190. Kinoshita, K. *Electrochemical Oxygen Technology.* New York: John Wiley & Sons Inc., **1992**.

191. Cheng, F.; Su, Y.; Liang, J.; Tao, Z.; Chen, J. MnO_2-based nanostructures as catalysts for electrochemical oxygen reduction in alkaline media. *Chem. Mater.* **2010**, *22*, 898–905.

192. Xiao, W.; Wang, D.; Lou, X. W. Shape-controlled synthesis of MnO_2 nanostructures with enhanced electrocatalytic activity for oxygen reduction. *J. Phys. Chem. C.* **2010**, *114*, 1694–1700.

193. Cao, Y.; Zheng, M.-S.; Cai, S.; Lin, X.; Yang, C.; Hu, W.; Dong, Q.-F. Carbon embedded α-MnO_2@graphene nanosheet composite: A bifunctional catalyst for high performance lithium oxygen batteries. *J. Mater. Chem. A.* **2014**, *2*, 18736–18741.

194. Salehi, M.; Shariatinia, Z. An optimization of MnO_2 amount in CNT-MnO_2 nanocomposite as a high rate cathode catalyst for the rechargeable Li-O_2 batteries. *Electrochim. Acta.* **2016**, *188*, 428–440.

195. Sumboja, A.; Ge, X.; Thomas Goh, F. W.; Li, B.; Geng, D.; Andy Hor, T. S.; Zong, Y.; Liu, Z. Manganese oxide catalyst grown on carbon paper as an air cathode for high-performance rechargeable zinc-air batteries. *ChemPlusChem* **2015**, *80*, 1341–1346.

196. Yang, Y.; Shi, M.; Li, Y.-S.; Fu, Z.-W. MnO_2-graphene composite air electrode for rechargeable Li-air batteries. *J. Electrochem. Soc.* **2012**, *159*, A1917–A1921.

197. Gong, K.; Yu, P.; Su, L.; Xiong, S.; Mao, L. Polymer-assisted synthesis of manganese dioxide/carbon nanotube nanocomposite with excellent electrocatalytic activity toward reduction of oxygen. *J. Phys. Chem. C.* **2007**, *111*, 1882–1887.

198. Liu, S.; Zhu, Y.; Xie, J.; Huo, Y.; Yang, H. Y.; Zhu, T.; Cao, G.; Zhao, X.; Zhang, S. Direct growth of flower-like δ-MnO_2 on three-dimensional graphene for high-performance rechargeable Li-O_2 batteries. *Adv. Energy Mater.* **2014**, *4*, 1301960.

199. Gao, R.; Li, Z.; Zhang, X.; Zhang, J.; Hu, Z.; Liu, X. Carbon-dotted defective CoO with oxygen vacancies: A synergetic deign of bifunctional cathode catalyst for Li-O2 batteries. *ACS Catal.* **2016**, *6*, 400–406.

200. Huang, B.-W.; Li, L.; He, Y.-J.; Liao, X.-Z.; He, Y.-S.; Zhang, W.; Ma, Z.-F. Enhanced electrochemical performance of nanofibrous CoO/CNF cathode catalyst for Li-O₂ batteries. *Electrochim. Acta* **2014**, *137*, 183–189.

201. Tan, P.; Shyy, W.; Zhao, T. S.; Zhu, X. B.; Wei, Z. H. A RuO₂ nanoparticle-decorated buckypaper cathode for non-aqueous lithium-oxygen batteries. *J. Mater. Chem. A.* **2015**, *3*, 19042–19049.

202. Jung, H.-G.; Jeong, Y. S.; Park, J.-B.; Sun, Y.-K.; Scrosati, B.; Lee, Y. J. Ruthenium-based electrocatalysts supported on reduced graphene oxide for lithium-air batteries. *ACS Nano* **2013**, *7*, 3532–3539.

203. Lee, Y.; Suntivich, J.; May, K. J.; Perry, E. E.; Shao-Horn, Y. Synthesis and activities of rutile IrO₂ and RuO₂ nanoparticles for oxygen evolution in acid and alkaline solutions. *J. Phys. Chem. Lett.* **2012**, *3*, 399–404.

204. Zhang, X.; Xiao, Q.; Zhang, Y.; Jiang, X.; Yang, Z.; Xue, Y.; Yan, Y.-M.; Sun, K. La₂O₃ doped carbonaceous microspheres: A novel bifunctional electrocatalyst for oxygen reduction and evolution reactions with ultrahigh mass activity. *J. Phys. Chem. C* **2014**, *118*, 20229–20237.

205. Ahn, C.-H.; Kalubarme, R. S.; Kim, Y.-H.; Jung, K.-N.; shin, K.-H.; Par, C.-J. Graphene/ doped ceria nano-blend for catalytic oxygen reduction in non-aqueous lithium-oxygen batteries. *Electrochim. Acta.* **2014**, *117*, 18–25.

206. Liu, X.; Park, M.; Kim, M. G.; Gupta, S.; Wang, X.; Wu, G.; Cho, J. High-performance non-spinel cobalt-manganese mixed oxide-based bifunctional electrocatalysts for rechargeable zinc-air batteries. *Nano Energy* **2016**, *20*, 315–325.

207. Lu, X.; Wang, G.; Zhai, T.; Yu, M.; Xie, S.; Ling, Y.; Liang, C.; Tong, Y. Li, Y. Stabilized TiN nanowire arrays for high-performance and flexible supercapacitors. *Nano Lett.* **2012**, *12*, 5376–5381.

208. He, P.; Wang, Y.; Zhou, H. Titanium nitride catalyst cathode in Li-air fuel cell with an acidic aqueous solution. *Chem. Commun.* **2011**, *47*, 10701–10703.

209. Wang, Y.; Ohnishi, R.; Yoo, E.; He, P. Kubota, J.; Domen, K. Nano- and micro-sized TiN as the electrocatalysts for ORR in Li-air fuel cell with alkaline aqueous electrolyte. *J. Mater. Chem.* **2012**, *22*, 15549–15555.

210. Chen, J.; Takanabe, K.; Ohnishi, R.; Lu, D.; Okada, S.; Hatasawa, H.; Morioka, H. M.; Antonietti, M.; Kubota, J.; Domen, K. Nano-sized TiN on carbon black as an efficient electrocatalyst for the oxygen reduction reaction prepared using an mgp-C₃N₄ template. *Chem. Commun.* **2010**, *46*, 7492–7494.

211. Li, F.; Ohnishi, R.; Yamada, Y.; Kubota, J.; Domen, K.; Yamada, A.; Zhou, H. Carbon supported TiN nanoparticles: An efficient bifunctional catalyst for non-aqueous Li-O₂ batteries. *Chem. Commun.* **2013**, *49*, 1175–1177.

212. Laoire, C. O.; Mukerjee, S.; Abraham, K. M.; Plichta, E. J.; Hendrickson, M. A. Influence of non-aqueous solvents on the electrochemistry of oxygen in the rechargeable lithium-air battery. *J. Phys. Chem. C.* **2010**, *114*, 9178–9186.

213. Laoire, C. O.; Mukerjee, S.; Abraham, K. M.; Plichta, E. J.; Hendrickson, M. A. Elucidating the mechanism of oxygen reduction for lithium-air battery applications. *J. Phys. Chem. C.* **2009**, *113*, 20127–20134.

214. Park, J.; Jun, Y.-S.; Lee, W.; Gerbec, J. A.; See, K. A.; Stucky, G. D. Bimodal mesoporous titanium nitride/carbon microfibers as efficient and stable electrocatalysts for Li-O₂ batteries. *Chem. Mater.* **2013**, *25*, 3779–3781.

215. Lyth, S. M.; Nabae, Y.; Moriya, S.; Kuroki, S.; Kakimoto, M. A.; Ozaki, J.-I.; Miyata, S. Carbon nitride as a nonprecious catalyst for electrochemical oxygen reduction. *J. Phys. Chem. C.* **2009**, *113*, 20148–20151.

216. Zhang, L.; Su, Z.; Jiang, F.; Yang, L.; Qian, J.; Zhou, Y.; Li, W.; Hong, M. Highly graphitized nitrogen-doped porous carbon nanopolyhedra derived from ZIF-8 nanocrystals as efficient electrocatalysts for oxygen reduction reactions. *Nanoscale* **2014**, *6*, 6590–6602.

217. Zhang, Y.; Mori, T.; Ye, J.; Antonietti, M. Phosphorus-doped carbon nitride solid: Enhanced electrical conductivity and photocurrent generation. *J. Am. Chem. Soc.* **2010**, *132*, 6294–6295.

218. Fu, X.; Hu, X.; Yan, Z.; Lei, K.; Li, F.; Cheng, F.; Chen, J. Template-free synthesis of porous graphitic carbon nitride/carbon composite spheres for electrocatalytic oxygen reduction reaction. *Chem. Commun.* **2016**, *52*, 1725–1728.

219. Ma, T. Y.; Ran, J.; Dai, S.; Jaroniec, M.; Qiao, S. Z. Phosphorus-doped graphitic carbon nitrides grown *in situ* on carbon-fiber paper: Flexible and reversible oxygen electrodes. *Angew. Chem. Int. Ed.* **2015**, *54*, 4646–4650.

220. Koo, B. S.; Lee, J. K.; Yoon, W. Y. Improved electrochemical performances of lithium-oxygen batteries with tungsten carbide-coated cathode. *Japan. J. Appl. Phys.* **2015**, *54*, 047101.

221. Luo, W.-B.; Chou, S.-L.; Wang, J.-Z.; Liu, H.-K. A B4C nanowire and carbon nanotube composite as a novel bifunctional electrocatalyst for high energy lithium oxygen batteries. *J. Mater. Chem. A.* **2015**, *3*, 18395–18399.

222. Huang, K.; Bi, K.; Xu, J. C.; Liang, C.; Lin, S.; Wang, W. J.; Yang, T. Z. et al. Novel graphite-carbon encased tungsten carbide nanocomposites by solid-state reaction and their ORR electrocatalytic performance in alkaline medium. *Electrochim. Acta.* **2015**, *174*, 172–177.

223. Wang, Y.-J.; Wilkinson, D. P.; Zhang, J. Noncarbon support materials for Polymer Electrolyte Membrane Fuel Cell electrocatalysts. *Chem. Rev.* **2011**, *111*, 7625–7651.

224. Lu, Y. C.; Yang, S. H. Probing the reaction kinetics of the charge reaction of non-aqueous Li-O₂ batteries. *J. Phys. Chem. Lett.* **2013**, *4*, 93–99.

225. Harding, J. R.; Lu, Y. C.; Tsukada, Y.; Yang, S. H. Evidence of catalyzed oxidation of Li₂O₂ for rechargeable Li-air battery applications. *Phys. Chem. Chem. Phys.* **2012**, *14*, 10540–10546.

226. Werheit, H.; Au, T.; Schmechel, R.; Shalamberidze, S. O.; Kalandadze, G. I.; Eristavi, A. M. IR-active phonons and structure elements of isotope-enriched boron carbide. *J. Solid State Chem.* **2000**, *154*, 79–86.

227. Lazzari, R.; Vst, N.; Besson, J. M.; Baroni, S.; Dal Corso, A. Atomic structure and vibrational properties of icosahedral B4C boron carbide. *Phys. Rev. Lett.* (1999): 83, 3230–3233.

228. Lyu, Z.; Zhang, J.; Wang, L.; Yuan, K.; Luan, Y.; Xiao, P.; Chen, W. CoS₂ nanoparticles-graphene hybrid as a cathode catalyst for aprotic Li-O₂ batteries. *RSC adv.* **2016**, *6*, 31739–31743.

229. Wu, J.; Dou, S.; Shen, A.; Wang, X.; Ma, Z.; Ouyang, C.; Wang, S. One-step hydrothermal synthesis of NiCo₂S₄-rGO as an efficient electrocatalyst for the oxygen reduction reaction. *J. Mater. Chem. A.* **2014**, *2*, 20990–20995.

230. Chen, W.; Lai, Y.; Zhang, Z.; Gan, Y.; Jiang, S.; Li, J. β-FeOOH decorated highly porous carbon aerogels composites as a cathode material for rechargeable Li-O₂ batteries. *J. Mater. Chem. A.* **2015**, *3*, 6447–6454.

231. Li, J.; Wen, W.; Zhou, M.; Guan, L.; Huang, Z. MWNT-supported bifunctional catalyst of β-FeOOH nanospindles for enhanced rechargeable Li-O₂ batteries. *J. Alloy. Compd.* **2015**, *639*, 428–434.

232. Hu, X.; Fu, X.; Chen, J. A soil/Vulcan XC-72 hybrid as a highly-effective catalytic cathode for rechargeable Li-O₂ batteries. *Inorg. Chem. Front.* **2015**, *2*, 1006–1010.

233. Lee, C. K.; Park, Y. J. Polyimide-wrapped carbon nanotube electrodes for long cycle Li-air batteries. *Chem. Commun.* **2015**, *51*, 1210–1213.

234. Yoo, E.; Zhou, H. Fe phthalocyanine supported by graphene nanosheet as catalyst in Li-air battery with the hybrid electrolyte. *J. Power Sources* **2013**, *244*, 429–434.

235. Wang, H.; Liang, Y.; Li, Y.; Dai, H. $Co_{1-x}S$-graphene hybrid: A high-performance metal chalcogenide electrocatalyst for oxygen reduction. *Angew. Chem. Int. Ed.* **2011**, *50*, 10969–10972.

236. Seabold, J. A.; Choi, K.-S. Efficient and stable photo-oxidation of water by a bismuth vanadate photoanode coupled with an iron oxyhydroxide oxygen evolution catalyst. *J. Am. Chem. Soc.* **2012**, *134*, 2186–2192.

237. Débart, A.; Bao, J.; Armstrong, G.; Bruce, P. G. An O_2 Cathode for rechargeable lithium batteries: The effect of a catalyst. *J. Power Sources* **2007**, *174*, 1177–1182.

238. Wang, X.; Li, Y. Rational synthesis of α-MnO_2 single-crystal nanorods. *Chem. Commun.* **2002**, 764–765.

239. Kraytsberg, A.; Ein-Eli, Y. The impact of nano-scaled materials on advanced metal-air battery systems. *Nano Energy* **2013**, *2*, 468–480.

240. Aetukuri, N. B.; McCloskey, B. D.; Carcía, J. M.; Krupp, L. E.; Viswanathan, V.; Luntz, A. C. Solvating additives drive solution-mediated electrochemistry and enhance toroid growth in non-aqueous Li-O_2 batteries. *Nat. Chem.* **2015**, *7*, 50–56.

241. Hu, X. F.; Han, X. P.; Hu, Y. X.; Cheng, F. Y.; Chen, J. ε-MnO_2 nanostructures directly grown on Ni foam: A cathode catalyst for rechargeable Li-O_2 batteries. *Nanoscale* **2014**, *6*, 3522–3525.

242. Su, C.; Yang, T.; Zhou, W.; Wang, W.; Xu, X.; Shao, Z. Pt/C-$LiCoO_2$ composites with ultralow Pt loadings as synergistic bifunctional electrocatalyst for oxygen reduction and evolution reactions. *J. Mater. Chem. A.* **2016**, *4*, 4516–4524.

243. Zhao, H.; Haili, W.; Liang, L.; Xu, H.; Fu, J. Preparation and characterization of Fe_3O_4-Pt/C electro-catalysts for zinc-air battery cathodes. *Adv. Mat. Res.* **2011**, *239–242*, 1184–1189.

244. Wu, Q.; Jiang, L.; Qi, L.; Wang, E.; Sun, G. Electrocatalytic performance of Ni modified MnO_x/C composites toward oxygen reduction reaction and their application in Zn-air battery. *Inter. J. Hydrogen Energ.* **2014**, *39*, 3423–3432.

245. Hu, J.; Shi, L.; Liu, Q.; Huang, H.; Jiao, T. Improved oxygen reduction activity on silver-modified $LaMnO_3$-graphene via shortens the conduction path of adsorbed oxygen. *RSC Adv.* **2015** *5*, 92096–92106.

246. Hu, F.-P.; Zhang, X.-G.; Xiao, F.; Zhang, J.-L. Oxygen reduction on Ag-MnO_2/SWNT and Ag-MnO_2/AB electrodes. *Carbon* **2005**, *43*, 2931–2936.

247. Zhai, X.; Yang, W.; Li, M.; Lv, G.; Liu, J.; Zhang, X. Noncovalent hybrid of $CoMn_2O_4$ spinel nanocrystals and polydiallyl (dimethylammonium chloride) functionalized carbon nanotubes as efficient electrocatalyst for oxygen reduction reaction. *Carbon* **2013**, *65*, 277–286.

248. Ma, H.; Wang, B. A bifunctional electrocatalyst α-MnO_2-$LaNiO_3$/carbon nanotube composite for rechargeable zinc-air batteries. *RSC Adv.* **2014**, *4*, 46084–46092.

249. Chen, D.; Chen, C.; Baiyee, Z. M.; Shao, Z.; Ciucci, F. Nonstoichiometric oxides as low-cost and highly-efficient oxygen reduction/evolution catalysts for low-temperature electrochemical devices. *Chem. Rev.* **2015**, *115*, 9869–9921.

250. Chen, F. Y.; Chen, J. Metal-air Batteries: From oxygen reduction electrochemistry to cathode catalysts. *Chem. Soc. Rev.* **2012**, *41*, 2172–2192.

251. Cao, R.; Lee, J.-S.; Liu, M.; Cho, J. Recent progress in non-precious catalysts for metal-air batteries. *Adv. Energy. Mater.* **2012**, *2*, 816–829.

252. Li, J.; Zou, M.; Chen, L.; Huang, Z.; Guan, L. An efficient bifunctional catalyst of Fe/Fe_3C carbon nanofibers for rechargeable Li-O_2 batteries. *J. Mater. Chem. A.* **2014**, *2*, 10634–10638.

253. Wang, X.; Liu, X.; Tong, C.-J.; Yuan, X.; Dong, W.; Lin, T.; Liu, L.-M.; Huang, F. An electron injection promoted highly efficient electrocatalyst of $FeNi_3$@GR@Fe-NiOOH for oxygen evolution and rechargeable metal-air batteries. *J. Mater. Chem. A.* **2016**, *4*, 7762–7771.

254. Mirzaeian, M.; Hall, P. J. Preparation of controlled porosity carbon aerogels for energy storage in rechargeable lithium oxygen batteries. *Electrochim. Acta.* **2009**, *54*, 7444–7451.
255. Shui, J.-L.; Karan, N. K.; Balasubramanian, M.; Li, S.-Y.; Liu, D.-J. Fe/N/C composite in Li-O$_2$ battery: Studies of catalytic structure and activity toward oxygen evolution reaction. *J. Am. Chem. Soc.* **2012**, *134*, 16654–16661.

4 Doped-Carbon Composited Bifunctional Electrocatalysts for Rechargeable Metal-Air Batteries

4.1 INTRODUCTION

Recently, developments related to material selection, preparation, and processing has continuously provided opportunities to increase MABs performance for next-generation energy devices. Based on the current state of technology, major efforts in RMAB development have been put into the progress of air-electrodes, in particular bifunctional electrocatalysts for both ORR and OER. Among the different types of MAB bifunctional electrocatalysts, the low-cost materials of doped-carbon composited with other elements or compounds have been extensively explored and have shown promising performances in terms of catalytic activity and stability.

4.1.1 COMPOSITES OF DOPED-CARBON WITH CARBON

Because the combination of two different carbon materials has been found to have comparable catalytic performance to the commercially available noble metal-based electrocatalysts for rechargeable metal-air batteries (RMABs), heteroatom(s)-doped carbons, as modified carbon materials, have been explored to compensate for the disadvantages of the other carbon material in preparing composites of heteroatom-doped carbon, with carbon as the advanced bifunctional electrocatalysts.

For example, using an injection chemical vapor deposition method, Park et al.[1] prepared a composite of ethylene diamine-derived nitrogen-doped carbon nanotubes (NCNT) with thermally reduced graphene oxide (TRGO) to investigate the electrocatalytic activities towards both ORR and OER for RMABs. They proposed that as a substrate material, TRGO could serve as a bridge to electrically connect discrete graphene sheets and thereby restore electrical conductivity. They believed the addition of NCNT onto graphene sheets could prevent the sheets from restacking during electrode fabrication and thus avoiding the loss of catalytic activities.[2,3] In a typical experimental synthesis, ethylene diamine (EDA) was used as a carbon source to prepare NCNT directly onto TRGO because EDA could offer high nitrogen contents and nitrogen configurations for EDA-derived NCNT,[4] favoring catalytic performance. In the analysis of morphology and structure of the TRGO/NCNT composite, a combination of SEM and TEM showed that the NCNTs were well distributed on the graphene sheet in which the NCNTs could contact several graphene sheets. The connection of NCNTs and graphene sheets could not only improve the electronic conduction but also form a porous structure between the graphene sheets. Based on BET measurements, the surface area of the TRGO/NCNT composite was 175.3 m^2 g^{-1}, which was sufficiently high for the uniform distribution of active sites on the carbon support, even though TRGO had a much higher surface area of 528.4 m^2 g^{-1}. Interestingly, the addition of NCNTs could result in the increased pore size in the TRGO/NCNT composite, suggesting that the NCNT grown on the surface of graphene sheets helped prevent the sheets from restacking and provided spaces between graphene sheets. XPS analysis showed the contents of carbon, oxygen, and nitrogen were 91.0%, 4.6%, and 3.9%, respectively. The electrochemical properties of the TRGO/NCNT composite were then evaluated using a RDE technique. In the CVs obtained in O$_2$-saturated 0.1 M KOH solution, it was seen that both TRGO/NCNT and TRGO exhibited only one oxidation peak. The peak of TRGO/NCNT had a higher current density and more positive potential than that of TRGO, suggesting that TRGO/NCNT possessed a higher ORR activity than TRGO. The LSV curves at 900 rpm further revealed that compared to TRGO, TRGO/NCNT presented 123 and 152 mV higher onset potential and half-wave potential, respectively. The comparison of TRGO and NCNT showed that NCNT had 75 and 92 mV improvements, respectively, in terms of onset and half-wave potentials. This clearly indicated that the addition of NCNT could contribute to the improvement of ORR activity as heteroatom-doping in the graphitic layer could act as ORR catalytic active sites.[5-7] TRGO/NCNT, following a four-electron transfer ORR mechanism, also had comparable ORR activities to those of state-of-the-art commercial 20wt% Pt/C. For the investigation of OER, it was found that at 1.0 V, the current density of TRGO/NCNT was ~27.8 mA cm^{-2}, which was ~6.5% and ~85% higher than NCNT and 20wt% Pt/C, respectively. After 200 continuous cycles, the residual current density of TRGO/NCNT was ~2.8 and ~7.6 times higher than those of NCNT and Pt/C, respectively. This suggested that TRGO/NCNT possessed higher and more durable OER activities than both TRGO and NCNT. This study demonstrated that the strategy of combining heteroatom-doped carbons with carbons of novel morphologies and structure to create bifunctional activities for ORR and OER should be an effective approach for RMABs. Similar to their study, Li et al.[8] developed a nanostructured, hierarchical composite cathode consisting of nitrogen-doped CNT (NCNT) arrays and a carbon paper for air-electrode of Li-air battery (LAB) with an aqueous catholyte. At 0.5 mA cm^{-2}, the tested discharging capacity and specific energy were ~710 mAh g^{-1} and ~2057 Wh kg^{-1},respectively, in which the LAB could reach a power density of ~10.4 mW cm^{-2}.

To solve the intrinsic hydrophobic characteristic of commercial carbon black (e.g., Ketjenblack) and develop advanced metal-free carbon-based composite bifunctional catalysts, N-doped carbon sheets derived from gelatin were grown on dispersed Ketjenblack nanoparticles to form the catalyst,[9] which had an irregularly stacked small sheet morphology with more exposed edge sites favorable to oxygen adsorption.[10,11] In a typical experiment, gelatin was the precursor for the nitrogen-doped carbon sheet. After a pyrolysis process, a composite of N-doped carbon sheets and Ketjenblack (NCS/KB) was obtained. In terms of different pyrolysis temperatures of 800°, 900°, and 1000°C, three corresponding samples labelled as NCS/KB-800, NCS/KB-900, and NCS/KB-1000, all had the content of Ketjenblack 13 times lower than that of gelatin. For comparison, a physical mixture of gelatin and Ketjenblack was used as the reference sample. The analysis of morphology by TEM images showed that the gelatin derived irregular carbon sheets grew randomly and tightly along aggregated Ketjenblack carbon with exposed graphitic edge sites, induced by the effective interactions between Ketjenblack and gelatin. When XPS was used to check the nitrogen chemical states of the catalyst, N-6 (pyridinic), N-5 (pyrrolic), N–G (graphitic), and N–X (oxidized nitrogen), species were found based on the deconvolution of N 1 s spectrum of NCS/KB-800 (see Figure 4.1 a–c). The pyridinic and graphitic nitrogen species were found to be the two predominant peaks. As the pyrolysis temperature in the synthesis increased from 800° to 1000°C, the ratio of graphitic (N–G) to pyridinic (N-6) increased from 0.98 to 1.43 and the overall nitrogen content decreased from 1.79% to 1.19 at%. In the investigation of ORR activity using a RDE technique, it was found that due to the strong interaction between gelatin derived NCS and Ketjenblack, all NCS/KB samples showed much better catalytic activity than the physical mixture of Ketjenblack and carbonized gelatin. In fact, the interaction between gelatin derived NCS and Ketjenblack was attributed to two factors: the improved electrical conductivity resulting from the addition of Ketjenblack, and N-doping in the carbon sheets. In detail, the comparison of the three NCS/KB composite catalysts revealed that NCS/KB-800 had a more negative onset potential, indicating a lower ORR

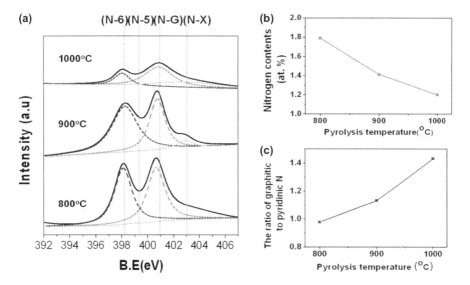

FIGURE 4.1 Chemical state of nitrogen species in NCS/K catalysts using XPS analysis. (a) N1 s XPS analysis of NCS/K-900. Labeled pyridinic nitrogen as (N-6), pyrrolic as (N-5), graphitic as (N–G), and oxidized nitrogen as (N–X). (b) Overall nitrogen content and (c) the ratio of graphitic N to pyridinic N depends on pyrolysis temperature. (From Nam, G. et al. *Nano Lett.* **2014**, *14*, 1870–1876.[9])

activity than those of the other two catalysts. When an assembled RZAB was run at 25 and 50 mA cm^{-2}, a considerably enhanced electrochemical improvement was observed for the NCS/KB-900 catalyst when compared to Pt/C catalysts. The peak power density of the battery with the NCS/KB-900 was about $\sim$193 mW cm^{-2}. This was slightly higher than that (about $\sim$188 mW cm^{-2}) with Pt/C, suggesting that NCS/KB-900 could be a potential catalyst for RZABs and that the compositing of NCS and KB should be a promising strategy to tune the hydrophobic characteristics of carbon.

Lin et al.[12] prepared a N-doped carbon/Ketjenblack (NC–KB) composite for RLABs by pyrolyzing nitrogen containing ionic liquids (IL) on Ketjenblack carbon. They believed that KB can provide porous structure for the transport of Li$^+$, O$_2$ and electrons, as well as offer a high surface area to distribute N-doped carbons derived from IL. It was demonstrated that the NC–KB composite could enhance the solvent pathway of Li$_2$O$_2$ growth, which altered the morphology of Li$_2$O$_2$ into loosely packed nanosheets to leave the passageway clear for Li$^+$/O$_2$ transport, resulting in a significantly improved discharge capacity ($\sim$9905 mAh g^{-1}). In particular, the pyridinic-N sites on the surface NC layer were revealed to be the main contributor of the solvent pathway for Li$_2$O$_2$ growth.

Kim et al.[13] designed and studied the composites of mesoporous carbon grown on N, S co-doped graphene nanosheets (MC/NSG) for rechargeable lithium-oxygen batteries (RLOBs). Both N and S, whose electronegativities and atomic sizes are higher than that of carbon, could modify the structure as well as the charge and spin densities of carbon, benefiting the electrocatalytic activity.[6,14] The selection of mesoporous carbon could provide favorable porous structures for the improvement of performance by accommodating insoluble Li$_2$O$_2$ formed during battery discharge.[15] Moreover, the growth of mesoporous carbon on graphene could prevent the restacking of graphene, resulting in a highly active surface and enhanced electronic conductivity as well.[16] In their synthesis, sucrose was used as the carbon source and thiourea was used as the source of both nitrogen and sulfur. For comparison, undoped mesoporous carbon on graphene (MC/G) was prepared using a similar procedure without thiourea. The combination of SEM, TEM, and HRTEM clearly revealed the presence of densely formed mesoporous structures on both sides of the graphene sheets and the occurrence of foam-like morphology in individual graphene sheets after high-temperature heat treatment. To understand the chemical nature and bonding configuration of the doped heteroatoms in the XPS analysis, the contents of N and S in the MC/NSG composite were measured to be $\sim$1.05 at% and $\sim$2.15 at%, respectively, confirming the successful doping of N and S. Deconvolution techniques for C 1 s, S 2p demonstrated the presence of C–S, C–N, and C–O bonds and the predominant doping of sulfur in the carbon framework. As for the three N species corresponding to N 1 s spectra, graphitic N was located in the carbon framework while pyridinic- and pyrrolic-N tended to stay at the edges or in the defect sites present in the basal plane and were bonded to two adjacent carbon atoms. Therefore, XPS results confirmed the successful doping of N and S in the carbon framework. Nitrogen adsorption/desorption measurements (BET measurements) for the MC/NSG showed the existence of many mesopores with an average pore size of 4 nm, associated with a high surface area of $\sim$1730 m^2 g^{-1} with a mesoporous surface area of $\sim$1032 m^2 g^{-1}.

Evaluation of ORR activity was conducted using RDE measurements at a rotation speed of 900 rpm in O_2-saturated 1 M LITFSI in TEGDME electrolyte at room temperature. For the voltammetric response at 3.02.2 V, the comparison of the onset potentials MC/NSG, MC/G, Ketjenblack carbon (KB) revealed a decreasing order for ORR activity: MC/NSG > MC/G > KB. Although both MC/NSG and MC/G catalysts had similar porous structures and other characteristics (e.g., surface area, porosity, size, etc.), MC/NSG demonstrated a better performance than MC. This should be attributed to the co-doping of N and S, which played a significant role in creating a favorite structure for performance. Meantime, compared to the referenced KB, the higher ORR activity of MC/G and MC/NSG resulting from their larger surface areas accommodating insoluble reaction products could improve electronic conductivity and create more defect sites on the carbon structure. Corresponding to ORR activity, the maximum power density of MC/NSG was ~26.4 W g^{-1} at 2.0 V, which was ~60% higher than that of KB (~16.4 W g^{-1}) and ~13% higher than that of GC (~22.9 W g^{-1}), further confirming the higher ORR performance of MC/NSG catalyst. In comparison with those of MC/G (~309 mV per decade) and KB catalysts (~314 mV per decade), the smallest slope (~299 mV per decade) in the Tafel plots further demonstrated that MC/NSG was the best ORR catalyst. For the OER activity evaluated using a galvanostatic intermittent titration technique (GITT), which combines transient and steady-state processes to determine the overpotential of the electrochemical system,[17,18] it was found that, compared to MC/G and KB catalysts, MC/NSG showed a lower overpotential during charging in each interval. The superior OER activity for MC/NSG was possibly due to the successful design of mesoporous carbons on graphene nanosheets in which they complemented each other. To further test the application of MC/NSG in RLOBs, electrochemical measurements were carried out using a Swagelok-type cell with a constant current density of 0.1 mA cm^{-2} (i.e., 100 mAg_{carbon}^{-1}). Compared to MC/G and KB cathodes, MC/NSG exhibited not only a higher discharge capacity, along with a higher efficiency, but also a lower overpotential, indicating a superior catalytic effect as an air-electrode for RLOB. A combination of SEM and XRD was used to monitor the morphology and crystalline phase for the reversibility of the MC/NSG air-electrode. Before cycling, SEM images of the MC/NSG electrode showed a macroporous morphology with a sponge-like sheet shape of the NSG (see Figure 4.2a), whereas after the fifth discharge, the electrode was deposited with ORR products (i.e., Li_2O_2) on the surface of the NSG (see Figure 4.2b). When recharging, these products disappeared (see Figure 4.2c). This reversible behavior was further confirmed by XRD analysis (see Figure 4.2d) in which the characteristic crystalline peaks during discharging corresponded to the Li_2O_2 product. For the MC/NSG composite, this reversibility resulted in an excellent cycling performance. After 100 cycles, there was no significant voltage profile change at the assembled RLOB with a specific capacity decreasing from 1000 to 500 mAh g^{-1} (see Figure 4.2e,f). It was demonstrated here that the design of a co-doped graphene and mesoporous carbon composite catalysts, including the control of porous morphology, could create promising performance RLOBs.

Zhu et al.[19] also prepared S, N co-doped carbon nanodots (S–N–C), and composited this S–N–C with one-dimensional (1D) cable-like multiwall carbon nanotubes by annealing to form a MWCNT@S–N–C composite. This composite was shown

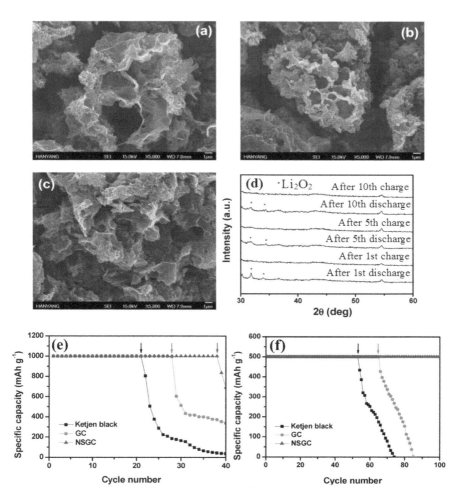

FIGURE 4.2   SEM images of the MC/NSG electrode (a) before cycling, (b) after the fifth discharge cycle, and (c) after the fifth charge cycle. (d) XRD diffraction patterns of the MC/NSG electrode analyzed at different charge states. (Modified from Kim, J.-H. et al. *J. Mater. Chem. A* **2015**, *3*, 18456–18465.[13])

to not only possess superior ORR activity and excellent durability comparable to commercial Pt/C, but also a higher OER activity compared to state-of-the-art IrO_2 catalysts, demonstrating its excellent overall bifunctional catalytic performance. Therefore, the combination of co-doped carbon materials and carbon, especially in a controlled porous structure is an effective strategy to develop high-performance carbon-based composite bifunctional non-metal catalysts for RMABs.

Other than the co-doping of S and N, Li et al.[20] fabricated and studied the composite of N and P co-doped graphene/carbon nanosheets (N,P–G/CNS) for bifunctional catalysts for both ORR and OER. In the composite design, more synergistic effects were expected between doped heteroatoms, such as N and P,[21,22] to contribute to the enhancement of catalytic performance. Graphene oxide (GO) nanosheets was selected not only as the precursor of graphene but also as the structure directing agent for the coating of N, P co-doped carbon during annealing. Their results showed that, compared to the corresponding single-heteroatom-doped counterpart, the N,P–G/CNS composite possessed much better catalytic activities towards both ORR and OER with a potential gap of 0.71 V between the OER potential at a current density of 10 mA cm^{-2} and the ORR potential at a current density of -3 mA cm^{-2}. This was attributed to the synergistic effects of doped N and P atoms, the high active sites on the surface of the N,P–G/CNS composite, the good conductivity of the incorporated graphene, and the unique porous structure for transportation of both gas and electrolyte, illustrating that N,P–G/CNS composites could be a promising bifunctional electrocatalyst for RMABs.

4.2   COMPOSITES OF DOPED-CARBON WITH OXIDES

4.2.1   COMPOSITES OF SINGLE ELEMENT-DOPED CARBON WITH OXIDES

4.2.1.1   Composites of Single Element-Doped Carbon and Perovskite Oxides

Perovskite oxides (with a regular formula of ABO$_3$) are promising in the application of bifunctional electrocatalysis, owing to their reported high specific catalytic activity (microamperes per square centimeter) toward both OER and ORR in alkaline media.[23–25] To further improve the catalytic activity, doped-carbon materials, used in the same way as un-doped carbon materials, have been used to enhance the electrically conductive pathway in electrochemical reactions by tailoring the morphology.[26] Moreover, doped carbon can act as an active ORR catalyst and also offer high surface areas to enhance the mass activities (milliamperes per milligram) of perovskite oxides.[27]

As a typical ABO$_3$-type perovskite oxide, lanthanum nickel oxide (LaNiO$_3$) nanoparticles were encapsulated by a high surface area network of N-doped CNT (NCNT) to form an intertwined core-corona structured LaNiO$_3$/NCNT composite bifunctional catalyst by Lee et al.[26] Based on the synergy between the two components, this unique composite morphology could not only enhance the charge transport property by providing rapid electron-conduction pathways but also facilitate the diffusion of hydroxyl and oxygen reactants through the highly porous framework. The obtained LaNiO$_3$/NCNT composite was highly tailored with nanosized perovskite nanoparticles to promote favorable NCNT growth directly on and around LaNiO$_3$ nanoparticles. In the two-step synthesis, La(NO$_3$)$_3$·6H$_2$O and Ni(NO$_3$)$_2$·6H$_2$O were used as the sources of La and Ni to prepare LaNiO$_3$ nanoparticle by a hydrothermal method (see Figure 4.3a). As in Figure 4.3a, the NCNT is grown with an injection chemical vapor deposition (CVD) method with 2.5wt% ferrocene (C$_{10}$H$_{10}$Fe) in which ethylenediamine (EDA) acts as the NCNT precursor solution (see Figure 4.3b). A combination of XRD, SEM, and TEM confirmed the phase and morphology of the LaNiO$_3$/NCNT composite and the N 1 s spectra in XPS clearly showed four nitrogen species of pyridine, pyrrolic, quaternary, and pyrrolidine, thus confirming the successful doping of N. The pyridinic and pyrrolic nitrogen configurations in NCNT should oversee ORR activity, and were combined with the OER active perovskite metal oxide to obtain highly efficient bifunctional catalysts.[28,29] For the C 1 s, a peak located

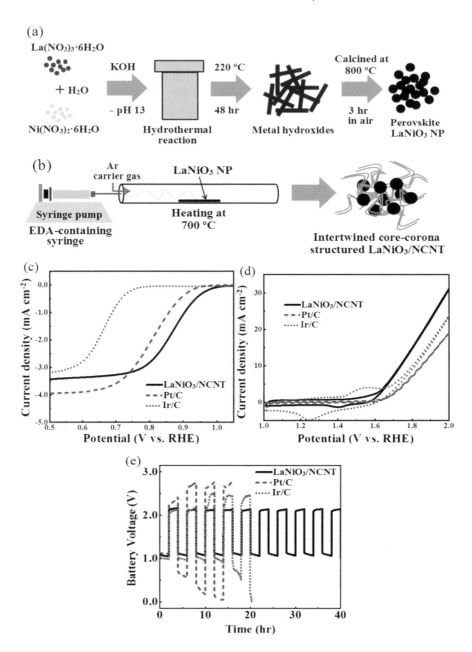

FIGURE 4.3 Schematic illustrations of (a) hydrothermal synthesis of LaNiO₃ nanoparticles (NP) and (b) injection CVD preparation of intertwined core-corona structured LaNiO₃/NCNT composites. (c) ORR and (d) OER polarization curves obtained at rotation speeds of 900 rpm with 10 and 50 mV s^{-1} scan rates, respectively, of LaNiO₃/NCNT, LaNiO₃, and NCNT. (e) Galvanostatic charge-discharge cycling of LaNiO₃/NCNT, Pt/C, and Ir/C obtained by utilizing realistic rechargeable zinc-air batteries. (From Lee, D. U. et al. *ACS Appl. Mater. Interfaces* **2015**, *7*, 902–910.[26])

at 284.1 eV was attributed to the sp2 configuration of the conjugated graphitic carbon making up the walls of NCNT. Half-cell testing was run to investigate both ORR and OER, and the comparison of ORR polarization curves at 900 rpm (see Figure 4.3c) shows that $LaNiO_3$/NCNT presents significant improvements over $LaNiO_3$ in terms of onset potential and half-wave potential, demonstrating the more ORR active sites from N-doping.[1,30] The half-wave potential of $LaNiO_3$/NCNT is superior to that of NCNT, suggesting that $LaNiO_3$/NCNT has an enhancement in ORR activity due to the synergistic effect between $LaNiO_3$ and NCNT. According to the measured electron transfer number, $LaNiO_3$/NCNT could give a four-electron reduction pathway[31] during ORR rather than a two-electron one. When the OER activity was analyzed, the OER current curves from 1.6 V (see Figure 4.3d) reveals that the OER current follows an increasing order of: NCNT < $LaNiO_3$ < $LaNiO_3$/NCNT, confirming the highest OER activity of the $LaNiO_3$/NCNT composite. This can be attributed to the addition of NCNT, which provides a conductive pathway for charge transfer, and the highly porous morphology of the composite benefits the better diffusion of active species such as evolved O_2 during the electrochemical reaction. In comparison with state-of-the-art ORR and OER catalysts, $LaNiO_3$/NCNT outperforms Pt/C and Ir/C, and showed an improved electrochemical stability after 500 continuous cycles. Interestingly, when a rechargeable zinc-air battery (RZAB) was tested using the three different air-electrodes, the $LaNiO_3$/NCNT electrode showed excellent cycle stability without any voltage decay during charging/discharging after 40 hours, while the operation of the Pt/C and Ir/C electrode were run for less than 15 and 20 hours, respectively, with significant voltage fading during charging/discharging (see Figure 4.3e). The improved charge and discharge properties of $LaNiO_3$/NCNT were attributed to the nanosized dimensions of $LaNiO_3$ nanoparticles, coupled with the prolific growth of encapsulation by highly porous NCNT networks, demonstrating the morphology-controlled $LaNiO_3$/NCNT composites being a potential carbon-based electrode catalyst for RZABs. The same group[32] also tested a NCNT/$LaNiO_3$ composite with non-uniform NCNT growth over the irregular surface of macrosized $LaNiO_3$ particles. In contrast to the intertwined core-corona structured $LaNiO_3$/NCNT composite,[26] the un-controlled morphology of the NCNT/$LaNiO_3$ composite resulted in a low interparticle interaction and limited electrocatalytic activity for both ORR and OER.

Using partial substitutions of B-sites in ABO_3-type perovskite oxides to improve the electronic structure and coordination chemistry, a perovskite lanthanum manganese cobaltite, $La(Co_{0.55}Mn_{0.45})_{0.99}O_{3-\delta}$ (LCMO), was designed to form a composite of LCMO nanorods and N-doped reduced graphene oxide (NrGO).[33] The LCMO with an e_g electron presented the 1% B-site vacancy,[23,24] favoring the achievement of OER activity[34] while the NrGO could establish an effective electrically percolating network and leverage its unique chemical properties for the achievement of both enhanced ORR and OER activity. In a typical two-step synthesis, LCMO was first prepared via a hydrothermal treatment route in which lanthanum nitrate hexahydrate, cobalt nitrate hexahydrate, and manganese nitrate hydrate acted as the La, Co, and Mn sources, respectively. LCMO was then dissolved into a self-made GO solution, followed by another hydrothermal method, so that the final LCMO/NrGO was obtained with 20wt% LCMO. For comparison, a physically mixed LCMO + NrGO composed of 20wt% LCMO and NrGO was used as a reference sample. When the structure characterization was measured using XRD, SEM, TEM, and HRTEM, it was found that in the SEM, TEM, and HRTEM images, the highly crystalline LCMO nanorods ($\sim$48.5 $\pm$ 7.2 nm) consisted of low-index and relatively isometric nanocrystals with grain sizes in the range of 19–28 nm, and XRD confirmed the crystalline orthorhombic structured perovskite phase of LCMO in the LCMO/NrGO composite. Based on the XPS spectra, it was found that the binding energies of O 1 s, C 1 s, and N 1 s of LCMO/NrGO were ca. 0.1–0.3 eV smaller than those of the two individual components. These matched the behavior of binding energies for La, Co, and Mn cations of LCMO/NrGO, resulting in molecular orbital sharing and the shift of the electron cloud from metallic cations to light elements. This also showed a covalent coupling between LCMO and NrGO in the LCMO/NrGO composite. In the investigation of ORR and OER using a three-electrode cell for electrochemical measurements, it was observed by the CV curves in O_2-saturated 0.1 M KOH solution that compared to LCMO and NrGO, LCMO/NrGO illustrated more positive ORR onset potentials. The ORR peak potential and current of LCMO/NrGO were 136 mV more positive and 6% larger than those of Pt/C. In the comparison using LSV curves, the half-wave potential of LCMO/NrGO (-0.184 V) was much more positive than those of LCMO and NrGO catalysts, the physically mixed LCMO + NrGO and Pt/C catalysts. This suggested that among all five tested samples, LCMO/NrGO possessed the best ORR activity due to the strong coupling between LCMO and NrGO. Based on the combined results from RDE and RRDE voltammograms, LCMO/NrGO dominantly followed a four-electron reaction mechanism in ORR. For OER, it was seen by the polarization curves that the OER onset potential of LCMO/NrGO was around 0.45 V, slightly higher than the 0.40 V of Ir/C and lower than LCMO + NrGO, LCMO, NrGO, and Pt/C catalysts, demonstrating that LCMO/NrGO possessed a better OER activity than LCMO + NrGO, LCMO, NrGO, and Pt/C catalysts and slightly worse OER activity than Ir/C. When the potential gap (ΔE) between the ORR potential at -3 mA cm^{-2} and the OER potential at 10 mA cm^{-2} was used to evaluate bifunctional activity, LCMO/NrGO had a gap of 0.960 V. This was lower than Ir/C ($\sim$1.086 V), demonstrating higher overall activity for LCMO/NrGO. The chronopotentiometric responses revealed a better durability for LCMO/NrGO than Pt/C and Ir/C. The enhanced bifunctional activity and durability of LCMO/NrGO demonstrated the synergistic covalent coupling of NrGO with the B-sites of LCMO perovskite oxides on electronic structure, providing a promising method for the development of low-cost, efficient, and durable electrocatalysts for RMABs.

Similar to LCMO/NrGO, co-doped $LaMnO_3$ (LMCO) was also used as a growth substrate for N-doped CNT to form a composite of LMCO/NrGO.[35] When used as an air-electrode in a RZAB, LMCO/NrGO was found to exhibit very stable charge and discharge voltages of 2.2 and 1.0 V, respectively, suggesting that LMCO/NrGO possessed excellent bifunctional activities towards both ORR and OER as well as strong stability. In addition, with the substitution and selection of B-site cation(s) in ABO_3 perovskite oxide, $LaTi_{0.65}Fe_{0.35}O_{3-\delta}$ (LTFO) was found to play a synergistic effect with N-doped carbon nanorods (NCNR) in LTFO/NCNR composites.[36]

Other than the partial substitution of B-site in perovskite oxides to be composited into a single heteroatom-doped carbon material, $La_{0.5}Sr_{0.5}Co_{0.8}Fe_{0.2}O_3$ (LSCF), a product from the substitution of both A-site and B-site in ABO_3-type perovskite oxides, was also studied as a composite catalyst with N-doped CNT (NCNT) for RLAB.[37] In the design, the introduction of Sr and Fe was not only expected to improve the catalytic activity of the perovskite oxide[38,39] but was also expected to produce stronger synergetic effects of the perovskite/NCNT composite than that of the previous NCNT/LaNiO$_3$ with uncontrolled morphology.[32] In a typical synthesis, a solution-based synthesis was first used to produce LSCF nanoparticle. Injection CVD method was then carried out to grow NCNT directly on the surface of LSCF nanoparticles to form a LSCF/NCNT composite. XRD, SEM, TEM, and EDX measurements confirmed that the NCNT-wrapped perovskite oxides could be found in structure and morphology, while TGA results showed that the LSCF/NCNT composite was composed of 78 wt% NCNTs and 32 wt% metal oxide. To evaluate the electrocatalytic activity of both ORR and OER, a half-cell test was conducted using a RDE technique with 0.1 M KOH solution. It was observed by LSV curves tested at 900 rpm in O_2-saturated 0.1 M KOH that compared to LSCF, LSCF/NCNT showed improvements in all catalytic features including onset potential and half-wave potential due to the synergy between NCNT and LSCF as well as N-doping offering more active ORR sites. The detailed comparison of the half-wave potential revealed that LSCF/NCNT displayed an improvement of 34 mV in comparison with NCNT, indicating that LSCF/NCNT possessed enhanced ORR performance due to the addition of LSCF. Based on the analysis of Koutecky-Levich plots at various rotation speeds, LSCF/NCNT was found to follow a pseudo-four-electron reduction pathway in ORR. In the OER polarization curves obtained at 0.6 V in O_2-saturated 0.1 M KOH solution, the observed OER currents of LSCF/NCNT were clearly higher than those of LSCF and NCNT samples. Particularly, at 1.0 V, the current density of LSCF/NCNT was 27% and 71% higher than the values for LSCF and NCNT samples, respectively, demonstrating that LSCF/NCNT possessed an enhanced OER activity, not only because of the addition of NCNT providing more conductive pathways, but also because of the synergistic effect between NCNT and LSCF favoring catalytic performance. Further durability tests with 500 continuous CVs in N_2-saturated 0.1 M KOH solution showed that the long-term bifunctional stability of LSCF/NCNT was greatly enhanced when compared to those of Pt/C and Ir/C. When an assembled RLAB was used to test the single-cell performance, the overpotential of LSCF/NCNT was 61 and 63 mV lower than those of LSCF and NCNT air-electrodes, respectively, confirming, once again, the enhanced ORR and OER activities of LSCF/NCNT. When cycle performances were tested for 10-minute interval at an applied current density of 0.5 mA cm^{-2}, LSCF/NCNT demonstrated a cycle stability for over 80 cycles. At 100 cycles, the end voltages of the discharge and charge were 2.86 and 3.91 V, respectively, which were very similar to the first cycle voltages of 2.90 and 3.84 V. Interestingly, another test of high energy cycle stability with 2 hours per cycle at an applied current density of 0.3 mA cm^{-2} showed that the voltage differences during discharging and charging between the 1st and 15th cycles were only 0.09 and 0.14 V, respectively. These results demonstrated the great potential of designed LSCF/NCNT in the application of RLABs due to the controlled morphology and composition, which could provide strong synergistic effects between LSCF and NCNT for the enhancement of bifunctional activity and durability. The same group[40] also studied LSCF/NRGO formed from an electrospun porous nanorod $La_{0.5}Sr_{0.5}Co_{0.8}Fe_{0.2}O_3$ combined with N-doped reduced graphene oxide (NRGO) as a high-performance and durable bifunctional catalyst. They found that LSCF/NRGO exhibited comparable ORR and superior OER performance to state-of-the-art Pt/C catalyst as well as better durability, highlighting the importance of controlling morphology in the development of high-performance heteroatom-doped carbon/perovskite oxide composite for RMAB bifunctional catalysts.

Aside from LSCF, $Ba_{0.5}Sr_{0.5}Co_{0.8}Fe_{0.2}O_{3-\delta}$ (BSCF) was also combined with a N-doped mesoporous carbon (NC) for the preparation of BSCF/NC composites for ORR and OER catalyst.[41] In the composite design, BSCF was selected due to its higher OER activity, which was at least an order of magnitude higher than that of iridium oxide.[24] It was expected the strong synergistic effect with the N-doped mesoporous carbon might result in the improvement of catalytic activity due to the evidenced interactions between BSCF and carbon.[42] Observed by electrochemical measurements, the BSCF/NC composite presented a more positive onset ORR potential and less onset OER potential than those composites made of BSCF and common commercial carbon materials such as acetylene black (AB) carbon, activated carbon (AC), and Vulcan XC-72 carbon. This enhanced catalytic activity for both ORR and OER resulted from the electronic interaction between BSCF and NC, coupled with its porous structure, high surface area, and uniform dispersion of BSCF.

4.2.1.2 Composites of Single Element-Doped Carbon with Spinel Oxides

The compositing between carbon and spinel oxides such as $CoFe_2O_4$,[43–45] $NiCo_2O_4$,[46] and Co_3O_4,[47] could result in the enhancement of catalytic activities of both ORR and OER due to the assisted fast electron transport and good interactions between oxide and carbon.[48] For carbon materials, the addition of N-doping can offer conducting pathways and facilitate the adsorption of reactants (e.g., O_2, OH^-). The interaction between the two components can also promote interfacial electron transfer between the catalyst surface and the reaction intermediates.[49,50]

Due to attractive characteristics such as abundant catalytic sites, environmental compatibility, and strong stability,[51,52] N-doped CNTs were selected to be combined with a ternary $ZnCo_2O_4$ spinel oxide to form a $ZnCo_2O_4$/N–CNT composite catalyst.[49] The merit integration of both the spinel-phase metal oxide and the N-doped CNTs through a controllable nanoscale strategy was expected to enhance the bifunctional catalytic performance. In the synthesis, CNTs were first oxidized and then acid-treated with 5% HCl solution to remove residual impurities. The obtained oxygen-containing functional groups on the CNTs could produce strong coupling with N dopants toward Zn^{2+} and Co^{2+} ions, anchoring and controlling the nucleation of $ZnCo_2O_4$ crystallites into highly dispersed and uniform quantum dots, as seen in Figure 4.4a. According to a typical one-step experimental route, $Co(OAc)_2 \cdot 4H_2O$ and $Zn(OAc)_2 \cdot 2H_2O$ were used as the sources of Co and Zn to form a mixture with acid-treated CNTs. After a one-pot hydrothermal route, $ZnCo_2O_4$ quantum dots were anchored onto N-doped CNTs to form $ZnCo_2O_4$/N–CNT composite. For comparison, Co_3O_4/N–CNT, IrO_2/N–CNT, and $ZnCo_2O_4$ were prepared according to the same route. XRD patterns of $ZnCo_2O_4$, $ZnCo_2O_4$/N–CNT and Co_3O_4/N–CNT, $ZnCo_2O_4$ showed a spinel phase with Zn^{2+} occupying the tetrahedral site and Co^{3+} occupying the octahedral site. In a detailed

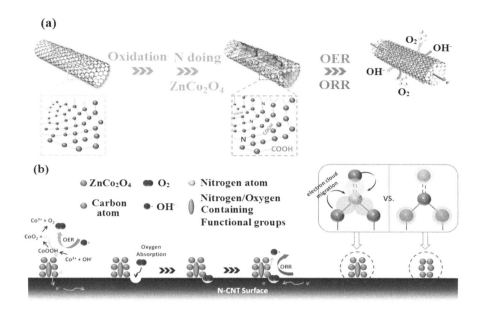

FIGURE 4.4 (a) Schematic illustration of the growth mechanism for the $ZnCo_2O_4$/N–CNT composite. (b) Schematic reaction mechanism of the OER and ORR processes catalyzed by the $ZnCo_2O_4$/N–CNT composite. (Modified from Liu, Z.-Q. et al. *Adv. Mater.* **2016**, *28*, 3777–3784.[49])

analysis of Raman spectra for pristine CNTs, oxidized CNTs, and $ZnCo_2O_4$/N–CNT, the intensity ratio (I_D/I_G) of the D band to G band was found to increase after the incorporation of $ZnCo_2O_4$, suggesting a partial degeneration of the graphitic lattice in the $ZnCo_2O_4$/N–CNT after the oxidization process. $ZnCo_2O_4$/N–CNT exhibited a lower I_D/I_G than that of the oxidized CNT, confirming that N-doping could decrease the quantity of defect sites in the carbon lattice with graphitic N filling up the vacancy of C. In SEM, TEM, and HR–TEM images coupled with an analysis of EDS, it was indicated that all the elements of Zn, Co, C, O, and N had uniform distributions in the $ZnCo_2O_4$/N–CNT composite in which $ZnCo_2O_4$ quantum dots (3.0–3.5 nm) covered the surface of N–CNT. To understand the synergistic interaction between $ZnCo_2O_4$ and N–CNT, XPS was used and revealed that after Zn^{2+} substitution of Co^{2+} in the tetrahedral sites, Co ions possessed a high valence state in $ZnCo_2O_4$. With the addition of N–CNT, Co 2p spectra shifted to lower binding energies due to the migration of the Co electron cloud induced by the strong electronegativity of C = N (especially pyridinic and pyrrodic nitrogen) and C = O on the N–CNT surface (See Figure 4.4b). This induction facilitated the oxidation of Co ions and, consequently, promoted the catalyst-electrolyte interfacial reversible chemisorption of oxygen species[53,54] while the population of Co ions with the higher valence state determined the enhancement of electrocatalytic activity.[55] In the investigation of OER using electrochemical measurements with a RDE technique, the LSV curves obtained in O_2-saturated 0.1 M KOH solution showed that compared to $ZnCo_2O_4$/N–CNT and Co_3O_4/N–CNT catalysts, $ZnCo_2O_4$ and N–CNT had inferior OER responses, demonstrating the coupling effects of the two components on catalytic activity. In addition, $ZnCo_2O_4$/N–CNT presented a lower Tafel slope ($\sim$70.6 mV decade^{-1}) than IrO_2/N–CNT ($\sim$74.5 mV decade^{-1}), Co_3O_4/N–CNT ($\sim$82.1 mV decade^{-1}), and $ZnCo_2O_4$ ($\sim$90.4 mV decade^{-1}), indicating that $ZnCo_2O_4$/N–CNT possessed excellent OER activity as confirmed by the analysis of tested electrochemical activity surface areas (ECSA) and charge transfer resistances. In a further chronoamperometry test, $ZnCo_2O_4$/N–CNT exhibited higher OER stability than the other catalysts. For ORR, the LSV curves in O_2-saturated 0.1 M KOH showed that the Tafel slope ($\sim$52.9 mV dec^{-1}) of $ZnCo_2O_4$/N–CNT was lower than those of Co_3O_4/N–CNT ($\sim$54.3 mV dec^{-1}), $ZnCo_2O_4$ ($\sim$81.6 mV dec^{-1}), and benchmark Pt/C catalysts, revealing that $ZnCo_2O_4$/N–CNT possessed more favorable ORR kinetics and could replace commercial noble metal catalysts for practical application. Furthermore, both the onset potential ($\sim$0.95 V) and half-wave potential ($\sim$0.87 V) of $ZnCo_2O_4$/N–CNT were superior to those of Co_3O_4/N–CNT ($\sim$0.95 V, $\sim$0.86 V) and $ZnCo_2O_4$ ($\sim$0.87 V, $\sim$0.84 V), and closer to that of benchmark Pt/C ($\sim$1.05 V, $\sim$0.87 V). Specifically, $ZnCo_2O_4$/N–CNT and Co_3O_4/N–CNT possessed the ORR electron transfer numbers of 3.8 and 3.6, respectively, which were close to that of Pt/C (3.9), demonstrating an efficient four-electron ORR process. It was concluded that the coupling of $ZnCo_2O_4$/N–CNT and Co_3O_4/N–CNT could enhance catalyst bifunctional activity, while the substitution of Zn in Co_3O_4 could result in stronger coupling, thus, higher catalytic performance for $ZnCo_2O_4$/N–CNT. Based on the promising ORR and OER activity of $ZnCo_2O_4$/N–CNT, a RZAB was used to test battery performance. During a long-term galvanostatic discharge process, the tested specific capacity of $ZnCo_2O_4$/N–CNT was $\sim$428.5 mAh g^{-1} corresponding to an energy density of $\sim$595.6 Wh kg^{-1}. This was comparable to that of Pt/C with a specific capacity of $\sim$548.4 mAh g^{-1} (i.e., 762.2 Wh kg^{-1}). When cycling tests were conducted with a charging current density of 0.1 A cm^{-2} and a discharging current density of 0.01 A cm^{-2} in a short interval (10 min per cycle), $ZnCo_2O_4$/N–CNT presented lower voltage fading for 17 cycles, demonstrating its potentiality in application of RMABs.

Using a simple hydrothermal method, Li et al.[56] also prepared two different ternary MFe_2O_4 (M = Co, Ni) spinel oxides coated on multi-walled carbon nanotubes (MFe_2O_4/CNT, M = Co, Ni). In addition to the enhanced ORR and OER activities, the two resulting MFe_2O_4/CNT composites also delivered better electrochemical performances during discharging/charging compared to pure Ketjenblack (KB) carbon when used as air-electrodes for RLOB.

Similar to N-doped CNTs, N-doped graphene was composited with different ternary spinel oxides such as $CoMn_2O_4$ (CMO)[57,58] and $LiMnO_4$ (LMO)[59] for advanced bifunctional catalysts under RMAB operating conditions. Prabu et al.[57] conducted an investigation on the composite of $CoMn_2O_4$ nanoparticles anchored on N-doped reduced graphene oxide (CMO/N–rGO) for bifunctional air-electrode for RZABs under ambient conditions.[57,58] Aside from CMO/N–rGO composites, a composite of $LiMnO_4$ nanoparticles dispersed on N-doped reduced graphene oxide nanosheet (LMO/N–rGO) was also synthesized and investigated as an air-electrode for aluminum-air batteries (Al–AB).[59] Although these two composites were synthesized using a similar modified Hummer's method,[60] the difference between them is the metal at the A-sites for the spinel oxides (regular formula: $A_xB_{3-x}O_4$): one was Co, and the other was Li. The selection of CMO with Co metal in the A-site possessed advantages such as low cost, high abundance, low toxicity, and multiple valence states,[57,58] while the LMO with Li metal in the A-site[59] was due to its stable three-dimensional tunnel structure, favoring the charge transfer, its cost-effectiveness, and low toxicity. In the XPS characterization of structure and morphology, these two composites had strong chemical coupling not only between CMO and NrGO but also between LMO and NrGO, which was shown to have an active effect on the enhancement of ORR and OER activities, as well as their durability. In the tested bifunctional activities of both ORR and OER, CMO/NrGO was superior to CMO/rGO, and LMO/NrGO was better than LMO/rGO. This demonstrated that N-doping could not only create more conductive pathways but also result in more active sites for both ORR and OER. Interestingly, when a constructed primary ZAB was discharged under ambient conditions (i.e., consuming air) to a 0.5 V limit with a current density of 20 mA cm^{-2}, CMO/NrGO with a four-electron transfer ORR mechanism outperformed CMO/rGO in discharge features such as plateau region, discharge time, and discharge capacity. The discharge capacity of CMO/NrGO was ~610 mAh g^{-1}, which was comparable to those of CoO/N−CNT[61] and 1D NiCo$_2$O$_4$-based materials[62] in primary ZABs. Meantime, when a constructed RZAB was tested under ambient conditions (i.e., consuming air) with short (10 minutes) and long intervals (1 and 2 hours) cycles for 200 cycles for both CMO/NrGO and CMO/rGO, the initial potential gap between charge and discharge was increased from 0.7 to 1.06 V after 200 cycles for CMO/NrGO, whereas CMO/rGO exhibited the increased values from 0.95 to 1.23 V, indicating better rechargeability of CMO/NrGO air-electrode. Table 4.1 displays the compared charge and discharge potential gaps of CMO/rGO and CMO/NrGO at the 1st and 100th cycles. Normally, compared to the use of pure oxygen in the battery, the use of air under ambient conditions results in an increased potential gap. These values are, however, more valuable for the future commercialization of CMO/NrGO. A further long-term cyclability (1- and 2-hour cycles) test with a current density of 15 mA cm^{-2} showed that CMO/NrGO exhibited higher round-trip efficiencies and lower overpotentials at both the 1st cycle and the 8th cycles than the commercial Pt/C and CMO/rGO, demonstrating its excellent stability. When LMO/NrGO and LMO/rGO were used as the air-electrode catalysts in an assembled Al-air battery under ambient conditions (i.e., consuming air) at 10 mA cm^{-2}, it was found at discharge durations over 8 hours, LMO/NrGO delivered a higher energy density of ~585 mAh g^{-1} than LMO/rGO (~568 mAh g^{-1}) while the decrease of discharge voltage of LMO/NrGO was 0.16 V, which was slightly lower than that of LMO/rGO (0.17 V). This result revealed that LMO/NrGO possessed better discharge performance owing to N-doping and the stronger coupling between LMO and NrGO.

Among binary spinel oxides, Co$_3$O$_4$ has become the most popular spinel oxide to combine with N-doped carbon for the exploration and development of advanced metal-free composite air-electrodes for RMABs. So far, several different carbon materials, including CNTs,[63] carbon nanowebs,[64] Vulcan XC-72,[65] and Ketjenblack,[66] have been selected to obtain composites of N-doped carbon with Co$_3$O$_4$ for bifunctional catalysts. Of the various carbon materials used, Ketjenblack is considered to be a typical carbon black with good electrical conductivity, high mesoporous area and porosity, high surface area, and stability. After N-doping, the N-doped Ketjenblack

TABLE 4.1

Summary of Charge and Discharge Potential Gap of CMO/rGO and CMO/ NrGO with a Current Density of 20 mA cm^{-2} by Consuming Pure Oxygen and Air under Ambient Conditions

	1st Cycle (V vs. Zn)			100th Cycle (V vs. Zn)		
	Charge Potential	Discharge Potential	C-D Potential Gap	Charge Potential	Discharge Potential	C-D Potential Gap
			O$_2$ Atmosphere			
CMO-rGO	1.90	1.19	0.71	1.99	0.98	1.01
CMO-N-rGO	1.78	1.08	0.70	1.90	1.04	0.86
			Air Atmosphere			
CMO-rGO	2.09	1.14	0.95	2.13	1.03	1.1
CMO-N-rGO	1.95	1.25	0.70	2.06	1.13	0.93

Source: Prabu, M. et al. *ACS Appl. Mater. Interfaces* **2014**, *6*, 16545–16555.[58]

(N–KB) can endow more electronic transportation and active sites favoring the improvement of catalytic performance. When Co$_3$O$_4$ nanoparticles were grown on N–KB to form a Co$_3$O$_4$/N–KB composite, Liu et al.[66] observed that the active synergy between N–KB and Co$_3$O$_4$ could result in the effective enhancement of catalytic performance in Al-air batteries. For comparison, a Co$_3$O$_4$/KB composite was prepared using the same method for reference. Depending on experimental controls, the content of Co$_3$O$_4$ was fixed to be 5wt%, 10wt%, and 20wt% in the Co$_3$O$_4$/N–KB composites. In the structure and morphology characterized using XRD and Raman spectra, the compared XRD spectra of N–KB and typical Co$_3$O$_4$/N–KB confirmed the crystalline spinel structure of Co$_3$O$_4$ with a crystalline size of 15.6 nm, as well as carbon graphitic and amorphous phases. Raman spectra found that the intensity ratio of the D band to G band (I$_D$/I$_G$) followed an increasing order: KB (I$_D$/I$_G$ = 1.347) < N–KB (I$_D$/I$_G$ = 1.373) < Co$_3$O$_4$/N–KB (I$_D$/I$_G$ = 1.389), suggesting that N-doping could result in more defect sites on carbon and the addition of Co$_3$O$_4$ could rebuild the carbon layer to form structural defects and exposed edge planes. SEM, TEM, HRTEM, and STEM showed that Co$_3$O$_4$ nanoparticles were uniformly distributed on the KB support and confirmed the uniform distributions of C, O, N, and Co elements in the Co$_3$O$_4$/N–KB composite. XPS was then used to investigate the synergistic interaction between Co$_3$O$_4$ and N–KB. The analysis of Co 2p, O 1 s, N 1 s, and C 1 s revealed the existence of electron coupling of Co^{3+} and Co^{2+} in Co$_3$O$_4$ and N–KB, which could result in enhanced electrocatalytic activity. To obtain the catalyzed ORR behavior, electrochemical measurements were carried out using a RDE technique. Based on the obtained LSV curves at 1600 rpm in O$_2$-saturated 0.1 M KOH, a comparison of the three different Co$_3$O$_4$/N–KB composites demonstrated that 5 wt% Co$_3$O$_4$/N–KB exhibited the best ORR activity. When ORR comparison of 5 wt% Co$_3$O$_4$/N–KB with Co$_3$O$_4$, KB, N–KB, and 5wt%Co$_3$O$_4$/KB catalysts was conducted, it was found that, compared to the very poor onset potentials of pure Co$_3$O$_4$

and KB catalysts, 5wt%Co_3O_4/KB gave higher positive onset potential, suggesting that the Co_3O_4 supported carbon material played an active role in improving ORR performance. N–KB also presented higher ORR activity than pure KB. Furthermore, the half-wave potential of 5wt% Co_3O_4/N–KB was ~0.79 V, which was higher than those of Co_3O_4 (~0.28 V), KB (~0.56 V), N–KB (~0.64 V), and Co_3O_4/KB (~0.72 V), and 40 mV less than that of commercial Pt/C, demonstrating the higher ORR performance of 5wt% Co_3O_4/N–KB compared to Co_3O_4/N–KB and the other catalysts. This result confirmed the N-doping and the synergistic coupling between Co_3O_4 and N–KB could create more active sites, thus favoring higher ORR activity. According to calculations from the LSV curves at different electrode rotating rates, it was revealed that, similar to Pt/C, Co_3O_4/N–KB and Co_3O_4/KB ran a direct four-electron transfer pathway in ORR. Finally, when assembled Al-air batteries with 6 M KOH solution were tested at a constant current discharge of 20 mA cm^{-2} for 15 hours, it was found that the average working voltage of Co_3O_4/N–KB was better than that of Co_3O_4/KB and similar to that of Pt/C. This showed that Co_3O_4/N–KB should be a promising catalyst with high efficiency and low cost for Al-air batteries even though further optimization is required.

Other than N-doping, P-doping can also improve the electronic conductivity of carbon and result in high plane edge exposure and charge redistribution, creating more active sites for catalytic activity.[67–69] For example, Cao et al.[69] anchored $MnCo_2O_4$ onto P-doped hierarchical porous carbon (MCO/P–HPC) for RLOBs.[69] It was found that P-doping could provide nucleation sites for metal oxides while the hierarchical porous structure could not only facilitate the electrolyte immersion and electron and ion transport, but also offer effective spaces for oxygen diffusion and O_2/Li_2O_2 conversion.[70–72] In a two-step synthesis, $C_{24}H_{20}PBr$ was used as the source of P to synthesize P-doped hierarchical porous carbon by a template method coupled with an electrostatic attraction process. $C_4H_6CoO_4 \cdot 4H_2O$ and $C_4H_6MnO_4 \cdot 4H_2O$ acted as the sources of Co and Mn for $MnCo_2O_4$, which was then directly anchored on P-doped hierarchical carbon through a hydrothermal method. TGA showed that the spinel $MnCo_2O_4$ possessed a content of 34 wt% while EDS revealed a P content of 2.10 at%, matching the results in ICP–MS. The mesopores of the MCO/P–HPC possessed two centers with diameters of 5.6–12 nm and 20–40 nm, respectively. XPS was used to check the synergistic interaction between P–HPC and MCO. The shifts binding energy of Co 2p, Mn 2p, and P 2p were associated with the covalent coupling between MCO and P–HPC (e.g., M–O–P), providing conductive pathways for charge transport and promoting electrochemical catalytic activity. To evaluate the electrochemical performance of MCO/P–HPC, a RLOB with 1 M LITFSI in TEGDME electrolyte was used for electrochemical measurements. At a current density of 200 mA g^{-1}, MCO/P–HPC gave a much higher discharge capacity of 13,150 mAh g^{-1} than MCO (~8950 mAh g^{-1}), P–HPC (~5850 mAh g^{-1}), and referenced acetylene black (~3850 mAh g^{-1}), indicating that MCO/P–HPC possessed superior bifunctional activity for both ORR and OER. Even at a high current density of 1000 mA g^{-1}, MCO/P–HPC could still present a high discharge capacity of 1728 mAh g^{-1}, demonstrating the excellent rate capability of the MCO/P–HPC catalyst. When cycling tests were conducted with a cutoff capacity of 1000 mAh g^{-1} at a current density of 200 mA g^{-1}, MCO/P–HPC based RLOB sustained 200 cycles with low overpotentials. The high capacity, enhanced rate capability, and excellent cycling stability with low overpotentials of MCO/P–HPC-based RLOB demonstrated the catalyst's coupling effects of MCO and P–HPC and P-doping as well as the hierarchical porous structure resulting in superior ORR and OER activities.

4.2.1.3 Composites of Single Element-Doped Carbon with Other Oxides

Although oxides such as manganese oxide[73-75] and Co_3O_4[76,77] do not exhibit the same level of catalytic performances as perovskite oxides do, researchers have found the enhanced bifunctional catalytic activities from their composites with single heteroatom-doped carbon. N-doped carbon has attracted more interest than S-doped carbon in the development and exploration of advanced bifunctional catalysts for RMABs, indicated by the widespread research of N-doping in various carbon materials for ORR and/or OER.

Among oxides, manganese oxides have been investigated as potential catalyst materials to enhance the performance of oxygen cathodes.[78] With different structure and morphologies, α-MnO_2 has presented reasonable capacity and catalytic activity, as well as stable cycling performances.[79-81] However, the performance still needs to be improved to meet the requirements for commercial application. Particularly, their catalytic activities were far below commercial Pt-based catalysts. To improve the catalytic activity, Zahoor et al.[75] explored an effective strategy to enhance the electrical conduction of α-MnO_2 using highly conductive carbon materials such as N-doped graphite nanofibers (N–GNFs) because of their large surface area, flexibility, chemical stability, as well as the ability to act as an excellent substrate to accommodate α-MnO_2. N-doping could also increase conductivity and create active sites for ORR. In their experimental route, $KMnO_4$ was used as the source of Mn to prepare α-MnO_2 nanorods grown on caterpillar-like N–GNFs (α-MnO_2/N–GNFs) composites. XRD identified the α-MnO_2 phase and graphite carbon while XPS confirmed the N-doping in the α-MnO_2/N–GNFs composite. In the investigation of morphology and surface area, TEM and SEM images showed that α-MnO_2 consisted of small nanorods (130–170 nm in diameter) grown on N–GNFs while the addition of N–GNFs resulted in the increase in the surface area of α-MnO_2 from 25.5 m^2 g^{-1} to 150 m^2 g^{-1}, indicated by BET measurements. N_2 adsorption/desorption isotherm showed an average pore size of 5.35 nm matching the mesoporous structure of α-MnO_2/N–GNFs. To investigate its electrochemical properties, α-MnO_2/N–GNFs was measured as an air-electrode in a Swagelok™ type cell with 1M LITFSI/TEGDME electrolyte in a potential window of 2–4.3 V. It was found by the first discharge-charge curves at 0.1 mA cm^{-2} that α-MnO_2/N–GNFs exhibited higher discharge capacity than α-MnO_2. When the current density was increased from 0.1 to 0.3 mA cm^{-2}, α-MnO_2/N–GNFs increased its capability from 2907 to 3260 mAh g^{-1}. This shows that α-MnO_2/N–GNFs had higher reversibility and rate capability than other reported MnO_2-based catalysts such as Pd/MnO_2,[82,83] α-MnO_2/CNT/CNF,[84] and α-MnO_2.[85] When cyclability was tested at a limited depth of discharge at 500 mAh g^{-1} in a controlled charge-discharge capacities for 50 cycles, α-MnO_2/N–GNFs exhibited much smaller and more stable discharge and charge potential difference than α-MnO_2, indicating that α-MnO_2/N–GNFs possessed a better rechargeability due to the enhanced ORR and OER activities. Moreover, according to the analysis during charging/discharging using

XRD and SEM techniques, the reversible adsorption/desorption of reactants/products of α-MnO$_2$/N–GNFs were the determination factors for the electrocatalytic activity.

Similar to α-MnO$_2$, Co$_3$O$_4$ was combined with carbon materials, such as partially graphitized carbon[76] and carbon nanofibers,[77] to induce the enhanced electron transfer and high efficiencies for both ORR and OER, which was due to the addition of carbon, offering abundant active sites and strong synergetic coupling within the composite. With the improved catalytic performances, both partially graphitized carbon and carbon nanofiber composited-Co$_3$O$_4$ catalysts further demonstrated that the addition of conducting carbon materials with sufficient mesoporous structures could be an effective strategy to improve the bifunctional activity, stability, and then the RMAB performance. In addition, two other oxides, Mn, Co co-substituted Fe$_3$O$_4$,[86] and non-spiel Co–Mn mixed oxide,[87] have also been reported to have important synergistic effects on the improvement of catalytic performances in RZABs when combined with carbon.

Replacing N-doping with S-doping, Gao et al.[88] studied MnO$_x$/S-doped graphitized carbon (MnO$_x$/S–GC) composites in which polyvinyl chloride (PVC) plastic, (NH$_4$)$_2$Fe(SO$_4$)$_2$, and KMnO$_4$ were used as the sources of C, S, and Mn, while (NH$_4$)$_2$Fe(SO$_4$)$_2$ also were used as a graphitization catalyst. They found that a synergistic coupling existed between MnO$_x$ and S–GC, and this coupling played an active role in the catalytic performance. In their investigation on the synergistic interaction between MnO$_x$ and S–GC using XPS, the downshift of Mn 2p was present due to the existence of S-doping and carbon in the MnO$_x$/S–GC composite when compared to that of pure MnO$_x$. Furthermore, the observation of S 2p spectra evidenced S-doping even though the S amount ($\sim$7.3%) was very low. In the evaluation of ORR activity of MnO$_x$/S–GC composite using a RDE technique, it was found that, based on ORR features such as onset potential, half-wave potential, and Tafel slope, the MnO$_x$/S–GC composite exhibited much higher ORR activity than S–GC, while being slightly lower than commercial Pt/C catalyst. This suggested the active coupling effect between MnO$_x$ and S–GC. Based on Koutecky-Levich plots collected at different electrode rotation rates, the calculated electron transfer number for MnO$_x$/S–GC was close to 4, demonstrating a four-electron oxygen reduction pathway. In fact, with the doping of S into graphitized carbon, charge properties such as electron density and distribution were changed due to its larger electronegativity, greater number of valence electrons, and larger atom radius.[6] Furthermore, the d-orbitals of S were easily polarized, resulting in easier adsorption/desorption with O$_2$ and H$_2$O$_2$, thus improving ORR and OER activities.[89] For the OER, the LVS curves at 2000 rpm in O$_2$-saturated KOH showed that MnO$_x$/S–GC presented a lower Tafel slope ($\sim$67 mV decade^{-1}) than S–GC ($\sim$97 mV decade^{-1}), suggesting that MnO$_x$/S–GC had a superior OER activity to Ir/C.[90] This also suggested that the addition of MnO$_x$ plays an effective role in the enhancement of OER process because MnO$_x$ had special crystal structures, nano-size particles, mixed valance states, and Mn–O bonds facilitating O$_2$ evolution.[90,91]

4.2.2 COMPOSITES OF DUAL ELEMENTS CO-DOPED CARBON WITH OXIDES

Recently, many studies have reported that combining oxides with conductive carbon materials can prevent oxide particle agglomeration, promote electron transfer, and result in preferable catalytic performances. With conductive carbon materials, doped carbon materials with heteroatoms (e.g., N, P, or S) can result in more active catalytic sites for catalyzing both ORR and OER. Compared to single heteroatom-doped carbon, dual heteroatoms co-doped carbon displays better bifunctional activity due to the synergetic interaction arising from the co-doping of more than one heteroatom in carbon.[92] Furthermore, this synergetic interaction that improves bifunctional activity and efficiency can become even stronger when dual heteroatoms co-doped carbons are combined with oxides. Having said this, few publications have focused on the composites of oxides and dual heteroatoms co-doped carbons as advanced bifunctional catalysts in RMABs.

Utilizing bacterial cellulose (BC) as the source of carbon nanofibers, Liu et al.[93] designed and studied a CoO@N/S–CNF composite with CoO nanoparticles embedded in N, S co-doped CNFs using a two-step method including a simple immersion step followed by direct pyrolysis. They indicated that as a typical biomass material consisting of interconnected 3D networks of native cellulose-I nanofibers with a high young's modulus and many hydrogen bonds,[94,95] BC could be used to form robust 3D CNF networks as anchoring sites to immobilize active CoO nanoparticles in the fabrication of the composite. In a typical experimental route, self-made BC aerogels were immersed into an aqueous solution containing $CoCl_2$, thiourea and urea with a molar ratio of 1:2:100 at room temperature for 3 days. Freeze-drying and annealing were then carried out to produce the final CoO@N/S–CNF product. For comparison, Co_3O_4@N–CNF samples prepared with $CoCl_2$, urea, and BC, and N/S–CNF samples obtained with urea, thiourea, and BC were used. For the characterization of structure and morphology, TEM and SEM images coupled with analysis of XRD spectra were used. In the cases of CoO@N/S–CNF and Co_3O_4@N–CNF, cylinder shaped CoO nanoparticles were found to be uniformly embedded in a 3D porous interconnected carbon framework in which the average size of CoO was much smaller than that of Co_3O_4; this suggested that the thiourea not only acted as a reducing agent but also prevented the agglomeration of CoO nanoparticles and decreased the nanoparticle size, possibly by the formed S-doping. It was believed that S-doping could reduce the size of active CoO particles, favoring activity, whereas N-doping could create defects, resulting in many active sites for catalytic activity.[96] Interestingly, the BET measurements revealed that, with small pore sizes, the formed CoO particles could result in the increase of surface area from 129 to ~377 $m^2 g^{-1}$. This high surface area could result in a large density of active sites and a more efficient mass transport, benefiting the enhancement of catalytic activities for both ORR and OER.[97] The Co 2p and O 1 s XPS spectra suggested that the existence of CoO in CoO@N/S–CNF composite had five distinguished species: C = C, C–S, C–N, C–O, and C = O. C1 s spectrum not only confirmed the successful doping of N and S into the carbon framework but also indicated that the doping amount of N was higher than that of S evidenced by the ratio of peak areas between C–S and C–N. Moreover, among the five N species including pyridinic N, pyrrolic N, quaternary N, oxidized N, and chemisorbed N obtained by the deconvolution of N1 s spectrum, pyridinic N, with a 43.5% content, was reported to have a positive effect on the ORR onset potential whereas other N species could improve the electrochemical performance of carbon by strengthening the initial chemical adsorption of oxygen.[98] The final S 2p spectrum indicated that S atoms were partially doped into the carbon network rather than a physical combination. To evaluate the electrocatalytic activities of ORR and OER, electrochemical measurements were run on a three-electrode electrochemical cell using RDE technique. For the CoO@N/S–CNT composite, the CV curves in O_2-saturated 0.1 M KOH solution presented a clear cathodic reduction peak at 0.740 V that was different from those two CV curves in air- and N_2-saturated 0.1 M KOH solutions, indicating a high electrocatalytic activity for ORR. In the LSV curves in O_2-saturated KOH solution, it was found that Co@N/S–CNT exhibited higher onset potentials than the other two samples of N/S–CNF and Co_3O_4@N–CNF. This suggested that Co@N/S–CNF possessed an improved ORR activity due to the incorporation of CoO nanoparticles, which played a positive effect on improving the ORR catalysis, and also due to the co-doping of N and S, which could contribute to the synergistic effect on the enhancement of ORR activity. Chronoamperometric responses in O_2-saturated 0.1 M KOH with a rotation rate of 1600 rpm showed that Co@N/S–CNF was possibly more stable than commercial Pt/C, as evidenced by the lower current loss (~20%) for Co@N/S–CNF as compared to Pt/C (~50%) after 10 hours of stability test. Koutecky-Levich plots revealed a four-electron transfer ORR pathway for Co@N/S–CNF. For the OER, the OER polarization curves in 0.1 M KOH solution at 1600 rpm showed that Co@N/S–CNF displayed a much lower overpotential and a higher OER current density than the other samples. At 10 mA cm^{-2}, the overpotential of CoO@N/S–CNF was ~1.55 V vs. RHE. This was much lower than those of Co_3O_4@N–CNF, N/S–CNF, and Pt/C catalysts, and close to that of RuO_2. The calculated Tafel slopes were 95.0, 96.7, 278.5, and 78.3 mV decade^{-1} for CoO@N/S–CNF, Co_3O_4@N–CNF, Pt/C, and RuO_2, respectively, confirming that CoO@N/S–CNF possessed high OER electrocatalytic performance. The calculated potential difference between the potential at 10 mA cm^{-2} for OER and −3 mA cm^{-2} for ORR[99] were 0.828 and 0.959 V for CoO@N/S–CNF and Co_3O_4@N–CNF, respectively, suggesting that CoO@N/S–CNF also possessed a better bifunctional activity than Co_3O_4@N–CNF. The above results demonstrated that CoO nanoparticles intimately coated on CNFs could facilitate electron transfer, favoring reaction kinetics and catalytic performance, whereas N, S-doped CNF networks provided good conductive pathways and porous channels to enhance mass transfer. Moreover, it could be inferred that the two synergistic interactions between CoO and N,S-doped CNF and between doped N and S atoms could significantly contribute to both ORR and OER.

To investigate the application of dual heteroatom co-doped carbons and oxides in hybrid LAB and to find a suitable substitution for state-of-the-art Pt/C, Li et al.[100] combined Co_3O_4 nanocrystals with O, N-doped carbon nanowebs (ON–CNW)[101] to synthesize a Co_3O_4/ON–CNW composite to act as a highly active ORR catalyst for LAB. They observed that the composite of ON–CNW and Co_3O_4 had a strong coupling effect, causing improved catalytic performance. Their XPS spectra of Co 2p and N 1 s revealed that Co_3O_4 could couple with O- and N-doped CNW. When an assembled LAB with the Co_3O_4/ON–CNW air-electrode was tested for 130 cycles at 0.5 mA cm^{-2}, only 0.016% decrease in efficiency per cycle was observed. This was better than the findings reported for N-doped mesoporous carbon (0.017% decrease per cycle for 100 cycles).[102] Although Co_3O_4/ON–CNW displayed a slightly lower initial discharge voltage (3.00 V) in comparison with Pt/C, the cycle life of Co_3O_4/ ON–CNW was superior to that of Pt/C (below 50 cycles), suggesting that Co_3O_4/ ON–CNW could provide a practical low-cost metal-free alternative to expensive Pt/C catalysts in hybrid LABs.

4.3 COMPOSITES OF DOPED-CARBON WITH METALS

Because carbon and metal(s) can produce strong synergistic effects on catalytic performance, heteroatom(s)-doped carbons have been composited with metal(s) to form composite catalysts for RMABs. Therefore, the doping of carbon with one or more non-metal heteroatoms,[103–109] such as N, P, O, and S, rather than physically mixing them together has proved much more efficient.

4.3.1 COMPOSITES OF SINGLE ELEMENT-DOPED CARBON AND METAL(S)

Although single heteroatom-doping can modify the electronic state of carbon, including its charge and spin density redistribution to for the composite catalysts, their ORR and OER activities are generally lower than those of state-of-the-art Pt/C and RuO_2. To further improve catalytic performance, single heteroatom-doped carbon has been combined with metal to form an effective composite in which non-metal heteroatom doped carbon can utilize a metal as a second dopant. The metal-doping tunes electronic properties and surface polarities, and creates more active sites for carbon materials, allowing for further improvements of the catalytic performances.[103,105,110,111]

The compositing between transition metal and N co-doped carbons (M/N–C) have exhibited comparable or even superior catalytic activity and durability to noble metal-based catalysts due to the synergistic effects of the carbon-metal interaction.[48,103] For example, using a noble metal like Pd as the second dopant, Crespiera et al.[104] fabricated mesoporous Pd, N-doped carbon nanofibers (PdN/CNF) through the electrospinning of a polyacrylonitrile/Pd(acetate)$_2$ mixture solution with a subsequent thermal treatment process. In a typical synthesis, electrospun polyacrylonitrile (PAN) nanofibers were converted into N-doped CNFs at a high temperature with stabilization and carbonization processes in which Pd^{2+} in the PAN nanofibers was reduced to Pd^0 and aggregated into Pd nanoparticles.[112] The size and distribution of Pd were strongly depended on the amounts of Pd(Ac)$_2$ and complexing silanes (i.e., N-[(aminoethyl)aminopropyl] trimethoxysilane, AEAPTS) in the precursor solution mixture. The final products were labelled as N/CNF, Pd2.5/N/CNF, and Pd5/N/CNF, respectively, corresponding to the amount of Pd(Ac)$_2$ used at 0, 2.5, and 5 mmol. For comparison, AEAPTS was processed in the same method to prepare Pd2.5A/N/CNF as a reference sample. XRD confirmed that the Pd phase possessed no impurities, TEM and HRTEM images of Pd2.5A/N/CNFs not only revealed the smaller size of Pd after the addition of AEAPTS but also showed the aggregation of Pd consisting of loosely packed metal nanoparticles, possibly resulting in a higher density of catalytic sites. According to the XRD patterns, the average particle sizes of Pd were calculated to be 27, 50, and 63 nm for Pd2.5A/N/ CNFs, Pd2.5/N/CNFs, and Pd5/N/CNFs, respectively, demonstrating that the increase of Pd(Ac)$_2$ content could result in the larger size of Pd. ICP–MS showed Pd contents of 7.06 wt%, 4.75 wt%, and 10.65 wt% for Pd2.5A/N/CNFs, Pd2.5/N/CNFs, and Pd5/N/

CNFs, respectively. Interestingly, BET measurements gave >100 m^2 g^{-1} the surface areas for all Pd/N/CNFs samples, which were significantly higher than that of N/CNFs (16 m^2 g^{-1}). Particularly, the addition of AEAPTS resulted in the partial conversion of micropores into mesopores, benefiting the distribution of active sites. When XPS was employed to examine the surface properties of the catalysts, the analysis of Pd 3d spectrum confirmed the presence of Pd(0) and PdO. The graphitic-N, ammonium-N, amines-N, and N–X species were found in the N 1 s spectra. The graphitic-N group, which could enhance ORR activity, was amplified with increasing amounts of Pd(Ac)$_2$, whereas the use of AEAPTS resulted in the increased amines-N groups. To evaluate the electrochemical performances of the catalysts, discharge-charge curves obtained between the open circuit potential (OCV) and 2.25 V at a current density of 20 mA g$^{-1}_{cathode}$, the obtained results showed that Pd2.5/N/CNFs could give the highest discharge capacity of 7828 mAh g^{-1} when compared with N/CNFs (6730 mAh g^{-1}), Pd5/N/CNFs (4941 mAh g^{-1}), and Pd2.5A/N/CNFs (3647 mAh g^{-1}) catalysts. Additionally, the tested galvanostatic charge curves revealed a superior coulombic efficiency ($\sim$95%) for Pd2.5/N/CNFs to that of N/CNFs, indicating that the addition of Pd could promote the decomposition of both Li$_2$O$_2$ and side products during the charging process. In a detailed investigation of different air-electrodes during charging-discharging using ex-situ FESEM and XRD, N/CNFs could only partially remove large-sized insulating discharge products ($\sim$1 μm) of Li$_2$O$_2$ and Li$_2$CO$_3$ particles on the surface of CNFs, resulting in a worse recharge ability. Pd2.5/N/CNFs, however, could totally decompose the Li$_2$O$_2$ product without the Li$_2$CO$_3$ product. For Pd5/N/CNFs, the larger amount of Pd resulted in broken fibers during electrode fabrication and the increment of Li$_2$O$_2$ ($>$1 μm) to form rougher sidewalls for Pd5/N/CNFs than for Pd2.5/N/CNFs. For Pd2.5A/N/CNFs, the coated Li$_2$O$_2$ product exhibited smaller particle sizes of $\sim$200 nm with a different morphology from those of Pd2.5/N/CNFs and Pd5/N/CNFs catalysts. The XRD also revealed that with Pd/N/CNFs electrode, no Li$_2$CO$_3$ was observed, and the oxidation of LiOH to Li$_2$O$_2$ occurred at lower charge potentials than that of Li$_2$CO$_3$, suggesting the easier decomposition of such products.[113] When the galvanostatic discharge and charge of the Li-O$_2$ cells were measured at a curtailed capacity of 200 mAh g^{-1} with an applied current of 20 mA g^{-1} cathode, it was found that with a round-trip cycle duration of 20 hours (10 per discharge), Pd2.5A/N/CNFs demonstrated a remarkably low charge potential of 3.2 V, whereas N/CNFs had limited stability with less than 20 cycles. Under the same conditions, Pd2.5/N/CNFs, Pd2.5A/N/CNFs, and Pd5/N/CNFs provided good cycling profiles for 73, 88, and 61 cycles, respectively, corresponding to 50–70 days of cell operation at 100% coulombic efficiency. The special morphology of the discharge products seen by ex-situ FESEM images could account for the easier decomposition of the peroxide, resulting in lower overpotentials for recharging and higher longevity. Furthermore, the nucleation of Li$_2$O$_2$ as the discharge product was found to be strongly dependent on the catalytic surface.[114,115] The particle size and distribution of Pd coupled with the content of N-doping are responsible for the alleviation of concomitant side reactions and the reduction of the accumulation of carbonate residuals, prolonging life-time and improving rechargeability of Pd2.5A/N/CNFs.

Non-noble metals such as Fe,[116–118] Co,[119–121] and Ni,[122] have been explored to replace Pd in the preparation of metal, N co-doped carbons (M/N–C) as air-electrodes for RMABs. Based on the proposed active sites of metal-nitrogen-carbon moieties (MeN_xC, Me = Co, Fe, Ni, Mn, etc., and normally x = 2 or 4), metal cations are often coordinated by pyridinic/pyrrolic nitrogen functionalities in the center or edge of a graphitic matrix.[123,124] In a study by Kattel and Wang,[125] the first principle density function theory (DFT) study for the ORR pathway on $Me-N_4$ catalytic clusters demonstrated that the active formation of $Me-N_4$ was inclined to stay at the edges of the graphitic pore. Furthermore, Co-based and Fe-based MeNCs were generally considered to be more active electrocatalysts than Ni-based MeNC. The activity was determined greatly by the structure and morphology of the nanocarbon materials, as well as the synergetic effects between carbon and dopants and between the two dopants.[126,127] As a typical example, Liu et al.[128] synthesized transition metal (Fe, Co, and Ni) nanoparticles encapsulated in N-doped CNTs (M/N–CNTs, M = Fe, Co, and Ni) via a solid–state thermal reaction followed by an annealing process. SEM and TEM images revealed that Co nanoparticles were encapsulated in graphitic layers and dispersed at the endpoint of N-doped CNTs in the typical Co/N–CNTs sample. Co 2p and N 1 s spectra in XPS clearly indicated the presence of $Co-N_x$ moieties in Co/N–CNTs through the coordination between Co and N_x.[129] To select optimal annealing temperatures, 700°, 800°, and 900°C were used to produce different Co/N–CNTs products labeled Co/N–CNTs-700, Co/N–CNTs-800, and Co/N–CNTs-900, respectively. In the comparison of electrocatalytic activity of these three Co/N–CNTs products, it was found that Co/N–CNTs-700 yielded the highest activities for both ORR and OER, perhaps because the Co/N–CNTs catalyst obtained at 700°C possessed the highest concentration of metal and N species[127] and the largest structural defects or microstructural rearrangement of the atom stemming from N-doping,[130] resulting in optimal conductivity and concentration of active species. When ORR and OER activities catalyzed by different metal-based N–CNTs catalysts were further compared, the calculated potential difference between the ORR (Potential at -3 mA cm^{-2}) and the OER (Potential at 10 mA cm^{-2})[99] from the polarization curves in O_2-saturated 0.1 M KOH solution at 1600 rpm were 0.78, 0.94, 1.09, 1.04, and 1.49 V for Co/N–CNTs, Fe/N–CNTs, Ni/N–CNTs, Pt/C, and RuO_2, respectively, showing that the catalytic activities of the three different metal-based N–CNTs catalysts were increased in the order of Ni/N–CNTs < Fe/N–CNTs < Co/N–CNTs. This suggested that Co nanoparticles played a significant role in both ORR and OER activities possibly due to its smaller size (3–4 nm as found in TEM images), favoring the fast electron transport between carbon and Co nanoparticles.[97] Importantly, for the synergistic catalytic effects between the N dopant and Co nanoparticles on Co/N–CNTs, accounting for the higher catalytic activity towards both ORR and OER, it was proposed that: (1) N-doping could produce more active sites due to its high electronegativity than C; (2) the N dopant could improve the electron donor-acceptor properties and enhance the bonding; (3) the Co could interact with N to form intimate bonds ($Co-N_x$), generating more active centers and thus improving catalytic performances; (4) the small sized Co nanoparticles favored the electron transfer from metal to carbon; (5) Co nanoparticles could essentially catalyze the formation of CNTs; and (6) the encapsulated Co nanoparticles could be protected from leaching to stabilize active sites.

Regarding metal, N co-doped carbon (i.e., M/N–C), two research groups investigated the possibility of metal-nitrogen-carbons as air-electrode catalysts for in RZABs[105] and RLOBs.[131] Li et al.[105] synthesized different metal-containing polydopamine (M–PDA) precursors via a self-polymerization of a mixture of transition metal salts and dopamine solution. After carbonization, three M/N–C catalysts labeled as Co–PDA–C, Ni–PDA–C, and Fe–PDA–C, respectively, were obtained; of the three, Co–PDA–C possessed much higher stability and catalytic activities for both ORR and OER in RZAB. Co–PDA–C was then steadily cycled up to 500 cycles (see Figure 4.5a) with only a slight increase (0.23 V) in the discharge-charge voltage gap with a discharge-charge current density of 2 mA cm^{-2} and 1 hour

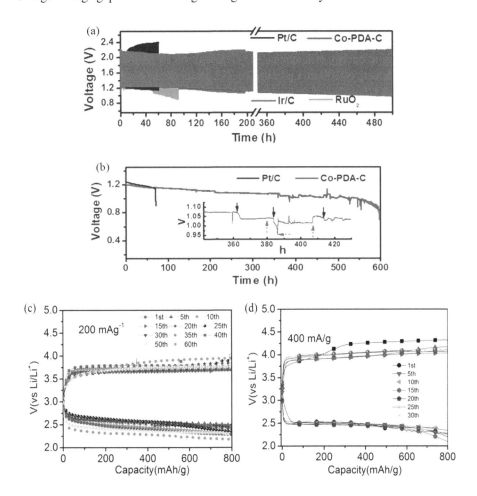

FIGURE 4.5 (a) Discharge-charge cycling performances of ZABs based on Co–PDA–C (red), Pt/C (black), Ir/C (blue) and RuO$_2$ (green) at 2 mA cm^{-2}; (b) Voltage profile of a Co–PDA–C based ZAB compared with a Pt/C based ZAB when fully discharged at a current density of 5 mA cm^{-2}, the inset shows the details of the selected period of the Co–PDA–C based ZAB (from 340 to 430 h); (c) cycling responses of Fe–N/C-based LOBs at 200 mA g^{-1} with a limited capacity of 800 mAh g^{-1}; and (d) cycling responses of Fe–N/C-based LOBs at 400 mA g^{-1} with a limited capacity of 800 mAh g^{-1}. (Modified from Li, B. et al. *Nanoscale* **2016**, *8*, 5067–5075.[105]; Guo, G. et al. *Nanotechnology* **2016**, *27*, 045401.[131])

per cycle. At a discharge current density of 5 mA cm^{-2}, the battery with Co–PDA–C as air-electrode could continuously discharge for more than 540 hours (see Figure 4.5b) with a discharge voltage above 1 V and a voltage drop rate of merely 0.37 mV h^{-1}. Meanwhile, two transition metal-nitrogen/carbon (M–N/C, M = Fe, Co) catalysts[131] were synthesized using environmentally-friendly histidine-tag-rich elastin protein beads, metal sulfate, and water-soluble carbon nanotubes, followed by post-annealing and acid leaching processes. The obtained Co–N/C air-electrode exhibited higher charge potentials than Fe–N/C when tested with non-aqueous RLOBs, suggesting that Fe–N/C possessed higher bifunctional activities for both ORR and OER. With a controlled capacity of 800 mAh g^{-1} at current densities of 200 or 400 mA g^{-1}, the Fe–N/C cathode showed stable charge voltages of ~3.65 or 3.90 V (see Figure 4.5c, d), corresponding to energy efficiencies of ~71.2% or 65.1%, respectively. Particularly, at a current density of 100 mA g^{-1}, the Fe–N/C cathode delivered a high capacity of ~12,441 mAh g^{-1}.

Metal, P co-doped carbon materials[106,132] have also been explored as novel electrocatalysts for air-electrodes of RMABs. For example, Zheng et al.[106] synthesized P and Co co-doped reduced-graphene (P–Co–rGO) via an electrostatic assembly followed by a pyrolysis process. Tetraphenylphosphonium bromide ($C_{24}H_{20}PBr$) and cobalt nitrate ($Co(NO_3)_2 \cdot 6H_2O$) were used as sources of P and Co in the synthesis. The hydrophilic anion groups of graphene (e.g., $-O^-$, $-COO^-$) resulted in the easy introduction of P and Co into the graphene network. For comparison, two single heteroatom-doped rGO samples of P–rGO and Co–rGO were also prepared using the same method for reference. In the characterization of microstructure and morphology, XRD patterns showed a narrow and sharp diffraction peak and a weak peak at 26° and 43°, respectively, corresponding to the (002) and (100) plane for rGO, confirming the reduction of graphene oxide induced by the graphite structure. In a detailed comparison of the (002) diffraction peak, the (002) peak became weaker and broader with increasing doping elements, suggesting more defect sites and disorders in the graphitic structure after the introduction of single and dual heteroatom(s) in the graphene, especially the introduction of P and Co. The (002) peak also shifted to a lower angle with increasing dopants, indicating the increased interlayer spacing between the graphene sheets. According to the results of atom force microscopy (AFM) measurements (see Figure 4.6a), P–Co–rGO was found to have five graphene layers with an average thickness of 3.5 nm. Furthermore, the ratio of intensities between the D band and G band (I_D/I_G, see Table 4.2) in Raman spectra was amplified with increasing dopants, implying a higher disorder in the graphitic structure and more defect sites in the P–Co–rGO. Further, XPS investigations for C 1 s, P 2p and Co 2p showed that C remained in the conjugated graphene system in the forms of C–P and C–O in which the three P species were P–C, P–O, and P–Co. Particularly, the Co 2p indicated the existence of Co^{2+} cations in P–Co–rGO without Co-based carbides, phosphides or Co metals. These XPS results not only confirmed the successful doping of P and Co, but also indicated the presence of Co–P moieties due to the possible bonding of Co^{2+} cations to P, favoring the enhancement of electrocatalytic performances.[103,105] Interestingly, the results of ICP–MS in Table 4.2 shows that, when compared to mono-doped graphene, P–Co–rGO exhibits an increased Co content and a decreased P content, demonstrating that the doping of P in the graphene lattice

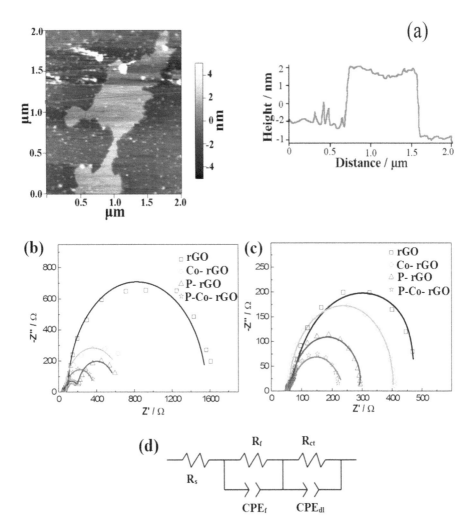

FIGURE 4.6 (a) AFM image of P–Co–rGO. Electrochemical impedance spectra of rGO, Co–rGO, P–rGO, and P–Co–rGO for ORR (b) and OER (c) with a rotating speed of 1600 rpm, (d) the corresponding equivalent circuits for the EIS of ORR and OER. (Modified from Zheng, X. et al. *RSC Adv.* **2016**, *6*, 64155–64164.[106])

can favor the co-doping of Co and the formation of Co–P moieties as evidenced by XPS, contributing to high catalytic activities and long-term stability for both ORR and OER. In electrochemical measurements in O_2-saturated 0.1 M KOH solution, P–Co–rGO followed a four-electron ORR transport mechanism and exhibited better ORR features, such as onset potential, half-wave potential, and Tafel slope, than rGO, P–rGO, or Co–rGO catalysts, indicating its excellent ORR activity. When compared to that of commercial Pt/C, the half-wave potential of P–Co–rGO only had a small negative shift (~12.8 mV). For the OER, the investigated LSV curves in N_2-saturated 0.1 M KOH revealed that P–Co–rGO showed a higher maximum

TABLE 4.2

The Ratio of I_D/I_G Based on Raman Spectra and the Content of Co and P of Co–rGO, P–rGO and P–Co–rGO Based on ICP–MS

	P (wt%)	Co (wt%)	I_D/I_G
rGO	–	–	1.11
Co–rGO	–	0.267	1.23
P–rGO	1.087	–	1.24
P–Co–rGO	0.639	2.990	1.28

Source: Modified from Zheng, X. et al. *RSC Adv.* **2016**, *6*, 64155–64164.[106]

current density ($\sim$29.07 mA cm^{-2}) than those of rGO, P–rGO, and Co–rGO catalysts, and was comparable to that of IrO$_2$ ($\sim$29.2 mA cm^{-2}). Furthermore, P–Co–rGO delivered the lowest potential ($\sim$1.62 V) at a current density of 10 mA cm^{-2}, which was very close to that of IrO$_2$ ($\sim$1.60 V). These results demonstrated that P–Co–rGO could be a promising bifunctional electrocatalyst for both ORR and OER with high activities. To further gain insight into the electrochemical properties of these heteroatom(s)-doped reduced graphene oxides, EIS was performed on a catalyst disk electrode at the onset potential of each catalyst in O$_2$-saturated electrolyte for ORR and at the half-wave potential in N$_2$-saturated electrolyte for OER under a rotating speed of 1600 rpm. Corresponding to the fitted Nyquist curves in Figure 4.6b and c, the equivalent circuits of ORR and OER in Figure 4.6d include R$_s$, R$_f$, R$_{ct}$, CPE$_f$ and CPE$_{dl}$, representing the electrolyte resistance, catalyst film resistance, charge transfer resistance, and constant phase elements of the catalyst layer and double layer respectively.[133] It can be seen in Figure 4.6b that each Nyquist curve is composed of two semicircles, in which the former is associated with the resistivity of the catalyst and the latter was the ORR taking place at the electrode-electrolyte interface. Compared to rGO, Co–rGO, and P–rGO catalysts, P–Co–rGO has the lowest R$_{ct}$, possibly due to the higher electrocatalytic ORR performance resulting from the chemical coupling of P–Co. Meantime, the span of the Nyquist curves in Figure 4.6c correspond to the magnitude of the charge transfer resistance associated with the OER process. P–Co–rGO displayed a smaller R$_{ct}$ than rGO, Co–rGO, and P–rGO catalysts, indicating lower charge transfer resistance and a faster OER process resulting from strong P–Co coupling.

4.3.2 COMPOSITES OF DUAL ELEMENTS CO-DOPED CARBON AND METALS

Recent studies have been carried out to achieve metal-based ternary-doped carbon catalysts with two other non-metal heteroatoms for ORR or OER.[107,134] Using a thermal co-decomposition of pre-synthesized polyaniline (PAn) nanofibers, cobalt chloride hexahydrate (CoCl$_2\cdot$6H$_2$O), and cyanamide (CM), Yang et al.[107] synthesized N–Co–O triply doped highly crystalline porous carbon materials for OER in acidic solution to evaluate its application in acid RLAB without side reactions with atmospheric CO$_2$.

In the synthesis, $CoCl_2 \cdot 6H_2O$, $Co(NO_3)_2 \cdot 6H_2O$, and $Co(CH_3COO)_2 \cdot 4H_2O$ were used as the sources of Co to obtain three N–Co–O triply doped carbons (TDC): TDC–Cl, TDC–NO$_3$, and TDC–Ac, respectively. In the characterization of morphology, phase, and doping, TEM images showed that TDC–Cl had an irregular porous structure (i.e., pore size above 10 nm) with significant morphology departure from doped carbon, whereas TDC–NO$_3$ and TDC–Ac were denser in structure with more unbleached Co nanoparticles and fewer pores in the carbon matrix due to the Co nanoparticles being encapsulated in the carbon layer. As evidenced by TGA results, the contents of Co for both TDC–NO$_3$ and TDC–Ac samples were 4.0wt% and 5.8wt%, respectively, both of which were higher than that of TDC–Cl. The tested BET surface area and pore volume of TDC–Cl were 235.9 m^2 g^{-1} and 0.27 cm^3 g^{-1}, respectively, which were higher than those of TDC–NO$_3$ (162.5 m^2 g^{-1}, 0.18 cm^3 g^{-1}) and TDC–Ac (182.6 m^2 g^{-1}, 0.22 cm^3 g^{-1}), confirming the larger amount of mesoporous structures for TDC–Cl. The quantitative analysis by XPS measurements revealed that, compared to TDC–NO$_3$ and TDC–Ac samples, TDC–Cl possessed a higher doped-Co content ($\sim$0.96 wt%) and a lower total-Co content ($\sim$2.1 wt%), indicating that Co achieved highly effective doping in the carbon matrix of TDC–Cl. For OER activity, LSV curves in O_2-saturated 0.5 M H_2SO_4 showed that at a 1.8 V potential cutoff, TDC–Cl delivered a higher current density of 6.8 mA cm^{-2} than TDC–NO$_3$ and TDC–Ac samples. When TDC–Cl was compared with Pt/C and Ir/C, the current density of TDC–Cl was about twice that of Pt/C, but much lower than that of Ir/C in acid solution. Furthermore, TDC–Cl showed a higher Tafel slope ($\sim$247 mV dec^{-1}) than Pt/C ($\sim$166 mV dec^{-1}) and Ir/C ($\sim$59 mV dec^{-1}). It was discussed that the activation for OER in an acidic solution was still higher for TDC–Cl than for Pt/C and Ir/C catalysts. To figure out what active sites are responsible for the improvement of OER in acidic solutions, XPS spectra of N 1 s, O 1 s and Co 2p demonstrated that graphitic-N and oxygen-bound cobalt species should be responsible for OER activity, whereas an abundance of both pyridinic-N and Co–N$_x$ sites commonly resulted in remarkable ORR activity. It was believed that the active sites were useful for guiding future research in developing high-performance composite cathodes in RMABs.

Using natural materials that are low-cost, environmentally-friendly, and able to be produced in large scale, Guo et al.[134] developed natural tea-leaf-derived, heteroatoms (N, P, and Fe) ternary-doped 3D hierarchically porous carbons (HDPC) for ORR applications. They discussed that tea leaves could be directly employed as precursors for carbon frameworks and for the heteroatom dopants responsible for the improvement in catalytic activity toward ORR.[135,136] In the research to investigate the effects of temperature on ORR, three pyrolysis temperatures of 700°, 800°, and 900°C were selected to prepare the corresponding HDPC-700, HDPC-800, and HDPC-900 samples. In the characterization of microstructure and morphology, SEM and TEM showed that HDPC-700 and HDPC-900 presented the interconnected macropores, whereas HDPC-800 had both macropores and mesopores, matching the results measured by BET and porosity. The BET surface area for HDPC-800 was $\sim$345.76 m^2 g^{-1}, which was higher than those of HDPC-700 ($\sim$130.77 m^2 g^{-1}) and HDPC-900 ($\sim$281.34 m^2 g^{-1}). Raman spectra revealed that the relative intensity ratio of the D and G bands (I_D/I_G) decreased with increasing pyrolysis temperatures, indicating an increase in the degree of ordered graphitic structures. For XPS, to check the doping of Fe, N, and P, the fractions of Fe–N_x and pyrrolic-N species were decreased with increasing pyrolysis temperatures, suggesting the decreased active sites when pyrolysis temperature was increased from 700° to 900°C. The P dopant also decreased with increasing temperatures. As reported, P-doping needed to be optimized for promoting catalytic activity toward ORR because high degrees of P-doping could cause large distortions in the hexagonal symmetry of the carbon framework, destroying the sp^2 carbon network and thus resulting in the decrease of ORR activity.[68] To evaluate the ORR activity, LSV curves were tested on a RDE in O_2-saturated 0.1 M KOH aqueous solution at 10 mV s^{-1} and 2000 rpm. The onset and half-wave potentials of HDPC-800 were 0.95, and 0.79 V, respectively. They were higher than those of HDPC-700 (0.91, 0.71 V) and HDPC-900 (0.89, 0.68 V), and close to that of commercial Pt/C (0.93, 0.77 V). This suggests that following a four-electron ORR transfer mechanism, HDPC-800 had a higher ORR activity than HDPC-700 and HDPC-900 and was comparable to commercial Pt/C. The high catalytic performance of HDPC-800 was attributed to the synergistic effect of heteroatom doping and 3D hierarchically porous structures through: (1) both N- and P-doping in carbon framework formed partially active sites for ORR; (2) newly formed Fe–N_x active sites favored ORR; and (3) 3D hierarchically porous structures composed of micropores and nanochannels benefited not only the transport and diffusion of oxygen molecules/electrolytes but also the distribution of heteroatoms and active sites. The same group[109] also studied novel heteroatoms (N, P, S and Fe) quaternary-doped carbon catalysts derived from Shewanella bacteria for ORR and found that the design and synthesis of multi-heteroatoms-doped carbons were a novel approach to developing advanced composite air-electrode materials for RMABs.

4.4   COMPOSITES OF DOPED CARBON AND NITRIDES

Based on that the combination of carbon and nitride can give enhanced electrocatalytic activities for both ORR and OER, researchers have developed the composites of heteroatom(s)-doped carbon and nitride as possible air-electrode materials for RMAB applications. Molybdenum (MoN), exhibiting high Pt-like electrocatalytic activities, electronic conductivity, and chemical stability, has become promising in the development of MoN-based composites as potential cathodes in RLOBs.[137,138]

To improve catalytic activity and build a rational design to obtain more active sites and mesoporous nanostructures, Zhang et al.[137] synthesized MoN/N-doped carbon nanosphere (MoN/N–C) composites though a hydrothermal method followed by ammonia annealing. In the synthesis, MoO_2 nanospheres were first synthesized with a one-pot hydrothermal treatment of the mixture containing ammonium heptamolybdate tetrahydrate, ethylene glycol, and polyvinylpyrrolidone (PVP). The as-prepared MoO_2 nanospheres were subsequently heated at 700°C under an ammonia atmosphere using cyanamide (NH_2CN) as the structure confinement agent to obtain well-crystallized MoN/N–C nanospheres. Both PVP and cyanamide acted as the sources of C. Characterization by SEM, TEM, and HR–TEM images showed that MoN/N–C consisted of uniform nanospheres with diameters of 50 ± 5 nm, in which each nanosphere-processed mesoporous structures resulted from the assembly of many primary nanoparticles (3–5 nm). The BET surface area of MoN/N–C was

~93 m^2 g^{-1} in which the pore sizes analyzed by BJH showed a range from 2 to 20 nm. XRD confirmed the MoN phase without any impurity. The oxidation states of the elements in MoN/N–C were investigated by XPS. The analysis of Mo 3d and N 1 s confirmed the formation of Mo–N bonds, and the combination of C 1 s and N 1 s evidenced not only by the successful doping of N but also the presence of pyridinic-N, pyrrolic-N, and graphitic-N. In the investigation of electrochemical properties, MoN/N–C presented higher ORR onset potential and higher ORR peak current than those of MoN, N–C and the physical mixture MoN + N–C, indicating the significant synergetic effects between MoN and N-doped C on the ORR activity. When further electrochemical measurements were conducted in an assembled RLOB, it was found that in the first charge-discharge profile, MoN/N–C exhibited lower overpotentials compared to super P cathodes. Furthermore, after 10 cycles, the capacity of super P dropped dramatically from 1150 to 280 mAh g^{-1} whereas MoN/N–C showed a stabilized capacity above 790 mAh g^{-1}. Particularly, under a controlled capacity of 400 mAh g^{-1}, MoN/N–C demonstrated a good cycling ability for over 30 cycles, demonstrating the application potential of MoN/N–C for OER and ORR in RLOBs.

Dong et al.[138] also synthesized and investigated molybdenum nitride/nitrogen-doped graphene nanosheets (MoN/NGS) as an alternative air-electrode for RLOBs. They used an in-situ synthesis route to obtain conductive MoN and then evenly coated it on nitrogen-doped graphene nanosheets. XRD was used to confirm the well-crystallite MoN phase (JCPDS No. 77-1999) without any impurities, SEM and TEM images showed that with a size range of 20–40 nm in diameter, the MoN nanoparticles were uniformly dispersed on thin layers of NGS. XPS revealed the existence of N-doping in graphene nanosheets due to the ammonia reduction of GO. Elemental analysis showed that the amount of NGS in MoN/NGS was 12 wt%, in which the total N amount was 2.2 wt%. According to electrochemical measurements using a RLOB, the MoN/NGS air-electrode exhibited a larger capacity with a significantly higher round-trip efficiency when compared with the NGS cathode, illustrating that the addition of MoN could result in the improvement of catalytic activity and thus the enhancement of battery performance. However, at a controlled capacity of 1100 mAh g^{-1} and a current density of 0.08 mA cm^{-2}, although MoN/NGS only ran 7 stable cycles with good reversibility, it did give a large specific capacity of 1050 mAh g^{-1}, suggesting that the combination of MoN and carbon could be a desirable strategy to develop alternative air-electrodes for RLOB.

4.5   OTHER DOPED CARBON-BASED COMPOSITES

Because transition metal-nitrogen/carbon (M–N/C) catalysts have attracted considerable attention as the most promising alternatives for both ORR and OER in RMABs, some derivative configurations are explored from the modification of M–N/C. Typically, with varied experimental conditions (e.g., high-temperature treatment and/or oxidizing environment), M–N/C catalysts tend to decompose and transform to metal/metal oxide/metal carbide-nitrogen-carbon structured catalysts in which the formed substrate-anchor configurations[103] tend to produce a synergetic host-guest electronic interaction. Additionally, the core-shell structure can exhibit a protective graphitic shell, resulting in a strongly coupled interface and thus improved activity and stability.[139–145] For instance, Ni et al.[141] designed and prepared in-situ N-doped mesoporous carbon-supported CoO@Co nanoparticles using ionic liquid (IL) 1-butyl-3-methylimidazolium tetrachlorocobalt ([BMlm]$_2$[CoCl$_4$]) as the precursor, with silica as the hard template. In their synthesis, N-doped mesoporous carbon-supported Co nanoparticles were first prepared after a calcination process under N$_2$ with a silica template. The catalyst was then exposed to air to obtain the final N-doped mesoporous carbon-supported CoO@Co product (i.e., Co@Co/C–IL) after Co was oxidized to CoO. Atomic absorption spectrometry (AAS) measurements measured the Co content to be 35.3 wt%. C–IL was prepared with [BMlm]Cl using the same method and then mixed with CoCl$_2$·6H$_2$O and treated to prepare a referenced mixture product (m–CoO@Co/C–IL) with the same Co wt% as that of Co@Co/C–IL. In the characterization of structure and morphology, BET measurements showed a surface area of 139.15 m^2 g^{-1} for CoO@Co/C–IL, despite the addition of Co in which the pore sizes of CoO@Co/C–IL were around 2 nm, confirming the presence of a mesoporous structure. TEM displayed that CoO@Co particles were irregularly shaped and possessed sizes ranging from 20–300 nm on the carbon. After XRD checked the crystal structure, XPS was used to analyze the surface properties of CoO@Co/C–IL and m–CoO@Co/C–IL. Based on N 1 s spectra, it was found that CoO@Co/C–IL only possessed three N species of pyridinic-N, quaternary-N, and Co–N$_x$, whereas the Co–N$_x$ species was not observed in m–CoO@Co/C–IL, confirming that the physical mixture m–CoO@Co/C–IL possessed no coupling between Co and N. Co 2p spectra confirmed two Co species in the form of Co^{2+} and Co–N$_x$, suggesting the existence of CoO on the surface of Co nanoparticles and the interaction between Co and N. Moreover, the recorded H$_2$-TPR curves in Figure 4.7a shows the reduction temperature of CoO@Co/C–IL in the range of 230–440°C, which is broader and lower than pure CoO and m–CoO@Co/C–IL, suggesting the possible interaction between CoO@Co and C–IL and therefore the acceleration of the reduction of CoO. When the catalytic activity of the catalyst was evaluated, it was observed that, compared to carbon electrodes such as XC-72 and C–IL, CoO@Co/C–IL exhibited higher peak current densities for both ORR and OER with similar ORR

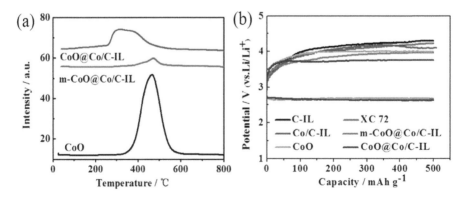

FIGURE 4.7   (a) H$_2$-TPR curves of CoO, m–CoO@Co/C–IL, and CoO@Co/C–IL; (b) capacity retention capability of Li-O$_2$ cells with XC-72, C–IL, and Co@Co/C–IL electrodes at various current densities. (Modified from Ni, W. et al. *J. Mater. Chem. A* **2016**, *4*, 7746–7753.[141])

onset potentials and lower OER onset potentials. This suggested that CoO@Co/C–IL possessed superior activities to the other two carbon samples. The discharge-charge curves of a tested RLOB showed that CoO@Co/C–IL possessed better rate capabilities with a higher coulombic efficiency than XC-72 and C–IL samples. Furthermore, as seen in Figure 4.7b, CoO@Co/C–IL in the range of 100–500 mAh g^{-1} presents lower charge potential than C–IL, XC72, Co/C–IL, m–CoO@Co/C–IL, CoO, and CoO@Co/C–IL samples despite at the same discharge potentials. The addition of Co alone was not effective enough for OER whereas the incorporation of CoO favored the improvement of catalytic activity. m–CoO@Co/C–IL exhibited a higher activity than pure CoO, suggesting that the amount of CoO was not the crucial factor in enhancing electrochemical performance. The comparison of CoO@Co/C–IL, m–CoO@Co/C–IL, and CoO showed that the enhanced catalytic activity of CoO@Co/C–IL should be resulted from the synergistic interaction between CoO@Co and C–IL and from the effects of N-doping. In the cycling measurements, CoO@Co/C–IL delivered a good cyclic performance with negligible decay after 55 cycles, whereas XC-72 and C–IL could only run 11 and 26 cycles, respectively. This research presented a new strategy in the development of modified M/N–C air-electrode catalysts for high-performance RLOBs.

In addition to the derivatives of designed metal-nitrogen/carbon catalysts, heteroatom(s)-doped carbon has also composited with other materials such as sulfide,[96,119,146] Cobalt-hydroxide,[147] and cobalt ferrite,[148] to obtain enhanced activities for both ORR and OER in RMAB applications. Compared to cobalt-hydroxide and cobalt ferrite, the composition of sulfide and heteroatom(s)-doped carbon has been studied by Geng et al.[146] as the air-electrodes for RZABs. They used thiourea as the single precursor to react with graphene oxide (GO) to obtain N, S co-doped graphene nanosheets (NS–GNS). In the reaction of thiourea with GO in the presence of Co^{2+}, the nucleation of CoSx nanoparticles was formed and grown into considerably larger sizes ($\sim$50 nm) to form the CoSx@NS–GNS catalyst. In the measurement of electrochemical performances in an assembled RZAB, the CoS$_x$@NS–GNS air-electrode was discharged and charged for 50 cycles at 1.25 mA cm^{-2} over a period of 50 hours with an almost constant discharge voltage of $\sim$1.23 V. For comparisons, commercial Pt/C after 50 cycles exhibited a discharge voltage loss of 11% and a clear increase in charge voltage. In a detailed comparison of 40 cycles, the difference between open cell voltage and the discharge potential of Pt/C ($\sim$0.240 V) was larger than CoS$_x$@NS–GNS ($\sim$0.151 V), suggesting that CoS$_x$@NS–GNS could perform better than a Pt-based catalyst and should be a potentially cost-effective alternative for RZABs.

Additionally, researchers have also explored heteroatom(s) doped carbon mixtures of two different carbon materials[149–151] and their further composites with oxide such as Co$_3$O$_4$[152] for both ORR and OER in the development of efficient composite bifunctional catalysts for RMABs.

4.6 CHAPTER SUMMARY

Based on literature data, this chapter reviewed the doped carbon-based composites as bifunctional catalysts for RMABs in terms of their material selection, synthesis strategy, structural characterization, and electrochemical performance. Their advantages induced by different components for enhancing RMAB performance are analyzed. Several challenges related to the catalyst properties, material preparation, and electrode/battery performance are also reviewed. The main points underscored are as the follows:

1. The composition of doped carbon and carbon can be used to strategically compensate for the disadvantages of pure carbon components to obtain designed structures, hybrid phase, and properties for ORR and OER catalysis. The synergetic interactions between different carbon components can favor not only electron transfer and mass transport but also improve catalytic activity and durability/stability.

2. Doped carbon-based composite catalysts can simultaneously possess two synergistic effects, leading to enhanced electrocatalytic performances of air-electrode for RMABs: one synergistic effect is the interaction between carbon and the other component, and the other effect is induced by the dopant(s) and the other component. For doped carbon-based composite catalysts, the interaction between dopant(s) and the other component can promote interfacial electron transfers. In catalyzing ORR and OER, dual-doping in doped carbon-carbon composite catalysts can result in additional synergies, that is the interaction between not only carbon and dopants but also between two dopants, induced by the coupling of more than one heteroatom in carbon.

3. The incorporation of metal (including noble metals and non-noble metals) into doped-carbon is realized by two strategies: (i) metal is used as a dopant; (ii) metal is used as a filler. The first strategy can introduce structural defects, tune electronic properties and surface polarities, and then create more active sites induced by the strong interaction between the doped metal and carbon. Because the incorporated metal is only used as a filler in the second strategy, it can serve as the active site to produce synergy with underlying carbon atoms, contributing to the fast transport of ion/electrons/oxygen and thereby the improvement of electrocatalytic activity and stability/durability.

4. The use of non-noble metals can increase catalytic sites by forming active complex structures (i.e., MeN_xC, Me = Co, Fe, Ni, Mn, etc., normally $x = 2$ or 4), which is induced by the strong interaction between the non-noble metal and carbon, favoring the enhancement of both ORR and OER activities as well as stability/durability.

5. Metal oxides have shown various interactions with doped carbon(s), leading to improved electrochemical performances of RMAB air-electrodes and good resistance to the corrosion of carbon. Particularly, perovskite oxides and spinel oxides can offer their intrinsic activities for ORR and OER, demonstrating further improvements of catalytic activity and stability/durability for both ORR and OER in RMAB.

6. Although nitrides can present Pt-like properties, electronic conductivity, chemical stability, new synthesis strategy, as well as advanced nanostructures are still needed in the development of doped carbon-nitride composites to

improve their bifunctional catalytic activity and durability/stability for RMAB applications.

REFERENCES

1. Park, H. W.; Lee, D. U.; Liu, Y.; Wu, J.; Nazar, L. F.; Chen, Z. Bi-functional N-doped CNT/graphene composite as highly active and durable electrocatalyst for metal air battery applications. *J. Electrochem. Soc.* **2013**, *160*, A2244–A2250.
2. Yu, A.; Park, H. W.; Davies, A.; Higgins, D. C.; Chen, Z.; Xiao, X. Free-standing layer-by-layer hybrid thin film of graphene-MnO2 nanotube as anode for lithium ion batteries. *J. Phys. Chem. Lett.* **2011**, *2*, 1855–1860.
3. Yoo, E.; Kim, J.; Hosono, E.; Zhou, H.-S.; Kudo, T.; Honma, I. Large reversible Li storage of graphene nanosheet families for use in rechargeable lithium ion batteries. *Nano Lett.* **2008**, *8*, 2277–2282.
4. Chen, Z.; Higgins, D.; Tao, H.; Hsu, R. S.; Chen, Z. Highly active nitrogen-doped carbon nanotubes for oxygen reduction reaction in fuel cell applications. *J. Phys. Chem. C* **2009**, *113*, 21008–21013.
5. Sheng, Z.-H.; Gao, H.-L.; Bao, W.-J.; Wang, F.-B.; Xia, X.-H. Synthesis of boron doped graphene for oxygen reduction reaction fuel cells. *J. Mater. Chem.* **2012**, *22*, 390–395.
6. Liang, J.; Jiao, Y.; Jaroniec, M.; Qiao, S. Z. Sulfur and nitrogen dual-doped mesoporous graphene electrocatalyst for oxygen reduction with synergistically enhanced performance. *Angew. Chem. Int. Ed.* **2012**, *51*, 11496–11500.
7. Sun, Y. Q.; Li, C.; Shi, G. Q. Nanoporous nitrogen doped carbon modified graphene as electrocatalyst for oxygen reduction reaction. *J. Mater. Chem.* **2012**, *22*, 12810–12816.
8. Li, Y.; Huang, Z.; Huang, K.; Carnahan, D.; Xing, Y. Hybrid Li-air battery cathodes with sparse carbon nanotube arrays directly grown on carbon fiber papers. *Energy Environ. Sci.* **2013**, *6*, 3339–3345.
9. Nam, G.; Park, J.; Kim, S. T.; Shin, D.; Park, N.; Kim, Y.; Lee, J.-S.; Cho, J. Metal-free Ketjenblack incorporated nitrogen-doped carbon sheets derived from gelatin as oxygen reduction catalysts. *Nano Lett.* **2014**, *14*, 1870–1876.
10. Morcos, I.; Yeager, E. Kinetic studies of the oxygen-peroxide couple on pyrolytic graphite. *Electrochim. Acta* **1970**, *15*, 953–975.
11. Qu, D. Investigation of oxygen reduction and activated carbon electrodes in alkaline solution. *Carbon* **2007**, *45*, 1296–1301.
12. Lin, H.; Liu, Z.; Mao, Y.; Liu, X.; Fang, Y.; Liu, Y.; Wang, D.; Xie, J. Effect of nitrogen-doped carbon/Ketjenblack composite on the morphology of Li_2O_2 for high-energy-density Li-air batteries. *Carbon* **2016**, *96*, 965–971.
13. Kim, J.-H.; Kannan, A. G.; Woo, H.-S.; Jin, D.-G.; Kim, W.; Ryu, K.; Kim, D.-W. A bi-functional metal-free catalyst composed of dual-doped graphene and mesoporous carbon for rechargeable lithium-oxygen batteries. *J. Mater. Chem. A* **2015**, *3*, 18456–18465.
14. Liu, Z.; Nie, H.; Yang, Z.; Zhang, J.; Jin, Z.; Lu, Y.; Xiao, Z.; Huang, S. Sulfur-nitrogen co-doped three-dimensional carbon foams with hierarchical pore structures as efficient metal-free electrocatalysts for oxygen reduction reactions. *Nanoscale* **2013**, *5*, 3283–3288.
15. Wang, Z.-L.; Xu, D.; Xu, J.-J.; Zhang, L.-L.; Zhang, X.-B. Graphene oxide gel-derived, free-standing hierarchically porous carbon for high-capacity and high-rate rechargeable $Li-O_2$ batteries. *Adv. Funct. Mater.* **2012**, *22*, 3699–3705.
16. Zhang, L.; Zhang, F.; Yang, X.; Long, G.; Wu, Y.; Zhang, T.; Leng, K. et al. Porous 3D graphene-based bulk materials with exceptional high surface area and excellent conductivity for supercapacitors. *Sci. Rep.* **2013**, *3*, 1408.

17. Kim, B. G.; Kim, H. J.; Back, S.; Nam, K. W.; Jung, Y.; Han, Y. K.; Choi, J. W. Improved reversibility in lithium-oxygen battery: Understanding elementary reactions and surface charge engineering of metal alloy catalyst. *Sci. Rep.* **2014**, *4*, 4225.
18. Cui, Z. H.; Guo, X. X.; Li, H. Equilibrium voltage and overpotential variation of non-aqueous Li-O_2 batteries using the galvanostatic intermittent titration technique. *Energy Environ. Sci.* **2015**, *8*, 182–1187.
19. Zhu, Q.; Lin, L.; Jiang, Y.-F.; Xie, X.; Yuan, C.-Z.; Xu, A.-W. Carbon nanotube/S–N–C nanohybrids as high performance bifunctional electrocatalysts for both oxygen reduction and evolution reactions. *New J. Chem.* **2015**, *39*, 6289–6296.
20. Li, R.; Wei, Z.; Gou, X. Nitrogen and phosphorus dual-doped graphene/carbon nanosheets as bifunctional electrocatalysts for oxygen reduction and evolution. *ACS Catal.* **2015**, *5*, 4133–4142.
21. Gong, X.; Liu, S.; Ouyang, C.; Strasser, P.; Yang, R. Nitrogen- and phosphorus-doped biocarbon with enhanced electrocatalytic activity for oxygen reduction. *ACS catal.* **2015**, *5*, 920–927.
22. Choi, C. H.; Chung, M. W.; Park, S. H.; Woo, S. I. Additional doping of phosphorus and/or sulfur into nitrogen-doped carbon for efficient oxygen reduction reaction in acidic media. *Phys. Chem. Chem. Phys.* **2013**, *15*, 1802–1805.
23. Suntivich, J.; Gasteiger, H. A.; Yabuuchi, N.; Nakanishi, H.; Goodenough, J. B.; Shao-Horn, Y. Design principles for oxygen-reduction activity on perovskite oxide catalysts for fuel cells and metal-air batteries. *Nat. Chem.* **2011**, *3*, 546–550.
24. Suntivich, J.; May, K. J.; Gasteiger, A.; Goodenough, J. B.; Shao-Horn, Y. A perovskite oxide optimized for oxygen evolution catalysis from molecular orbital principles. *Science* **2011**, *334*, 1383–1385.
25. Hardin, W. G.; Slanac, D. A.; Wang, X.; Dai, S.; Johnston, K. P.; Stevenson, K. J. Highly active, nonprecious metal perovskite electrocatalysts for bifunctional metal-air battery electrodes. *J. Phys. Chem. Lett.* **2013**, *4*, 1254–1259.
26. Lee, D. U.; Park, H. W.; Park, M. G.; Ismayilov, V.; Chen, Z. Synergistic bifunctional catalyst design based on perovskite oxide nanoparticles and intertwined carbon nanotubes for rechargeable zinc-air battery applications. *ACS Appl. Mater. Interfaces* **2015**, *7*, 902–910.
27. Hardi, W. G.; Mefford, J. T.; Slanac, D. A.; Patel, B. B.; Wang, X. Q.; Dai, S.; Zhao, X.; Ruoff, R. S.; Johnston, K. P.; Stevenson, K. J. Tuning the electrocatalytic activity of perovskites through active site variation and support interactions. *Chem. Mater.* **2014**, *26*, 3368–3376.
28. Zhu, S.; Chen, Z.; Li, B.; Higgins, D.; Wang, H.; Li, H.; Chen, Z. Nitrogen-doped carbon nanotubes as air cathode catalysts in zinc-air battery. *Electrochim. Acta* **2011**, *56*, 5080–5984.
29. Lee, D. U.; Park, H. W.; Higgins, D.; Nazar, L.; Chen, Z. Highly active graphene nanosheets prepared via extremely rapid heating as efficient zinc-air battery electrode material. *J. Electrochem. Soc.* **2013**, *160*, F910–F915.
30. Li, Y.; Wang, J.; Li, X.; Liu, J.; Geng, D.; Yang, J.; Li, R.; Sun, X. Nitrogen-doped carbon nanotubes as cathode for lithium-air batteries. *Electrochem. Commun.* **2011**, *13*, 668–672.
31. Kim, J. G.; Kim, Y.; Noh, Y.; Kim, W. B. $MnCo_2O_4$ nanowires anchored on reduced graphene oxide sheets as effective bifunctional catalysts for Li-O_2 battery cathodes. *ChemSusChem* **2015**, *22*, 1752–1760.
32. Chen, Z.; Yu, A.; Higgins, D.; Li, H.; Wang, H.; Chen, Z. Highly active and durable core-corona structured bifunctional catalysts for rechargeable metal-air battery application. *Nano Lett.* **2012**, *12*, 1946–1952.
33. Ge, X.; Thomas Goh, F. W.; Li, B.; Anday Hor, T. S.; Zhang, J.; Xiao, P.; Wang, X.; Zong, Y.; Liu, Z. Efficient and durable oxygen reduction and evolution of a hydrothermally synthesized $La(Co_{0.55}Mn_{0.45})_{0.99}O_{3-\delta}$ nanorod/graphene hybrid in alkaline media. *Nanoscale* **2015**, *7*, 9046–9054.

34. Kim, J.; Yin, X.; Tsao, K.-C.; Fang, S.; Yang, H. $Ca_2Mn_2O_5$ as oxygen-deficient perovskite electrocatalyst for oxygen evolution reaction. *J. Am. Chem. Soc.* **2014**, *136*, 14646–14649.

35. Lee, D. U.; Par, M. G.; Park, H. W.; Seo, M. H.; Ismayilov, V.; Ahmed, R.; Chen, Z. Highly active Co-doped $LaMnO_3$ perovskite oxide and N-doped carbon nanotube hybrid bi-functional catalyst for rechargeable zinc-air batteries. *Electrochem. Commun.* **2015**, *60*, 38–41.

36. Prabu, M.; Ramakrishnan, P.; Ganesan, P.; Manthiram, A.; Shanmugam, S. $LaTi_{0.65}Fe_{0.35}O_{3-\delta}$ nanoparticle-decorated nitrogen-doped carbon nanorods as an advanced hierarchical air electrode for rechargeable metal-air batteries. *Nano Energy* **2015**, *15*, 92–103.

37. Park, H. W.; Lee, D. U.; Park, M. G.; Ahmed, R.; Seo, M. H.; Nazar, L. F.; Chen, Z. Perovskite-nitrogen-doped carbon nanotube composite as bifunctional catalyst for rechargeable lithium-air batteries. *ChemSusChem* **2015**, *8*, 1058–1065.

38. Garcia, E. M.; Taroco, H. A.; Matencio, T. M.; Domingues, R. Z.; dos Santos, J. A. F. Electrochemical study of $La_{0.6}Sr_{0.4}Co_{0.8}Fe_{0.2}O_3$ during oxygen evolution reaction. *Int. J. Hydrogen Energ.* **2012**, *37*, 6400–6406.

39. Singh, R. N.; Lal, B. High surface area lanthanum cobaltate and its A and B sites substituted derivatives for electrocatalysis of O_2 evolution in alkaline solution. *Int. J. Hydrogen Energ.* **2002**, *27*, 45–55.

40. Park, H. W.; Lee, D. U.; Zamani, P.; Seo, M. H.; Nazar, L. F.; Chen, Z. Electrospun porous nanorod perovskite oxide/nitrogen-doped graphene composite as a bi-functional catalyst for metal air batteries. *Nano Energy* **2014**, *10*, 192–200.

41. Wang, J.; Zhao, H.; Gao, Y.; Chen, D.; Chen, C.; Saccoccio, M.; Ciucci, F. $Ba_{0.5}Sr_{0.5}Co_{0.8}Fe_{0.2}O_{3-\delta}$ on N-doped mesoporous carbon derived from organic waste as a bi-functional oxygen catalyst. *Int. J. Hydrogen Energ.* **2016**, *41*, 10744–10754.

42. Fabbri, E.; Nachtegaal, M.; Cheng, X.; Schmidt, T. J. Superior bifunctional electrocatalytic activity of $Ba_{0.5}Sr_{0.5}Co_{0.8}Fe_{0.2}O_{3-\delta}$/carbon composite electrodes: Insight into the local electronic structure. *Adv. Energy Mater.* **2015**, *5*, 1402033.

43. Li, P.; Ma, R.; Zhou, Y.; Chen, Y.; Zhou, Z.; Liu, G.; Liu, Q.; Peng, G.; Liang, Z.; Wang, J. In situ growth of spinel $CoFe_2O_4$ nanoparticles on rod-like ordered mesoporous carbon for bifunctional electrocatalysis of both oxygen reduction and oxygen evolution. *J. Mater. Chem. A* **2015**, *3*, 15598–15606.

44. Bian, W.; Yang, Z.; Strasser, P., Yang, R. A $CoFe_2O_4$/graphene nanohybrid as an efficient bi-functional electrocatalyst for oxygen reduction and oxygen evolution. *J. Power Sources* **2014**, *250*, 196–203.

45. Liu, S.; Bian, W.; Yang, Z.; Tian, J.; Jin, C.; Shen, M.; Zhou, Z.; Yang, R. A facile synthesis of $CoFe_2O_4$/biocarbon nanocomposites as efficient bi-functional electrocatalysts for the oxygen reduction and oxygen evolution reaction. *J. Mater. Chem. A* **2014**, *2*, 18012–18017.

46. Lee, D. U.; Kim, B. J.; Chen, Z. One-pot synthesis of a mesoporous $NiCo_2O_4$ nanoplatelet and graphene hybrid and its oxygen reduction and evolution activities as an efficient bi-functional electrocatalyst. *J. Mater. Chem. A* **2013**, *1*, 4754–4762.

47. Liu, Y.; Higgins, D. C.; Wu, J.; Fowler, M.; Chen, Z. Cubic spinel cobalt oxide/multi-walled carbon nanotube composites as an efficient bifunctional electrocatalyst for oxygen reaction. *Electrochem. Commun.* **2013**, *34*, 125–129.

48. Liang, Y.; Li, Y.; Wang, H.; Zhou, J.; Wang, J.; Regier, T.; Dai, H. Co_3O_4 nanocrystals on graphene as a synergistic catalyst for oxygen reduction reaction. *Nat. Mater.* **2011**, *10*, 780–786.

49. Liu, Z.-Q.; Cheng, H.; Li, N.; Ma, T. Y.; Su, Y.-Z. $ZnCo_2O_4$ quantum dots anchored on nitrogen-doped carbon nanotubes as reversible oxygen reduction/evolution electrocatalysts. *Adv. Mater.* **2016**, *28*, 3777–3784.

50. Liang, Y.; Wang, H.; Diao, P.; Chang, W.; Hong, G.; Li, Y.; Gong, M. et al. Oxygen reduction electrocatalyst based on strongly coupled cobalt oxide nanocrystals and carbon nanotubes. *J. Am. Chem. Soc.* **2012**, *134*, 15849–15857.

51. Hao, L.; Zhang, S.; Liu, R.; Ning, J.; Zhang, G.; Zhi, L. Bottom-up construction of triazine-based frameworks as metal-free electrocatalysts for oxygen reduction reaction. *Adv. Mater.* **2015**, *27*, 3190–3195.

52. Unni, S. M.; Bhange, S. N.; Illathvalappil, R.; Mutneja, N.; Patil, K. R.; Kurungot, S. Nitrogen-induced surface area and conductivity modulation of carbon nanoform and its function as an efficient metal-free oxygen reduction electrocatalyst for anion-exchange membrane fuel cells. *Small* **2015**, *11*, 352–360.

53. Tung, C. W.; Hsu, Y. Y.; Shen, Y. P.; Zheng, Y.; Chan, T. S.; Sheu, H. S.; Cheng, Y. C.; Chen, H. M. Reversible adapting layer produces robust single-crystal electrocatalyst for oxygen evolution. *Nat. Commun.* **2015**, *6*, 8106.

54. Wei, Z.; Wang, J.; Mao, S.; Su, D.; Jin, H.; Wang, Y.; Xu, F.; Li, H.; Wang, Y. In situ-generated Co^0-Co_3O_4/N-doped carbon nanotubes hybrids as efficient and chemoselective catalysts for hydrogenation of nitroarenes. *ACS Catal.* **2015**, *5*, 4783–4789.

55. Cheng, H.; Su, Y. Z.; Kuang, P. Y.; Chen, G. F.; Liu, Z. Q. Hierarchical $NiCo_2O_4$ nanosheet-decorated carbon nanotubes towards highly efficient electrocatalyst for water oxidation. *J. Mater. Chem. A* **2015**, *3*, 19314–19321.

56. Li, J.; Zou, M.; Wen, W.; Zhao, Y.; Lin, Y.; Chen, L.; Lai, H.; Guan, L.; Huang, Z. Spinel MFe_2O_4 (M = Co, Ni) nanoparticles coated on multi-walled carbon nanotubes as electrocatalysts for Li-O_2 batteries. *J. Mater. Chem. A* **2014**, *2*, 10257–10262.

57. Prabu, M.; Ramakrishnan, P.; Shanmugam, S. $CoMn_2O_4$ nanoparticles anchored on nitrogen-doped graphene nanosheets as bifunctional electrocatalyst for rechargeable zinc-air battery. *Electrochem. Commun.* **2014**, *41*, 59–63.

58. Prabu, M.; Ramakrishnan, P.; Nara, H.; Momma, T.; Osaka, T., Shanmugam, S. Zinc-air battery: Understanding the structure and morphology changes of graphene-supported $CoMn_2O_4$ bifunctional catalysts under practical rechargeable conditions. *ACS Appl. Mater. Interfaces* **2014**, *6*, 16545–16555.

59. Liu, Y.; Li, J.; Li, W.; Li, Y.; Chen, Q.; Liu, Y. Spinel $LiMn_2O_4$ nanoparticles dispersed on nitrogen-doped reduced graphene oxide nanosheets as an efficient electrocatalyst for aluminum-air battery. *Int. J. Hydrog. Energ.* **2015**, *40*, 9225–9234.

60. Hummers, W. S.; Offeman, R. E. Preparation of graphitic oxide. *J. Am. Chem. Soc.* **1958**, *80*, 1339–1339.

61. Li, Y.; Gong, M.; Liang, Y.; Feng, J.; Kim, J. E.; Wang, H.; Hong, G.; Zhang, B.; Dai, H. Advanced zinc-air batteries based on high-performance hybrid electrocatalysts. *Nat. Commun.* **2013**, *4*, 1805.

62. Prabu, M.; Ketpang, K.; Shanmugam, S. Hierarchical nanostructured $NiCo_2O_4$ as an efficient bifunctional non-precious metal catalyst for rechargeable zinc-air batteries. *Nanoscale* **2014**, *6*, 3173–3181.

63. Lee, D. U.; Park, M. G.; Park, H. W.; Seo, M. H.; Wang, X.; Chen, Z. Highly active and durable nanocrystal-decorated bifunctional electrocatalyst for rechargeable zinc–air batteries. *ChemSusChem* **2015**, *8*, 3129–3138.

64. Liu, S.; Li, L.; Ahn, H. S.; Manthiram, A. Delineating the roles of Co_3O_4 and N-doped carbon nanoweb (CNW) in bifunctional Co_3O_4/CNW catalysts for oxygen reduction and oxygen evolution reactions. *J. Mater. Chem. A* **2015**, *3*, 11615–11623.

65. An, T.; Ge, X.; Andy Hor, T. S.; Thomas Goh, F. W.; Geng, D.; Du, G.; Zhan, Y.; Liu, Z.; Zong, Y. Co_3O_4 nanoparticles grown on N-doped Vulcan carbon as a scalable bifunctional electrocatalyst for rechargeable zinc-air batteries. *RSC Adv.* **2015**, *5*, 75773–75780.

66. Liu, K.; Zhou, Z.; Wang, H.; Huang, X.; Xu, J.; Tang, Y.; Li, J.; Chu, H.; Chen, J. N-doped carbon supported Co_3O_4 nanoparticles as an advanced electrocatalyst for the oxygen reduction reaction in Al-air batteries. *RSC Adv.* **2016**, *6*, 55552–55559.

67. Zhang, M.; Dai, L. Carbon nanomaterials as metal-free catalysts in next generation fuel cells. *Nano Energy* **2012**, *1*, 514–517.

68. Liu, Z.-W.; Peng, F.; Wang, H.-J.; Yu, H.; Zheng, W.-X.; Yang, J. Phosphorus-doped graphite layers with high electrocatalytic activity for the O_2 reduction in an alkaline medium. *Angew. Chem., Int. Ed.* **2011**, *50*, 3257–3261.

69. Cao, X.; Wu, J.; Jin, C.; Tian, J.; Strasser, P.; Yang, R. $MnCo_2O_4$ anchored on P-doped hierarchical porous carbon as an electrocatalyst for high-performance rechargeable Li-O_2 batteries. *ACS Catal.* **2015**, *5*, 4890–4896.

70. Zhang, Z.; Bao, J.; He, C.; Chen, Y. N.; Wei, J. P.; Zhou, Z. Hierarchical carbon-nitrogen architectures with both mesopores and macrochannels as excellent cathodes for rechargeable Li-O_2 batteries. *Adv. Funct. Mater.* **2014**, *24*, 6826–6833.

71. Xu, J.-J.; Wang, Z.-L.; Xu, D.; Zhang, L.-L.; Zhang, X.-B. Tailoring deposition and morphology of discharge products towards high-rate and long-life lithium-oxygen batteries. *Nat. Commun.* **2013**, *4*, 2438.

72. Xu, J.-J.; Wang, Z.-L.; Xu, D.; Men, F.-Z.; Zhang, X.-B. 3D ordered macroporous $LaFeO_3$ as efficient electrocatalyst for Li-O_2 batteries with enhanced rate capability and cyclic performance. *Energy Environ. Sci.* **2014**, *7*, 2213–2219.

73. Chen, Z.; Yu, A.; Ahmed, R.; Wang, H.; Li, H.; Chen, Z. Manganese dioxide nanotube and nitrogen-doped carbon nanotube based composite bifunctional catalyst for rechargeable zinc-air battery. *Electrochim. Acta* **2012**, *69*, 295–300.

74. Park, H. W.; Lee, D. U.; Nazar, L. F.; Chen, Z. Oxygen reduction reaction using MnO_2 nanotubes/nitrogen-doped exfoliated graphene hybrid catalyst for Li-O_2 battery applications. *J. Electrochem. Soc.* **2013**, *160*, A344–A350.

75. Zahoor, A.; Christy, M.; Jang, H.; Nahm, K. S.; Lee, Y. S. Increasing the reversibility of Li-O_2 batteries with caterpillar structured α-MnO_2/N–GNF bifunctional electrocatalysts. *Electrochim. Acta* **2015**, *157*, 299–306.

76. Li, G.; Wang, X.; Fu, J.; Li, J.; Park, M. G.; Zhang, Y.; Lui, G.; Chen, Z. Pomegranate-inspired design of highly active and durable bifunctional electrocatalysts for rechargeable metal-air batteries. *Angew. Chem. Int. Ed.* **2016**, *55*, 4977–4982.

77. Li, B.; Ge, X.; Goh, T.; Andy Hor, T. S.; Geng, D.; Du, G.; Liu, Z.; Zhang, J.; Liu, X.; Zong, Y. Co_3O_4 nanoparticles decorated carbon nanofiber mat as binder-free air-cathode for high performance rechargeable zinc-air batteries. *Nanoscale* **2015**, *7*, 1830–1838.

78. Kinoshita, K. *Electrochemical Oxygen Technology*. New York: John Wiley & Sons Inc., **1992**.

79. Song, K.; Jung, J.; Heo, Y.; Lee, Y. C.; Cho, K.; Kang, Y. M. α-MnO_2 nanowire catalysts with ultra-high capacity and extremely low overpotential in lithium–air batteries through tailored surface arrangement. *Phys. Chem. Chem. Phys.* **2013**, *15*, 20075–20079.

80. Zahoor, A.; Jang, H. S.; Jeong, J. S.; Christy, M.; Hwang, Y.; Nahm, K. S. A comparative study of nanostructured α and δ MnO_2 for lithium oxygen battery application. *RSC Adv.* **2014**, *4*, 8973–8977.

81. Hu, X.; Cheng, F.; Han, X.; Zhang, T.; Chen, J. Oxygen bubble-templated hierarchical porous ε-MnO_2 as a superior catalyst for rechargeable Li-O_2 batteries. *Small* **2015**, *11*, 809–813.

82. Thapa, A. K.; Saimen, K.; Ishihara, T. Pd/MnO_2 air electrode catalyst for rechargeable lithium/air battery. *Electrochem. Solid-State Lett.* **2010**, *13*, A165–A167.

83. Thapa, A. K.; Ishihara, T. Mesoporous α-MnO_2/Pd catalyst air electrode for rechargeable lithium–air battery. *J. Power Sources* **2011**, *196*, 7016–7020.

84. Zhang, G. Q.; Zheng, J. P.; Liang, R.; Zhang, C.; Wang, B.; Au, M.; Hendrickson, M.; Plichta, E. J. α-MnO_2/carbon nanotube/carbon nanofiber composite catalytic air electrodes for rechargeable lithium-air batteries. *J. Electrochem. Soc.* **2011**, *158*, A822–A827.

85. Débart, A.; Bao, J.; Armstrong, G.; Bruce, P. G. An O_2 Cathode for rechargeable lithium batteries: The effect of a catalyst. *J. Power Sources* **2007**, *174*, 1177–1182.

86. Zhan, Y.; Xu, C.; Lu, M.; Liu, Z.; Lee, J. Y. Mn and Co co-substituted Fe_3O_4 nanoparticles on nitrogen-doped reduced graphene oxide for oxygen electrocatalysis in alkaline solution. *J. Mater. Chem. A.* **2014**, *2*, 16217–16223.

87. Liu, X.; Park, M.; Kim, M. G.; Gupta, S.; Wang, X.; Wu, G.; Cho, J. High-performance non-spinel cobalt-manganese mixed oxide-based bifunctional electrocatalysts for rechargeable zinc-air batteries. *Nano Energy* **2016**, *20*, 315–325.

88. Gao, Y.; Zhao, H.; Chen, D.; Chen, C.; Ciucci, F. In situ synthesis of mesoporous manganese oxide/sulfur-doped graphitized carbon as a bifunctional catalyst for oxygen evolution/reduction reactions. *Carbon* **2015**, *94*, 1028–1036.

89. Wohlgemuth, S.-A.; White, R. J.; Willinger, M. G.; Titirici, M.-M.; Antonietti, M. A one-pot hydrothermal synthesis of sulfur and nitrogen doped carbon aerogels with enhanced electrocatalytic activity in the oxygen reduction reaction. *Green Chem.* **2012**, *14*, 1515–1523.

90. Gorlin, Y.; Jaramillo, T. F. A bifunctional nonprecious metal catalyst for oxygen reduction and water oxidation. *J. Am. Chem. Soc.* **2010**, *132*, 13612–13614.

91. Ramirez, A.; Hillebrand, P.; Stellmach, D.; May, M. M.; Bogdanoff, P.; Fiechter, S. Evaluation of MnO_x, Mn_2O_3, and Mn_3O_4 electrodeposited films for the oxygen evolution reaction of water. *J. Phys. Chem. C* **2014**, *118*, 14073–14081.

92. Choi, C. H.; Park, S. H.; Woo, S. I. Binary and ternary doping of nitrogen, boron, and phosphorus into carbon for enhancing electrochemical oxygen reduction activity. *ACS Nano* **2012**, *6*, 7084–7091.

93. Liu, T.; Guo, Y.-F.; Yan, Y.-M.; Wang, F.; Deng, C.; Rooney, D.; Sun, K.-N. CoO nanoparticles embedded in three-dimensional nitrogen/sulfur co-doped carbon nanofiber networks as a bifunctional catalyst for oxygen reduction/evolution reactions. *Carbon* **2016**, *106*, 84–92.

94. Yano, H.; Sugiyama, J.; Nakagaito, A. N.; Nogi, M.; Matsuura, T.; Hikita, M.; Handa, K. Optically transparent composites reinforced with networks of bacterial nanofibers. *Adv. Mater.* **2005**, *17*, 153–155.

95. Wu, Z. Y.; Li, C.; Liang, H. W.; Chen, J. F.; Yu, S. H. Ultralight, flexible, and fire-resistant carbon nanofiber aerogels from bacterial cellulose. *Angew. Chem. Int. Ed.* **2013**, *52*, 2925–2929.

96. Ganesan, P.; Prabu, M.; Sanetuntikul, J.; Shanmugam, S. Cobalt sulfide nanoparticles grown on nitrogen and sulfur co-doped graphene oxide: An efficient electrocatalyst for oxygen reduction and evolution reactions. *ACS Catal.* **2015**, *5*, 3625–3637.

97. Su, Y.; Zhu, Y.; Jiang, H.; Shen, J.; Yang, X.; Zou, W.; Chen, J.; Li, C. Cobalt nanoparticles embedded in N-doped carbon as an efficient bifunctional electrocatalyst for oxygen reduction and evolution reactions. *Nanoscale* **2014**, *6*, 15080–15089.

98. Sheng, Z.-H.; Shao, L.; Chen, J.-J.; Bao, W.-J.; Wang, F.-B.; Xia, X.-H. Catalyst-free synthesis of nitrogen-doped graphene via thermal annealing graphite oxide with melamine and its excellent electrocatalysis. *ACS Nano* **2011**, *5*, 4350–4358.

99. Chen, D.; Chen, C.; Baiyee, Z. M.; Shao, Z.; Ciucci, F. Nonstoichiometric oxides as low-cost and highly-efficient oxygen reduction/evolution catalysts for low-temperature electrochemical devices. *Chem. Rev.* **2015**, *115*, 9869–9921.

100. Li, L.; Liu, S.; Manthiram, A. Co_3O_4 nanocrystals coupled with O- and N-doped carbon nanoweb as a synergistic catalyst for hybrid Li-air batteries. *Nano Energy* **2015**, *12*, 852–860.

101. Li, L.; Manthiram, A. O- and N-doped carbon nanowebs as metal-free catalysts for hybrid Li-air batteries. *Adv. Energy Mater.* **2014**, *4*, 1301795.

102. Li, L.; Chai, S.-H.; Dai, S.; Manthiram, A. Advanced hybrid Li-air batteries with high-performance mesoporous nanocatalysts. *Energy Environ. Sci.* **2014**, *7*, 2630–1302636.

103. Tang, C.; Zhang, Q. Can metal-nitrogen-carbon catalysts satisfy oxygen electrochemistry? *J. Mater. Chem. A* **2016**, *4*, 4998–5001.

104. Crespiera, S. M.; Amantia, D.; Knipping, E.; Aucher, C.; Aubouy, L.; Amici, J.; Zeng, J.; Francia, C.; Bodoardo, S. Electrospun Pd-doped mesoporous carbon nanofibers as catalysts for rechargeable Li-O$_2$ batteries. *RSC Adv.* **2016**, *6*, 57335–57345.
105. Li, B.; Chen, Y.; Ge, X.; Chai, J.; Zhang, X.; Andy Hor, T. S.; Du, G.; Liu, Z.; Zhang, H.; Zong, Y. Mussel-inspired one-pot synthesis of transition metal and nitrogen co-doped carbon M/N–C as efficient oxygen catalysts for Zn-air batteries. *Nanoscale* **2016**, *8*, 5067–5075.
106. Zheng, X.; Yang, Z.; Wu, J.; Jin, C.; Tian, J.-H.; Yang, R. Phosphorus and cobalt co-doped reduced graphene oxide bifunctional electrocatalyst for oxygen reduction and evolution reactions. *RSC Adv.* **2016**, *6*, 64155–64164.
107. Yang, S.; Zhan, Y.; Li, J.; Lee, J. Y. N–Co–O triply doped highly crystalline porous carbon: An acid-proof nonprecious metal oxygen evolution catalyst. *ACS Appl. Mater. Interfaces* **2016**, *8*, 3535–3542.
108. Zeng, X.; You, C.; Leng, L.; Dang, D.; Qiao, X.; Li, X.; Li, Y.; Liao, S.; Adzic, R. R. Ruthenium nanoparticles mounted on multielement co-doped graphene: An ultra-high-efficiency cathode catalyst for Li-O$_2$ batteries. *J. Mater. Chem. A* **2015**, *3*, 11224–11231.
109. Guo, Z.; Ren, G.; Jiang, C.; Lu, X.; Zhu, Y.; Jiang, L.; Dai, L. High performance heteroatoms quaternary-doped carbon catalysts derived from Shewanella bacteria for oxygen reduction. *Sci. Rep.* **2015**, *5*, 17064.
110. Wu, G.; More, K. L.; Johnston, C. M.; Zelenay, P. High-performance electrocatalysts for oxygen reduction derived from polyaniline, iron, and cobalt. *Science* **2011**, *22*, 443–17447.
111. Yang, R.; Bonakdarpour, A.; Bradley Easton, E.; Stoffyn-Egli, P.; Dahn, J. R. Co–C–N oxygen reduction catalyst prepared by combinatorial magnetron sputter deposition. *J. Electrochem. Soc.* **2007**, *154*, A275–A282.
112. Huang, J. S.; Wang, D. W.; Hou, H. Q.; You, T. Y. Electrospun palladium nanoparticle-loaded carbon nanofibers and their electrocatalytic activities towards hydrogen peroxide and NADH. *Adv. Funct. Mater.* **2008**, *18*, 441–448.
113. Zhang, T.; Zhou, H. S. A reversible long-life lithium-air battery in ambient air. *Nat. Commun.* **2013**, *4*, 1817–1824.
114. Lu, Y.-C.; Gallant, B. M.; Kwabi, D. G.; Harding, J. R.; Mitchell, R. R.; Whittingham, M. S.; Shao-Horn, Y. Lithium-oxygen batteries: Bridging mechanistic understanding and battery performance. *Energy Environ. Sci.* **2013**, *6*, 750–768.
115. Gallant, B. M.; Mitchell, R. R.; Kwabi, D. G.; Zhou, J.; Zuin, L.; Thompson, C. V.; Shao-Horn, Y. Chemical and morphological changes of Li-O$_2$ battery electrodes upon cycling. *J. Phys. Chem. C* **2012**, *116*, 20800–20805.
116. Wu, M.; Tang, Q.; Dong, F.; Wang, Y.; Li, D.; Guo, Q.; Liu, Y.; Qiao, J. The design of Fe, N-doped hierarchically porous carbons as highly active and durable electrocatalysts for a Zn-air battery. *Phys. Chem. Chem. Phys.* **2016**, *18*, 18665–18669.
117. Wang, J.; Wu, H.; Gao, D.; Miao, S.; Wang, G.; Bao, X. High-density iron nanoparticles encapsulated within nitrogen-doped carbon nanoshell as efficient oxygen electrocatalyst for zinc-air battery. *Nano Energy* **2015**, *13*, 387–396.
118. Zhou, M.; Yang, C.; Chan, K.-Y. Structuring porous iron-nitrogen-doped carbon in a core/shell geometry for the oxygen reduction reaction. *Adv. Energy Mater.* **2014**, *4*, 1400840.
119. Cao, X.; Zheng, X.; Tian, J.; Jin, C.; Ke, K.; Yang, R. Cobalt sulfide embedded in porous nitrogen-doped carbon as a bifunctional electrocatalyst for oxygen reduction and evolution reactions. *Electrochim. Acta* **2016**, *191*, 776–783.
120. Thippani, T.; Mandal, S.; Wang, G.; Ramani, V. K.; Kothandaraman, R. Probing oxygen reduction and oxygen evolution reactions bifunctional non-precious metal catalysts for metal-air batteries. *RSC Adv.* **2016**, *6*, 71122–71133.

121. Song, J.; Zhu, C.; Fu, S.; Song, Y.; Du, D.; Lin, Y. Optimization of cobalt/nitrogen embedded carbon nanotubes as an efficient bifunctional oxygen electrode for rechargeable zinc-air batteries. *J. Mater. Chem. A* **2016**, *4*, 4864–4870.

122. Lin, X.; Lu, X.; Huang, T.; Liu, Z.; Yu, A. Binder-free nitrogen-doped carbon nanotubes electrodes for lithium-oxygen batteries. *J. Power Sources* **2013**, *242*, 855–859.

123. Li, Q.; Cao, R.; Cho, J.; Wu, G. Nanostructured carbon-based cathode for non-aqueous lithium-oxygen batteries. *Phys. Chem. Chem. Phys.* **2014**, *16*, 13568–13582.

124. Chen, Z.; Higgins, D.; Yu, A.; Zhang, L.; Zhang, J. A review on non-precious metal electrocatalysts for PEM fuel cells. *Energy Environ. Sci.* **2011**, *4*, 3167–3192.

125. Kattel, S.; Wang, G. A density functional theory study of oxygen reduction reaction on Me–N4 (Me = Fe, Co, or Ni) clusters between graphitic pores. *J. Mater. Chem. A* **2013**, *1*, 10790–10797.

126. Deng, D.; Yu, L.; Chen, X.; Wang, G.; Jin, L.; Pan, X.; Deng, J.; Sun, G.; Bao, X. Iron encapsulated within pod-like carbon nanotubes for oxygen reduction reaction. *Angew. Chem. Int. Ed.* **2013**, *52*, 371–375.

127. Zou, X.; Huang, X.; Goswami, A.; Silva, R.; Sathe, B. R.; Mikmeková, E.; Asefa, T. Cobalt-embedded nitrogen-rich carbon nanotubes efficiently catalyze hydrogen evolution reaction at all pH values. *Angew. Chem. Int. Ed.* **2014**, *53*, 4372–4376.

128. Liu, Y.; Jiang, H.; Zhu, Y.; Yang, X.; Li, C. Transition metals (Fe, Co, and Ni) encapsulated in nitrogen-doped carbon nanotubes as bi-functional catalysts for oxygen electrode reactions. *J. Mater. Chem. A* **2016**, *4*, 1694–1701.

129. Wang, Z.; Xiao, S.; Zhu, Z.; Long, X.; Zheng, X.; Lu, X.; Yang, S. Cobalt-embedded nitrogen doped carbon nanotubes: A bifunctional catalyst for oxygen electrode reactions in a wide pH range. *ACS Appl. Mater. Interfaces* **2015**, *7*, 4048–4055.

130. Nørskov, J. K.; Rossmeisl, J.; Logadottir, A.; Lindqvist, L.; Kitchin, J. R.; Bligaard, T.; Jónsson, H. Origin of the overpotential for oxygen reduction at a fuel-cell cathode. *J. Phys. Chem. B* **2004**, *108*, 17886–17892.

131. Guo, G.; Yao, X.; Ang, H.; Tan, H.; Zhang, Y.; Guo, Y.; Fong, E.; Yan, Q. Using elastin protein to develop highly efficient air cathodes for lithium–O_2 batteries. *Nanotechnology* **2016**, *27*, 045401.

132. Li, Q.; Yuan, F.; Yan, C.; Zhu, J.; Sun, J.; Wang, Y.; Ren, J.; She, X. Germanium and phosphorus co-doped carbon nanotubes with high electrocatalytic activity for oxygen reduction reaction. *RSC Adv.* **2016**, *6*, 33205–033211.

133. Cao, X. C.; Jin, C.; Lu, F. L.; Yang, Z. R.; Shen, M.; Yang, R. Z. Electrochemical properties of $MnCo_2O_4$ spinel bifunctional catalyst for oxygen reduction and evolution reaction. *J. Electrochem. Soc.* **2014**, *161*, H296–H300.

134. Guo, Z.; Xiao, Z.; Ren, G.; Xiao, G.; Zhu, Y.; Dai, L.; Jiang, L. Natural tea-leaf-derived, ternary-doped 3D porous carbon as a high-performance electrocatalyst for the oxygen reduction reaction. *Nano Research* **2016**, *9*, 1244–1255.

135. Li, Y.; Zhao, Y.; Cheng, H.; Hu, Y.; Shi, G.; Dai, L.; Qu, L. Nitrogen-doped graphene quantum dots with oxygen-rich functional groups. *J. Am. Chem. Soc.* **2012**, *134*, 15–18.

136. Cao, R.; Thapa, R.; Kim, H.; Xu, X.; Gyu Kim, M.; Li, Q.; Park, N.; Liu, M.; Cho, J. Promotion of oxygen reduction by a bio-inspired tethered iron phthalocyanine carbon nanotube-based catalyst. *Nat. Commun.* **2013**, *4*, 2076.

137. Zhang, K.; Zhang, L.; Chen, X.; He, X.; Wang, X.; Dong, S.; Gu, L.; Liu, Z.; Huang, C.; Cui, G. Molybdenum nitride/N-doped carbon nanospheres for lithium–O_2 battery cathode electrocatalyst. *ACS Appl. Mater. Interfaces* **2013**, *5*, 3677–3682.

138. Dong, S.; Chen, X.; Zhang, K.; Gu, L.; Zhang, L.; Zhou, X.; Li, L. et al. Molybdenum nitride based hybrid cathode for rechargeable lithium–O_2 batteries. *Chem. Commun.* **2011**, *47*, 11291–11293.

139. Zang, Y.; Zhang, H.; Zhang, X.; Liu, R.; Liu, S.; Wang, G.; Zhang, Y.; Zhao, H. Fe/Fe$_2$O$_3$ nanoparticles anchored on Fe–N-doped carbon nanosheets as bifunctional oxygen electrocatalysts for rechargeable zinc-air batteries. *Nano Research* **2016**, *9*, 2123–2137.

140. Lai, Y.; Chen, W.; Zhang, Z.; Qu, Y.; Gan, Y.; Li, J. Fe/Fe$_3$C decorated 3-D porous nitrogen-doped graphene as a cathode material for rechargeable Li-O$_2$ batteries. *Electrochim. Acta* **2016**, *191*, 733–742.

141. Ni, W.; Liu, S.; Fei, Y.; He, Y.; Ma, X.; Lu, L.; Deng, Y. CoO@Co and N-doped mesoporous carbon composites derived from ionic liquids as cathode catalysts for rechargeable lithium-oxygen batteries. *J. Mater. Chem. A* **2016**, *4*, 7746–7753.

142. Zhang, X.; Liu, R.; Zang, Y.; Liu, G.; Wang, G.; Zhang, Y.; Zhang, H.; Zhao, H. Co/CoO nanoparticles immobilized on Co–N-doped carbon as trifunctional electrocatalysts for oxygen reduction, oxygen evolution and hydrogen evolution reactions. *Chem. Commun.* **2016**, *52*, 5946–5949.

143. Xiao, J.; Chen, C.; Xi, J.; Xu, Y.; Xiao, F.; Wang, S.; Yang, S. Core-shell Co@Co$_3$O$_4$ nanoparticles-embedded bamboo-like nitrogen-doped carbon nanotubes (BNCNTs) as a highly active electrocatalyst for the oxygen reduction reaction. *Nanoscale* **2015**, *7*, 7056–7064.

144. Wu, Z.-Y.; Xu, X.-X.; Hu, B.-C.; Liang, H.-W.; Lin, Y.; Chen, L.-F.; Yu, S.-H. Iron carbide nanoparticles encapsulated in mesoporous Fe–N-doped carbon nanofibers for efficient electrocatalysis. *Angew. Chem. Int. Ed.* **2015**, *54*, 8179–8183.

145. Jiang, H.; Yao, Y.; Zhu, Y.; Liu, Y.; Su, Y.; Yang, X.; Li, C. Iron carbide nanoparticles encapsulated in mesoporous Fe–N-doped graphene-like carbon hybrids as efficient bifunctional oxygen electrocatalysts. *ACS Appl. Mater. Interfaces* **2015**, *7*, 21511–21520.

146. Geng, D.; Ding, N.-N.; Andy Hor, T. S.; Chien, S. W.; Liu, Z.; Zong, Y. Cobalt sulfide nanoparticles impregnated nitrogen and sulfur co-doped graphene as bifunctional catalyst for rechargeable Zn-air batteries. *RSC Adv.* **2015**, *5*, 7280–7284.

147. Kim, J. E.; Lim, J.; Lee, G. Y.; Choi, S. H.; Maiti, U. N.; Lee, W. J.; Lee, H. J.; Kim, S. O. Subnanometer cobalt-hydroxide-anchored N-doped carbon nanotube forest for bifunctional oxygen catalyst. *ACS Appl. Mater. Interfaces* **2016**, *8*, 1571–1577.

148. Kashyap, V.; Singh, S. K.; Kurungot, S. Cobalt ferrite bearing nitrogen-doped reduced graphene oxide layers spatially separated with microporous carbon as efficient oxygen reduction electrocatalyst. *ACS Appl. Mater. Interfaces* **2016**, *8*, 20730–20740.

149. Higgins, D. C.; Hoque, M. A.; Hassan, F.; Choi, J.-Y.; Kim, B.; Chen, Z. Oxygen reduction on graphene-carbon nanotube composites doped sequentially with nitrogen and sulfur. *ACS Catal.* **2014**, *4*, 2734–2740.

150. Tian, G.-L.; Zhao, M.-Q.; Yu, D.; Kong, X.-Y.; Huang, J.-Q.; Zhang, Q.; Wei, F. Nitrogen-doped graphene/carbon nanotube hybrids: *In situ* formation on bifunctional catalysts and their superior electrocatalytic activity for oxygen evolution/reduction reaction. *Small* **2014**, *10*, 2251–2259.

151. Zhang, Y.; Jiang, W.-J.; Zhang, X.; Guo, L.; Hu, J.-S.; Wei, Z.; Wan, L.-J. Engineering self-assembled N-doped graphene-carbon nanotube composites towards efficient oxygen reduction electrocatalysts. *Phys. Chem. Chem. Phys.* **2014**, *16*, 13605–13609.

152. Lu, X.; Chan, H. M.; Sun, C.-L.; Tseng, C.-M.; Zhao, C. Interconnected core-shell carbon nanotube-graphene nanoribbon scaffolds for anchoring cobalt oxides as bifunctional electrocatalysts for oxygen evolution and reduction. *J. Mater. Chem. A* **2015**, *3*, 13371–13376.

5 Noncarbon-Based Bifunctional Electrocatalysts for Rechargeable Metal-Air Batteries

5.1 INTRODUCTION

Previously, in Chapters 2–4, we discussed carbon-based composite bifunctional electrocatalysts for air-electrodes of metal-air batteries (MABs). In literature, there are also many researches on non-carbon-based bifunctional composite electrocatalysts for air-electrodes of MABs. The major focus is on developing low-cost, high active, and stable catalytic materials towards both the oxygen reduction reaction (ORR, the discharge reaction) and the oxygen evolution reaction (OER, the charge reaction). In this chapter, non-carbon based composite bifunctional electrocatalysts will be reviewed in terms of their material selection, synthesis strategy, structural characterization, and electrochemical performance. Their advantages induced by different components for enhancing MABs performance will be discussed, and several challenges related to the catalyst properties, material preparation, and electrode/battery performance are also analyzed.

5.1.1 COMPOSITES OF DIFFERENT METALS

Little research has been conducted on single metals as bifunctional air-electrodes for RMABs; this is probably due to the disadvantages of materials such as low surface area, low active sites, low stability, and/or inadequate bifunctional activities. Therefore, single metals are often composited with other materials such as another metal, oxide, carbon, or modified carbon. In this subsection, we will discuss composites of different metal(s) as bifunctional composite catalysts for RMABs.

Ru could be combined with Ni foam to directly form a Ru-Ni composite air-electrode material for MABs, according to Liao et al.[1] The formed porous Ru nanoparticles encapsulated on a 3D network of Ni foam (Ru@Ni) could provide more void spaces to support L_2O_2 and O_2 transfer as well as to supply sufficient electrons for the electro-catalytic reaction. The synthesis process was a facile and tunable galvanic replacement reaction. In their experiment, Ni foam acted as both the current collector and the substrate, while the porous Ru nanoparticles were made from $RuCl_3$. The Ru

157

content in the catalyst was measured to be 0.3 wt%, indicating a low usage of Ru in the electrode. According to TEM and HRTEM images coupled with XRD results, the porous Ru nanoparticles consisted of polycrystalline Ru ($\sim$5 nm) were homogeneously distributed on the pristine Ni skeleton. BET results showed that the surface area of the Ru-coating layer was 58.3 m^2 g_{Ni}^{-1} whereas the pristine Ni skeleton was 1.1 m^2 g_{Ni}^{-1}. Further BJH methods showed a broad distribution of mesoporous ranging from 8–40 nm, confirming the porous structure of Ru. An assembled rechargeable lithium-oxygen battery (RLOB) was used to evaluate the catalyst performance, which was run in an O_2 atmosphere at 200 mA g_{Ni}^{-1} with a controlled capacity of 1000 mAh g_{Ni}^{-1}. Based on the discharge-charge process monitored by TEM, ex-situ SEM, and an impedance test (as shown in Figures 5.1a–f), the Li_2O_2 uniformly covers the porous structure of Ru during discharge and grows up to secondary aggregates, increasing the resistance of the Ru@Ni electrode. During a full recharge, all Li_2O_2 products disappear and the highly porous nature is restored, resulting in a reversion of resistance of the Ru@Ni electrode. These observations indicate that the porous Ru@Ni cathode is able to catalyze both ORR and OER reactions in RLOBs. The CV curves obtained in the electrolyte of TEGDME and $LiClO_4$ under O_2 atmosphere showed that at 2.0–4.3 V, the referenced pristine Ni had no peaks corresponding to the catalytic activity towards either ORR or OER, whereas the porous Ru@Ni not only showed typical ORR currents as the potential went down during the cathodic scan but also presented three current peaks related to OER during the anodic scan, evidencing the significant catalytic activity of porous Ru. The ORR peaks were a result of the reduction of O_2 to Li_2O_2 via a sequence of intermediate steps[2,3] whereas the oxidation peaks were associated with the delithiation and bulk oxidation of Li_2O_2.[4,5] When the discharge-charge performance of the RLOB was compared using Ru@Ni and pristine Ni cathodes, it was revealed that pristine Ni foam had almost no specific capacity in the discharge or charge process, suggesting that Ni foam could only act as a stable current collector in the battery.[6–9] For the Ru@Ni air-electrode, the discharge-charge overpotentials of the RLOB was almost stable when current density was increased from 400 to 800 mA g_{Ni}^{-1}, demonstrating a good rate performance. When cycling tests were run with a controlled capacity of 1000 mAh g_{Ni}^{-1} and a current density of 200 mA g_{Ni}^{-1}, the Ru@Ni electrode could run 100 cycles without any variation in discharge and charge, exhibiting high cycling stability. The discharge potential was nearly in agreement with the theoretical reaction potential of Li to Li_2O_2 evidenced by a very small polarization value (ca. 0.15 V), again demonstrating the excellent ORR activity of porous Ru. They claimed the excellent performance was based on several factors: (1) good distribution of porous Ru on the Ni foam current collector with high adherence and electronic conductivity, (2) 3D network of the porous Ru@Ni electrode, favoring the permeation of O_2 and electrolyte in the three-phase area, and (3) good suppression of side reactions that often take place with high-voltage cycling. This research provides an important insight into the design and construction of porous carbon- and binder-free cathodes for RLOBs.

Compared to Ru@Ni, a Pd nanolayer covered Ni foam catalyst[7] was supported on a Super-P carbon black as an air-electrode for rechargeable lithium-air battery (RLAB) in ambient or simulated air was also explored and exhibited improved energy output, round-trip efficiency, and cyclability.

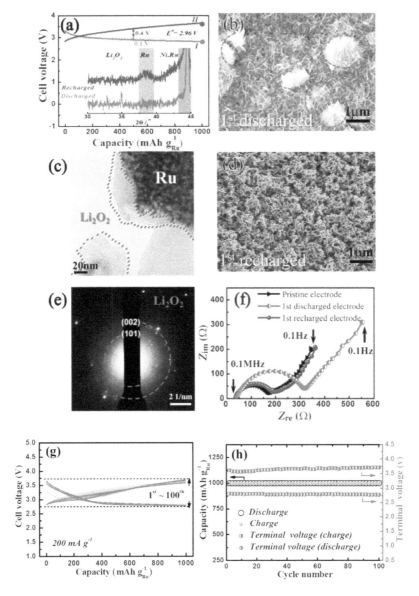

FIGURE 5.1 (a) Discharge-charge profile of the Li-O$_2$ battery with the Ru@Ni cathode at 200 mA g$_{Ru}^{-1}$ with a limited capacity of 1000 mAh g$_{Ru}^{-1}$. The inset is the XRD patterns of the cathode Ru@Ni after discharge and charge, corresponding to the points I and II respectively. (b, d) SEM images of the Ru@Ni cathode after the first discharge and charge, corresponding to the point I and II in (a). (c) TEM image and (e) SAED pattern of the discharged Ru@Ni cathode at the end of the 1st cycle. (f) Electrochemical impedance spectra of the battery with the Ru@Ni cathode at different states in the 1st cycle. (g) Discharge and charge profiles of the Li-O$_2$ battery at 200 mA g$_{Ru}^{-1}$ for 100 cycles. (h) Cycling stability and the variation of the terminal charge and discharge voltages with cycle number for the Li-O$_2$ battery at 200 mA g$_{Ru}^{-1}$ with a limited capacity of 1000 mAh g$_{Ru}^{-1}$. (Modified from Liao, K. et al. *Chem. Sus. Chem.* **2015**, 8, 1429–1434.[1])

Similar to single Ru and Pd above, a bimetal Ag-Cu alloy was also distributed on a Ni foam and then studied as air-electrode material for rechargeable zin-air battery (RZAB) by ZABs in Wu et al.[10] and Jin et al.[11] Normally, Ag is a competitive ORR catalyst with reasonable catalytic activities for ORR in alkaline electrolyte. Ag-Cu alloy should give more efficient ORR activity than that of pure Ag according to first-principles calculations.[12] In the synthesis, Ag-Cu nanoalloys with different atomic ratios of Ag to Cu of 90:10, 50:50, 75:25, and 25:75, respectively, were deposited on Ni foam using pulsed laser deposition (PLD) to form $Ag_{90}Cu_{10}$@Ni, $Ag_{50}Cu_{50}$@Ni, $Ag_{75}Cu_{25}$@Ni, and $Ag_{25}Cu_{75}$@Ni catalysts. HRTEM, SAED, and bright-field transmission electron microscopy revealed that the crystalline Ag-Cu nanoalloy particles with an average size of 2.58 nm were embedded in the amorphous Cu films. In XPS, Ag 3d shifted to higher binding energies, whereas Cu 2p moved to lower binding energies, demonstrating the effects of the Ag-Cu alloy. In the LSV curves collected in O_2-saturated KOH solution, $Ag_{90}Cu_{10}$@Ni, $Ag_{50}Cu_{50}$@Ni, and $Ag_{25}Cu_{75}$@Ni had similar ORR current peaks to pure Ag. However, the peak current was increased with increasing Cu amounts from 0 to 50 and then decreased with continuously increasing Cu amount. This suggested that $Ag_{50}Cu_{50}$@Ni had the maximum peak current, and that Cu doping needed to be optimized in obtaining the most effective catalytic activity of Ag-Cu catalysts. For ORR kinetics at different rotation rates using RDE, $Ag_{90}Cu_{10}$@Ni, $Ag_{75}Cu_{25}$@Ni, and $Ag_{50}Cu_{50}$@Ni catalysts at 1600 rpm exhibited more positive onset potential than pure Ag, demonstrating the enhanced ORR activities of the Ag-Cu catalyst due to Cu-doping into Ag. From Koutecky-Levich plots, the calculated electron transfer numbers were 3.4, 4, 4, 3.9, and 3.1 for Ag, $Ag_{90}Cu_{10}$@Ni, $Ag_{75}Cu_{25}$@Ni, $Ag_{50}Cu_{50}$@Ni, and $Ag_{25}Cu_{75}$@Ni catalysts, respectively. As for OER, the LSV curves showed that all Ag-Cu catalysts could deliver higher OER current densities and lower onset potentials than pure Ag and Ni foams, confirming the OER activity of Ag was also enhanced by alloying with Cu. When primary ZABs were fabricated with $Ag_{50}Cu_{50}$@Ni and $Ag_{90}Cu_{10}$@Ni, the $Ag_{50}Cu_{50}$@Ni electrode displayed a higher power density of $\sim$86.5 mW cm^{-2} at 100 mA cm^{-2}, possessed a higher discharge voltage stability, and became more stable than the $Ag_{90}Cu_{10}$@Ni electrode. When the charge-discharge performance of a home-made RZAB was examined, $Ag_{50}Cu_{50}$@Ni air-electrode displayed a low polarization potential and experienced no significant degradation after 200 cycles. The round-trip efficiency for $Ag_{50}Cu_{50}$@Ni was $\sim$50% at a current density of 20 mA cm^{-2}. The RZAB gave a round-trip efficiency of $\sim$53.9% for 100 cycles during the charge-discharge. This work provides a strategy to develop cost-effective metal-based bifunctional catalyst for RMAB applications.

To enhance the fast transport of O_2 and electrolytes as well as the accommodation of solid reaction products such as Li_2O_2 during the discharging-charging process of RLOBs, a hierarchical nanoporous Au-Ag alloy (h-NPG-AuAg) catalyst was synthesized via a two-step dealloying method by Guo et al.[13] in which commercial $Au_{15}Ag_{85}$ (at%) alloy was twice leached away in a 1 M HNO_3 solution to form h-NPG-AuAg. SEM and EDS revealed pore sizes of $\sim$20 nm with residual Ag at $\sim$70 at%. The BET surface area and porosity of h-NPG-$Au_{89}Ag_{11}$ were $\sim$82.9 m^2 g^{-1} and $\sim$81%, respectively. This was higher than those of as-prepared np-$Au_{30}Ag_{70}$ ($\sim$27.4 m^2 g^{-1} and $\sim$50%) and coarsened np-$Au_{30}Ag_{70}$ ($\sim$59.8 m^2 g^{-1} and $\sim$50%), suggesting that the second dealloying played a strong effect on the improvement of porosity and surface area, benefiting both the transfer of oxygen and electrolyte and the formation and decomposition of Li_2O_2. To examine electrocatalytic performance of the catalysts, an assembled RLOB was used. It was found that, due to a larger surface area and higher porosity, h-NPG-$Au_{89}Ag_{11}$ exhibited higher ORR activity than as-prepared np-$Au_{30}Ag_{70}$ and coarsened np-$Au_{30}Ag_{70}$. Furthermore, it was observed that the ORR activity was not related to the amount of silver because the residual Ag only resided in the interior of the gold ligaments and did not participate in air-electrode reactions.[14] However, due to low charging performances, h-NPG-$Au_{89}Ag_{11}$ had poor cycling retention and failed after tens of cycles. This was probably due to its deep nanoporous channels which retained Li_2O_2 and resulted in slow kinetics. With the assistance of a redox mediator tetrathiafulvalene (TTF), the hierarchical NPG continued to produce a highly enhanced reversible capacity and long cycling lifetime at low charge-discharge overpotentials and high current densities. They indicated the optimization of nanoporous morphology played an important part in the exploration of high performance catalyst through generation of hierarchical porous structures.

5.2   COMPOSITES OF OXIDE AND METAL

5.2.1   Metal Oxides

Normally, under extremely oxidative conditions at high potentials (e.g., >1.4 V vs. RHE), conventional carbon materials suffer from rapid oxidation, which leads to significant performance degradation in carbon-based catalysts.[15,16] Metal oxide catalysts, however, are inherently more stable in oxidizing environments.[17,18] Metal oxides, in particular highly efficient double perovskite oxides and oxygen-deficient perovskite oxides, have higher activity than state-of-the-art IrO_2 catalysts in alkaline media.[19–21] Although metal oxide catalysts do not fall into the categories of our composite catalysts in this chapter, it is necessary to briefly discuss some nanostructured oxide catalysts and their recent developments and advances in electrolysis for both ORR and OER in RMAB applications.

Perovskite oxides are generally denoted as ABO_3 in which A sites are rare earth or alkali metal ions and B sites are transition metal ions. To obtain expected physicochemical and catalytic properties, the substitution of A-site and/or B-site metal cations, and the generation of oxygen deficiency/vacancy play large effects on electronic structures and coordination chemistry.[16] For example, Du et al.[22] introduced oxygen defects in $CaMnO_3$ to obtain nonstoichiometric $CaMnO_{3-\delta}$ ($0 < \delta \leq 0.5$) (see Figure 5.2a–d) for bifunctionally ORR and OER electrocatalysts. In their experiment, the oxide nonstoichiometry was adjusted by controlling the thermal reduction of pristine perovskite microspheres and nanoparticles synthesized from thermal decomposition of two carbonate precursors (i.e., $CaMn(CO_3)_2$ and $(NH_4)_2CO_3$) and the Pechini route. XPS analysis and electrochemical measurements in 0.1 M KOH solution showed that in addition to the respectable catalytic stability, $CaMnO_{3-\delta}$ also exhibited catalytic activity which was dependent on not only the concentration of oxygen deficiency but also the oxidation state of Mn in $CaMnO_{3-\delta}$.

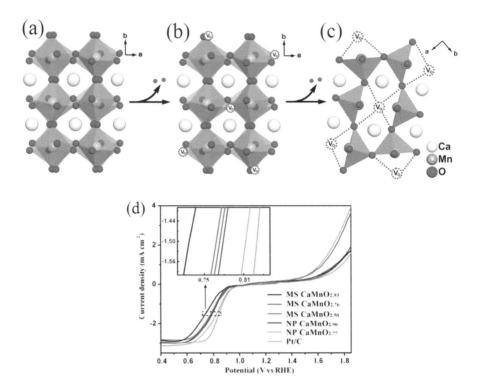

FIGURE 5.2 (a–c) Schematic representation of the crystal frameworks of pristine and oxygen-deficient perovskite oxides showing a transition from stoichiometric $CaMnO_3$ (a) to nonstoichiometric $CaMnO_{2.76}$ (b) and $CaMnO_{2.5}$ (c). (d) Bifunctional OER/ORR electrocatalytic activities of $CaMnO_{3-\delta}$ microspheres (MS), nanoparticles (NP), and the comparative Pt/C in O_2-saturated alkaline solution. The inset enlarges the ORR curves in the half-wave potential region. (From Du, J. et al. *Inorg. Chem.* **2014**, *53*, 9106–9114.[22])

Particularly, the nonstoichiometric $CaMnO_{3-\delta}$ with a δ of ~0.25 and an average Mn valence close to 3.5 delivered the highest ORR activity (36.7 A g^{-1} at 0.70 V vs. RHE, with an onset potential of 0.96 V), that was comparable to that of benchmarked Pt/C (see Figure 5.2d). Furthermore, the moderate oxygen-defective $CaMnO_{3-\delta}$ with a δ close to 0.25 also favored OER. Their density functional theory (DFT) studies and electrical conductivity measurements revealed that the high catalytic performance could be attributed to the improved intrinsic electrical conductivity that favored oxygen activation.

Furthermore, perovskite oxides such as $La_{0.6}Ca_{0.4}CoO_3$,[23] $La_{0.9}Sr_{0.1}CoO_3$,[24] $LaNi_{1-x}Fe_xO_3$ (x = 0, 0.1, 0.2, and 0.6),[25] $LaNi_xM_{1-x}O_3$ (M = Co, Mn, Fe, and Ni; x = 0.25 or 0.5),[26] $Ba_{0.5}Sr_{0.5}Co_{0.8}Fe_{0.2}O_3$,[27] and $Ba_{0.5}Sr_{0.5}Co_xFe_{1-x}O_{3-\delta}$ (x = 0.2 and 0.8);[28] derived from the substitution of A-site and/or B-site metal cations, also exhibited enhanced catalytic activities as well as efficient electronic and structural properties. A recent study by Petrie et al.[29] also demonstrated that like noble metal films, perovskite $LaNiO_3$ films used as a typical transition-metal oxide thin film could significantly improve both catalytic ORR and OER activities due to the strain-induced splitting of e_g orbitals and the corresponding orbital asymmetry at the surface.

As another candidate for highly electroactive oxides, spinel-type transition metal oxides with a general formula of AB_2O_4 have attracted attention due to their well-known spinel structures consisting of closely packed arrays of A^{2+} and B^{3+} cations occupying part or all of the tetrahedral and octahedral sites.[30,31] The cations in spinel-type transition metal oxides can exist in more than one oxidation state between the tetrahedral and octahedral sites. The type, distribution, and compositions of the cations can determine physical, chemical, and electrochemical properties. To develop spinel oxides as alternative low-cost bifunctional electrocatalysts, Cheng et al.[32] synthesized crystalline $Co_xMn_{3-x}O_4$ spinel oxides and studied their electrocatalytic performance. In their experiment, two crystalline $Co_xMn_{3-x}O_4$ spinel oxides, CoMnO-P and CoMnO-B, were prepared through a room-temperature synthesis route using NaH_2PO_2 and $NaBH_4$ as reductants. For comparison, two shape-controlled Co-Mn-O spinel oxides, tetragonal $CoMn_2O_4$ (HT-T-spinel) and cubic Co_2MnO_4 (HT-C-spinel) were obtained using a traditional high-temperature method. In electrochemical measurements run in 0.1 M KOH solution, four Co-Mn-O spinel oxides were found to exhibit high-efficiency electrocatalytic ORR following the decreasing order: CoMnO-P > CoMnO-B > HT-C-spinel > HT-T-spinel. The room-temperature spinel oxides displayed higher OER activities than the corresponding high-temperature samples in which the tetragonal HT-T-spinel oxide resulted in a two-fold OER activity than the cubic HT-C-spinel oxide, revealing phase-dependent catalytic activities toward ORR and OER, as confirmed in their first-principle theoretical studies. They believed that $Co_xMn_{3-x}O_4$ nanoparticles could produce considerable catalytic activity for both ORR and OER because of their high surface areas and abundant defects. Although these catalysts have similar spinel structures, they were reported to have different catalytic activities for ORR and OER. One report by Pletcher et al.[33] compared the electrochemical activities of spinel Co_3O_4 and $NiCo_2O_4$ oxides in alkaline media. The electrochemical measurements in O_2-saturated 1 M KOH solution showed that $NiCo_2O_4$ possessed lower ORR activities than Pt black, although the OER activity of Pt black was lower than those of $NiCo_2O_4$ and Co_3O_4. In addition, the activities of $NiCo_2O_4$ for both ORR and OER were higher than those of Co_3O_4, suggesting that $NiCo_2O_4$ could be a potential bifunctional catalyst.

Due to the importance of exploring cost-effective non-noble metal catalysts with desirable activities for both ORR and OER, other catalytic oxides were also investigated, including nanostructured MnO_2[34] and Ruddlesden-Popper-type $La_{n+1}Ni_nO_{3n+1}$ oxides.[35] Using different methods, Meng et al.[34] synthesized and examined several different structures of manganese oxides. The electrocatalytic properties measured in 0.1 M KOH showed that both the ORR and OER activity were associated with the crystallographic structures, and followed a decreasing order of α-MnO_2 > amorphous MnO_2 > β-MnO_2 > δ-MnO_2. At 10 mA cm^{-2}, the overpotential of α-MnO_2 electrode was ca. 490 mV, close to Ir/C (~380 mV), indicating its excellent OER activity. Under a current density of 5 mA cm^{-2}, α-MnO_2 also exhibited higher stability in comparison with amorphous MnO_2, β-MnO_2, and δ-MnO_2. Furthermore, at 3 mA cm^{-2}, α-MnO_2 displayed a potential of ~760 mV, which is close to that of commercial 20 wt%Pt/C ($\sim$860 mV), indicating that α-MnO$_2$ possessed strong O$_2$ adsorption capabilities and thus high ORR activity. Meantime, Yu et al.[35] designed and developed Ruddlesden-Popper-type La$_{n+1}$Ni$_n$O$_{3n+1}$ oxides with layered structures in which n-LaNiO$_3$ perovskite layers were sandwiched between two LaO rock-salt layers, with all the layers stacked along the c-axis. They observed the layered structure could accommodate extra oxygen in the LaO rock-salt layer, favoring the transport of oxygen.[36–38] Their results revealed that the Ni-O bond length and the hyperstoichiometric oxides in the rock-salt layers were correlated with ORR activities, while the OER activity was determined by the content of OH$^-$ on the surface of the catalyst. This research of layered-structured materials provides a guideline to develop new bifunctional electrocatalysts for improving ORR and OER performances in RMABs.

Although the combination of two different metal oxides can be considered a strategy in developing novel bifunctional catalysts to avoid carbon corrosion in RMABs, few studies have reported composite catalysts of two different metal oxides for high-performance electrocatalysts of both ORR and OER, possibly due to the drawbacks of the oxides such as low surface area, poor electronic conduction, and insufficient catalytic performance. One typical example is binary Co$_3$O$_4$/Co$_2$MnO$_4$ metal oxides[39] with an ORR-active Co$_2$MnO$_4$ component. At an optimal ratio of 1:2.7 for Co$_3$O$_4$/2.7Co$_2$MnO$_4$, the difference in potentials for OER (at 10 mA cm^{-2}) and ORR (at -3 mA cm^{-2}) was 1.09 V, indicating lower bifunctional activity than that of Ni$_{0.4}$Co$_{2.6}$O$_4$ ($\sim$0.96 V). Therefore, the combined oxides for catalysts still require conducting carbon to form effective bifunctional composite catalysts in the practical application of RMABs.

5.2.2 COMPOSITES OF PEROVSKITE OXIDE AND METAL

To reduce carbon corrosion and improve RMAB performance, the combination of oxide and metal has been adopted as an efficient strategy in developing high-performance carbon-free composite bifunctional catalysts. For example, perovskite oxides have been combined with different metals to improve their functional electrocatalysis for both ORR and OER. To date, there has been several metals which have been tested, including Pt,[40] Ag,[41] and Cu[42] in the exploration and development of perovskite oxide-metal composite catalysts.

Pt is a well-known metal catalyst for ORR in the application of polymer electrolyte membrane fuel cells (PEMFCs).[43–45] Therefore, researchers have extensively studied various composites of perovskite oxides with Pt as possible bifunctional catalysts. To improve catalytic activities, especially OER activity and solve the issue of carbon corrosion, Han et al.[40] studied a new porous Pt/CaMnO$_3$ nanocomposite material by taking advantages of the synergistic effect of Pt and CaMnO$_3$. Their synthesis included three steps: (1) Pechini synthesis of porous CaMnO$_3$ support, (2) pyrolysis deposition of Pt clusters on porous CaMnO$_3$ support to produce a mid-product of Pt/CaMnO$_3$, and (3) hydrogenation treatment of Pt/CaMnO$_3$ under a flowing 5% H$_2$/Ar gas for 2 hours to obtain the final H-Pt/CaMnO$_3$ product. In the characterization of structure and morphology, SEM and TEM images showed that Pt clusters with sizes of around 1.0 nm were supported on the interconnected porous

CaMnO$_3$ nanoparticles. ICP-MS revealed the Pt content to be 8.06 wt% in H-Pt/CaMnO$_3$. XPS measurements for CaMnO$_3$, Pt/CaMnO$_3$ and H-Pt/CaMnO$_3$ found that the Pt 4f slightly shifted to higher binding energies relative to bulk Pt, suggesting that Pt electronic structure was modified by the metal oxide.[46,47] Meantime, the analysis of Mn 2p and Mn 3 s indicated that hydrogenation introduced oxygen defects existed in H-Pt/CaMnO$_3$, resulting in a lower Mn valence than that of Pt/CaMnO$_3$. Furthermore, O 1 s spectra indicated the ratio of surface oxygen to lattice oxygen followed an increasing order: CaMnO$_3$ < Pt/CaMnO$_3$ < H-Pt/CaMnO$_3$; this suggests stronger interactions between hybrids and the adsorbed oxygen-containing species due to hydrogenation. In electrochemical measurements, the LSVs in O$_2$-saturated 0.1 M KOH solution showed that the two composite catalysts were more active, as evidenced by higher half-wave potentials than pristine CaMnO$_3$, and that H-Pt/CaMnO$_3$ outperformed the unhydrogenated Pt/CaMnO$_3$, indicated by lower overpotentials and larger reduction currents. Although the ORR performance of H-Pt/CaMnO$_3$ was still lower than commercial Pt/C, the tested chronoamperometric responses at 0.8 V in O$_2$-saturated 0.1 M KOH showed a better stability of such a H-Pt/CaMnO$_3$ catalyst. Interestingly, a combination of 30 wt% carbon and 70 wt% H-Pt/CaMnO$_3$ exhibited comparable ORR to that of commercial Pt/C. Based on the analysis of Koutecky-Levich curves, the calculated electron transport numbers suggested that both H-Pt/CaMnO$_3$ and Pt/CaMnO$_3$ followed an apparent 4e$^-$ ORR pathway. For OER, it was found that, compared to CaMnO$_3$ and Pt/CaMnO$_3$, H-Pt/CaMnO$_3$ displayed a lower overpotential and a greater OER current, demonstrating higher OER activity. When the bifunctional activity was evaluated using the charge-discharge potentials between the OER potential (at 10 mA cm^{-2}) and the ORR potential (half-wave potential of the ORR), the potential difference of H-Pt/CaMnO$_3$ was 1.01 V, showing that H-Pt/CaMnO$_3$ had a potential as a bifunctional catalyst.

To further enhance the activity of Ag, Park et al.[41] employed an oxophilic LaMnO$_3$ as the support material to produce bifunctional catalytic activity. They observed that efficient water activation could be a key factor in determining ORR activity in alkaline media because the proton could be easily generated by water activation (xH$_2$O + M → M − [OH]$_x$ + xH$^+$ + xe$^-$) after the incorporation of LaMnO$_3$, favoring the reduction of O$_2$ to H$_2$O.[48] In their synthesis, La(CH$_3$COO)$_3$·1.5H$_2$O and Mn(CH$_3$COO)$_2$·4H$_2$O were used as the sources of La and Mn to prepare LaMnO$_3$ after a co-precipitation process. AgNO$_3$ was then reduced using NaBH4 as a reducing agent to prepare Ag/LaMnO$_3$. For comparison, Ag/C samples were also synthesized with the same method. In TEM images, Ag had an average size below 20 nm in Ag/C and 20–30 nm in Ag/LaMnO$_3$; this was possibly attributed to the higher BET surface areas and larger porous sizes of carbon than that of LaMnO$_3$, resulting in smaller Ag sizes on the carbon than on the LaMnO$_3$. Electrochemical measurements using a RDE technique in O$_2$-saturated 0.1 M KOH solution showed that Ag/LaMnO$_3$ exhibited better onset potentials, suggesting its higher ORR activity than Ag/C. Furthermore, Ag/LaMnO$_3$ possessed a comparable limiting current (∼5.5 mA cm^{-2}) to Pt/C (∼5.62 mA cm^{-2}), showing a nearly four-electron pathway, while the limiting current of Ag/C was below 5 mA cm^{-2}, indicating a considerable portion of the ORR progressed via the two-electron pathway leading to peroxide formation. Moreover, in the comparison of the ligand and strain effects on ORR, the strain effect was found to be negligible while the ligand effect, resulted from the charge transfer from Mn to Ag, was apparent and led to the enhancement of both oxygen activation on Ag and water activation on Mn, and then an improved ORR catalytic activity of Ag/LaMnO$_3$. This research demonstrated the efficient oxophilic nature and bifunctional effect between catalyst and support for synthesizing ORR electrocatalysts in alkaline media for RMAB applications.

Cu has also been combined with perovskites to form new bifunctional composite catalysts for RMABs. Yang et al.[42] fabricated oxo perovskite Sr$_{0.95}$Ce$_{0.05}$CoO$_{3-\delta}$ (SCCO) particles loaded with copper nanoparticles (SCCO-Cu) and studied its activity in aqueous solution for RLAB. They observed that the deposition of Cu on the surface of SCCO could improve catalytic activity, resulting in high battery performances. Here, Cu(COOH)$_2$·H$_2$O was selected as the source of Cu. Their SEM measurements showed Cu nanoparticles with a size range of tens to several hundred nanometers on the surface of SCCO particles. The Cu loading was 20 wt% based on the analysis of ICP-MS. To evaluate the catalytic performance of SCCO-Cu, electrochemical measurements in a RLAB found that the difference between discharge and charge potentials for Vulcan XC-72, SCCO, and SCCO-Cu catalysts were 1.51, 1.11, and 0.98 V, respectively, indicating that the catalytic activity followed a decreasing order of SCCO-Cu > SCCO > Vulcan XC-72. The round-trip efficiencies of SCCO and SCCO-Cu were 72.3% and 75.2%, respectively, which were much higher than that of Vulcan XC-72 (~58.8%). SCCO-Cu exhibited a better performance than SCCO because the addition of metallic Cu nanoparticles not only improved the conductivity of SCCO but also catalyzed ORR at its surface. Further, long-term discharge-charge curves showed that, at a testing current density of 0.05 mA cm^{-2}, SCCO-Cu catalysts presented a stable performance over 115 hours, corresponding to 1500 mAh g^{-1}. This demonstrated that SCCO-Cu composites could provide a new way to develop efficient, low-cost bifunctional catalysts for RMABs.

5.2.3 COMPOSITES OF SPINEL OXIDE AND METAL

Up to now, two kinds of metals, Ag[49] and Ni,[50–59] have been combined with spinel oxides to form composites of spinel oxide-metal in the research and development of high-performance bifunctional catalysts for RMAB applications. Compared to the composites between perovskite and metal, compositing spinel oxide and metal, in particular that of spinel oxide with Ni, have been found to play an important role in improving performances when they are used as air-electrode materials for RMABs.

In alkaline electrolytes, Ag is very active for ORR, following a four-electron pathway, while Co$_3$O$_4$ spinel oxides are active for OER. Based on this, Amin et al.[49] carried out a strategy of combining an ORR catalyst with an OER one to obtain an Ag-Co$_3$O$_4$ composite for a bifunctional catalyst of ORR and OER. In their experiment, Co$_3$O$_4$ nanoparticles were mixed with commercial Ag (Ag311) to form Ag311 + Co$_3$O$_4$ (x wt%) composite in which the Co$_3$O$_4$ content (x wt%) was adjusted to be 0 wt%, 5 wt%, 10 wt%, 20 wt%, 30 wt%, and 50 wt%, respectively. The TEM images of Ag311 + Co$_3$O$_4$ (30 wt%) showed that Co$_3$O$_4$ (~50 nm) nanoparticles covered large Ag particles (~2 μm size). In the investigation of electrocatalytic activity using

RRDE measurements, the polarization curves in O_2-saturated 0.1 M KOH solution showed that Ag311 + Co_3O_4 (10 wt%) outperformed the other composites in terms of the onset potential ($\sim$0.91 V at 200 μA cm^{-2}) and half-wave potential (0.79 V), suggesting that among all Ag311 + Co_3O_4 (x wt%) composites, Ag311 + Co_3O_4 (10 wt%) had the highest ORR activity and followed a four-electron pathway. Further, RRDE measurements revealed that Ag311 + Co_3O_4 (10 wt%) had negligible peroxide species formation, and its ORR overpotential was 70 mV lower than that of Ag and only ca. 80 mV higher than that of commercial 20 wt% Pt/C. In an investigation of OER by differential electrochemical mass spectrometry (DEMS) using a dual thin layer flow-through cell,[60] the CV and mass spectrometric cyclic voltammograms (MSCV) of catalysts such as Ag311, Co_3O_4, Ag311 + Co_3O_4 (10 wt%), and glassy carbon (GC) electrodes showed that for pure Ag 311, the oxygen evolution around 1.73 V remained much lower than that of Ag311 + Co_3O_4 (10 wt%), possibly because the mixed catalyst (i.e., Ag311 + Co_3O_4 (10 wt%)) could result in oxide formation.

To avoid the extemporaneous parasitic reactions of carbon with Li_2O_2 resulted in the formation of resistive carbonate and, thus, large overpotentials during OER and poor cyclability of RLOBs.[61,62] Kalubarme et al.[53] studied carbon-free $MnCo_2O_4@$ Ni as an oxygen-electrode in which Co and Mn oxides, a typical spinel oxide, could provide potential activities for both ORR and OER. In a hydrolysis and hydrothermal synthesis, $MnSO_4 \cdot 6H_2O$ and $(CH_3COO)_2Co$ were used as sources of Mn and Co to obtain the direct growth of ternary spinel $MnCo_2O_4$ nanorod arrays on metallic Ni foam substrate. This Ni foam was slightly etched in an ultrasonic bath containing a mixture of HCl and HNO_3 to remove surface oxide layers before use. XRD confirmed the presence of well-crystalline $MnCo_2O_4$ with spinel structures (Fd-3 m) in which manganese and cobalt were distributed over both octahedral and tetrahedral sites. The $MnCo_2O_4$ nanorods were then oriented and assembled in a radial form from the center to the surface of the micro-spherical superstructure, akin to a chestnut bur-like structure. The formed spherical structures had sizes in the range of 1–5 μm. When ORR and OER activities of the freestanding $MnCo_2O_4@$Ni electrode (labeled as FSMCO) were assessed in 1 M LITFSI-TEGDME electrolyte using a RLOB, the ORR onset potential and ORR peak current density for the FSMCO electrode were higher than those of KB and the physically blended mixture of KB and $MnCo_2O_4$ (1:1) (KB-MCO), indicating that the FSMCO electrode had a higher ORR activity than other two electrodes. The OER was found to be associated with the decomposition of lithium oxides for the KB, KB-MCO, and FSMCO electrodes. The anodic linear sweep voltammograms showed a lower peak potential and a higher peak current density of the FSMCO electrode than those of the KB and KB-MCO, demonstrating that, in non-aqueous electrolytes, the FSMCO catalyst could give higher OER activity. The first tested discharge-charge curves on the RLOB showed that the FSMCO electrode displayed a lower overpotential than the other two electrodes. Furthermore, the discharge capacity of FSMCO was $\sim$10,520 mAh g^{-1}, which was higher than those of KB ($\sim$3830 mAh g^{-1}) and KB-MCO ($\sim$8650 mAh g^{-1}). It was claimed that the highly porous FSMCO could supply more three-phase (oxygen/catalyst/electrolyte phases) reaction active sites to decompose Li_2O_2, and that the chestnut bur-like superstructures could offer more active sites and enough transmission paths for O_2 and Li$^+$ ions, resulting in the fast kinetics of both ORR

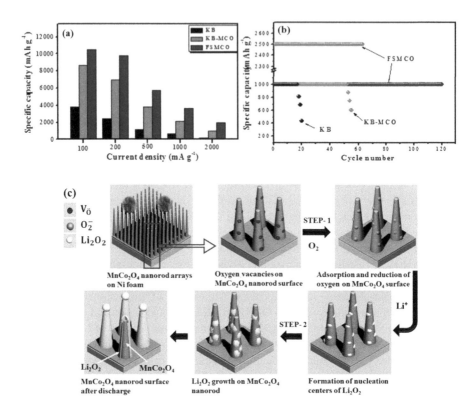

FIGURE 5.3 (a) Discharge capacities as a function of applied current. (b) Cyclability measured using the limited capacity discharge mode for the Li-O$_2$ cell containing oxygen electrodes composed of only KB, KB-MCO, and FSMCO, respectively. (c) Schematic illustration of the nucleation and growth of Li$_2$O$_2$ on the MnCo$_2$O$_4$ nanorod array. (Modified from Kalubarme, R. S. et al. *Sci. Rep.* **2015**, *5*, 13266.[53])

and OER. When the rate capability of the RLOB was examined, FSMCO retained higher capacity at different current densities of 100, 200, 500, 1000, 2000 mA g^{-1}, respectively, in comparison with the KB and KB-MCO electrodes (see Figure 5.3a). At a higher current density of 2000 mA g^{-1}, FSMCO had a capacity of 2015 mAh g^{-1}. The tested long-term cyclability showed that FSMCO had a good reversibility up to the 10th cycle, maintaining a capacity of over 4000 mAh g^{-1}, while KB-MCO showed a fast capacity fading. RLOB test (see Figure 5.3b) with a specific current of 500 mA g^{-1} under a controlled discharge-charge capacity mode of 1000 mAh g^{-1}, FSMCO ran for 119 cycles, showing higher cycle life than KB-MCO ($\sim$55 cycles) and KB ($\sim$17 cycles). It was believed that this excellent battery performance should be attributed to the integrated design of the hierarchical porous MCO as well as the good connection between the MnCo$_2$O$_4$ nanorods and the Ni substrate for increasing electron transport. For physical characterization, the field-emission scanning electron microscopy (FE-SEM), TEM, XRD and Raman spectra were used to examine the electrodes after full discharge and charge. It was observed that the FSMCO catalyst without carbon could affect the morphology and crystallinity of the discharge product of Li_2O_2 simultaneously, thus improving the reversibility through increasing contact areas between the Li_2O_2 and the catalyst. The growth process of Li_2O_2 follows two steps (see Figure 5.3c): (1) the nucleation sites were formed on the surface of $MnCo_2O_4$ nanorods, and (2) Li_2O_2 grew laterally to completely cover the nanorods. The morphological evolution of Li_2O_2 on the FSMCO electrode was highly reversible during the discharge-charge process, resulting in the FSMCO electrode producing higher reversibility and cyclability than the KB and KB-MCO ones.

Instead of ternary spinel $MnCo_2O_4$, Wu et al.[51] also used binary spinel Co_3O_4 to form $Co_3O_4@Ni$ composites in which hierarchical mesoporous/macroporous Co_3O_4 ultrathin nanosheets were directly grown on Ni foam via a simple hydrothermal reaction. Their assembled RLOB tests showed that under a current density of 100 mA g^{-1}, the morphology-controlled $Co_3O_4@Ni$ catalyst exhibited significantly improved electrochemical performances with a capacity of 11,882 mAh g^{-1} during the initial discharge, as well as a good cycling stability (more than 80 cycles at 200 mA g^{-1} with the capacity limited to 500 mAh g^{-1}). Furthermore, the charge voltage after 5 cycles was significantly reduced to circa 3.7 V, which was lower than most of the reported data (almost above 4.0 V).[63-65] The better electrochemical performance of $Co_3O_4@Ni$ catalyst could be ascribed to its hierarchical structures with macropores to supply enough space for Li_2O_2 storage and O_2 diffusion, and mesopores to provide sufficient catalyst active sites for both ORR and OER. Their results demonstrated the importance of structure- and morphology-control in the development of carbon-free oxide-metal composite catalysts for RMAB applications.

It was also reported by Lee et al.[57] that a spinel Co_3O_4 nanowire array could be combined with stainless steel mesh as an air-electrode material in RZABs. Their results demonstrated that, due to the direct coupling of Co_3O_4 and SS, the Co_3O_4/SS composite electrode exhibited remarkable electrochemical longevity, as evidenced by an extended cycling of 600 hours with charge and discharge potential retentions of 97% and 94%, respectively.

5.2.4 COMPOSITES OF OTHER OXIDE AND METAL

To solve the issue of carbon corrosion and afford high catalytic activity for ORR/OER, as well as to suppress harmful parasitic reactions during the charge-discharge process,[66,67] other oxides than those discussed in above subsections were also composited with metal(s), such as Ag,[68] Ni,[69] Cu,[70] or Pd,[71] to form new carbon-free oxide-metal composite bifunctional catalysts for RMABs.

To avoid Ag catalyst deactivation induced by the aggregation of nanoparticles, and to develop carbon-free bifunctional catalysts, Yang et al.[68] anodized the electrodeposited Ag-Sn alloys to form nanoporous Ag-embedded SnO_2 layers (AgSn NPL). They conducted a detailed analysis on the tested cyclic voltammograms (CVs) and linear sweep voltammetry (LSV). The ordered nanoporous layer was confirmed to have three advantages in enhancing the electrocatalytic performance: (1) nanopores had an average size of ~10 nm; (2) nanopores were ordered and interconnected; and (3) only a trace amount of Ag was embedded into the SnO_2 matrix, thereby avoiding the aggregation of Ag nanoparticles and reducing the cost of the electrocatalyst.

XPS showed that AgSn NPL consisted of Sn, Ag, and O with Ag/Sn in a 1.7% atomic ratio, corresponding to a Ag loading of $\sim$3.8 μg cm^{-2}. Further XPS spectra showed the presence of Ag 3d at 368.2 eV, which was related to pure Ag rather than Ag oxides,[72] as confirmed by XRD. The Sn3d spin-orbit peaks of SnO$_2$ were observed at 486.7 and 495. eV,[73] respectively. When the bifunctional electrocatalytic activity of AgSn NPL was examined through electrochemical measurements, the tested CVs (see Figure 5.4a) at a potential range of 0.2–1.1 V showed a prominent cathodic ORR peak for AgSn NPL with an onset potential of $\sim$0.87 V (vs RHE) in O$_2$-saturated 0.1 M NaOH solution. Additionally, CV in a wide potential range of 0.3–1.8 V (See Figure 5.4b) demonstrated the electrode reactions by a pair of redox reaction peaks (represented by $Ag_2O + 2OH^- \leftrightarrows 2AgO + H_2O + 2e^-$, labeled as A2 at 1.6 V vs RHE and C2 (1.3 V vs RHE, corresponding to the formation and reduction of AgO, respectively), and an additional pair of redox reaction peaks represented by $2Ag + 2OH^- \leftrightarrows Ag_2O + H_2O + 2e^-$, labeled as A1 at 1.28 V vs RHE and C1 (0.96 V vs RHE, corresponding to the formation and reduction of Ag$_2$O, respectively). The presence of a shoulder (1.22 V vs RHE) close to A1 was ascribed to AgOH

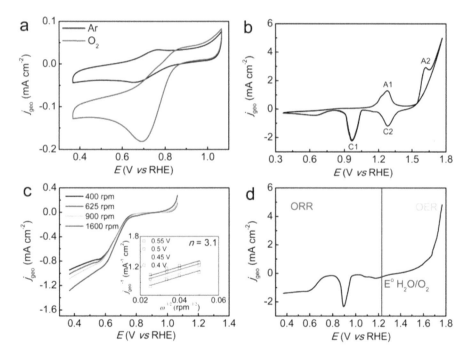

FIGURE 5.4 Bifunctional electrocatalytic performance of AgSn NPL as the oxygen electrode measured at scan rate of 5 mV s^{-1}. (a) ORR performance tested by CV in Ar (black line) and O$_2$ (red line). (b) CV tested in a wide potential range to demonstrate the ORR and OER performance. (c) LSV tested by rotating disk electrode at rotation rates from 400 to 1600 rpm shows the ORR performance. The inset is Koutecky-Levich (K-L) plots at different overpotentials. (d) LSV performed at 900 rpm shows bifunctional ORR and OER performance. (From Yang, Y. et al. *ACS Appl. Mater. Interfaces* **2015**, 7, 20607–20611.[68])

intermediates.[53] From the LSVs recorded by RDE from 400 to 1600 rpm (See Figure 5.4c), the electron-transfer number of AgSn NPL was calculated to be ~3.1, indicating that AgSn NPL mainly followed a four-electron transfer pathway (with oxygen being directly reduced to water). The bifunctional oxygen electrode activity of AgSn NPL was also found in the LSV at 900 rpm (See Figure 5.4d). During cycling tests, 2000 cycles for AgSn NPL showed the increased catalytic currents for both ORR and OER, suggesting that the porous layer of AgSn NPL was gradually activated to give a good performance for long-term utilization without decay.

To form efficient carbon-free bifunctional catalysts without carbon corrosion during charge-discharge, Ni and Cu have also been composited with oxides to obtain CoO/Ni[69] and Co_3O_4/Cu[70] composites for MAB applications. CoO/Ni was synthesized by building freestanding CoO nanowire arrays on Ni foam. This could offer enormous free volume and surface areas for the accommodation of discharge products and the confinement of side reactions during the battery cycling process. Co_3O_4/Cu was prepared via a hydrothermal method in which the formed flower-like Co_3O_4 microsphere particles were loaded with Cu nanoparticles, providing an open mesoporous structure, exposing more Co^{3+} species, and increasing the dispersion of other active Cu components. In the comparison of CoO/Ni and Super-P carbon for electrochemical properties using an assembled aprotic RLOB with 1.0 M LITFSI/TEGDME solution, it was found that at a current density of 100 mA g^{-1} with a cutoff specific capacity of 500 mAh g^{-1}, the CoO air-electrode had a significantly better cycling stability, as evidenced by higher discharge voltages (>2.6 V) and lower charge voltages (<4.4 V). After cycling over 500 hours, the discharge and charge voltages were above 2.5 and below 4.5 V, respectively, which were comparable to the previous results of Co_3O_4 nanowire air-electrode at 100 mA g^{-1} with 500 mAh g^{-1}.[74] This result suggested the mesoporous nanowire array architecture and stable surface activity of CoO/Ni air-electrode could provide positive effects on the fast transport of oxygen and electrolytes and the beneficial accommodation of discharge products in its mesopores for the improvement of catalytic activities for both ORR and OER. Meanwhile, when the ORR and OER activities of Co_3O_4/Cu were evaluated using an aqueous RLAB in the discharge and charge processes, compared to Vulcan XC-72, Co_3O_4/Cu exhibited higher discharge voltages and lower charge voltages, as well as a smaller difference between the discharge and charge voltages, indicating its higher ORR and OER activities. At a current density of 0.05 mA cm^{-2}, Co_3O_4/Cu could deliver stable performances over 110 hours, corresponding to 1250 mAh g^{-1}, demonstrating its promising potential for RMAB applications.

To further modify and optimize the design of oxide-metal composites for developing high-performance carbon-free bifunctional catalysts in RMABs, Leng et al.[71] designed and fabricated Co_3O_4 nanowire clusters on a nickel foam (NF) substrate and then decorated these clusters with Pd nanoparticles using a pulse electrodeposition method to obtain nickel foam-supported Pd/Co_3O_4 (Pd/Co_3O_4/NF) air-electrode for RLOBs. It was believed that these well-designed 3D porous architecture clusters modified with Pd nanoparticles could offer high surface areas, favoring the storage of discharge products, and that the homogeneous distribution of Pd nanoparticles could enhance the conductivity of the nanowires as well as the ORR/OER activities. Their TEM and SEM images clearly revealed a similar flower-like morphology for Pd/Co$_3$O$_4$/NF with Pd particle sizes of $\sim$10 nm, and that the flower-like shapes of Co$_3$O$_4$ were composed of nanowires $\sim$5 µm in length. In the investigation of electrocatalytic activity of Pd/Co$_3$O$_4$/NF, the CVs between 2.0 and 4.5 V in O$_2$-saturated 1 M LiN(CF$_3$SO$_2$)$_2$/TEGDME solution showed that, compared to pristine Co$_3$O$_4$/NF, Pd/Co$_3$O$_4$/NF displayed both higher ORR onset potentials and peak currents, implying higher ORR activity was due to the improved conductivity by the addition of Pd. For the OER process, the decomposition voltage of the discharge products started earlier and the current density catalyzed by Pd/Co$_3$O$_4$/NF was much greater than the reference Co$_3$O$_4$/NF catalyst, indicating Pd/Co$_3$O$_4$/NF not only possessed higher OER activity but also had better abilities to catalyze discharge products. The galvanostatic discharge-charge curves of RLOB showed that the specific discharge capacities of Pd/Co$_3$O$_4$/NF and pristine Co$_3$O$_4$/NF electrodes at a current density of 0.05 mA cm^{-2} were 1842.7 and 1550.9 mAh g^{-1}, respectively. These values were larger than previous Pt/Co$_3$O$_4$ and Co$_3$O$_4$ nanostructures (1371 and 1348 mAh g^{-1}, respectively),[75] demonstrating the advantages of these flower-like nanowire structures. It was also revealed that the presence of Pd not only decreased the polarization of the discharge-charge process but also facilitated the decomposition of insulating products upon recharge. During cycling tests at 0.1 mA cm^{-2} with a capacity of 300 mAh g^{-1}, Pd/Co$_3$O$_4$/NF remained stable for 70 cycles, which was better than that of pristine Co$_3$O$_4$/NF, which remained stable for only 40 cycles, indicating the enhanced stability of Pd/Co$_3$O$_4$/NF. When the air-electrodes were monitored during discharge-charge using a combination of SEM, EIS, and XPS, it was found that Pd/Co$_3$O$_4$/NF tended to deal with the discharge products more completely than pristine Co$_3$O$_4$/NF, resulting in a better rechargeability. With the use of carbon-free electrodes, some byproducts (i.e., Li$_2$CO$_3$) were discovered in XPS due to electrolyte instability and decomposition.[76,77] This research,[71] using the advantages of conducting metals and the 3D porous architecture of oxides to not only enhance the bifunctional electrolysis of both ORR and OER but also reduce the corrosion from discharge products, resulted in enhanced battery performances and provided an approach to develop high-performance carbon-free catalysts for RMABs.

5.3   OTHER NONCARBON-BASED COMPOSITES

As discussed above, significant efforts have been devoted to the design and preparation of noncarbon bifunctional catalysts for RMABs because carbon materials are not stable in the presence of the discharge intermediates and products during electrochemical charge-discharge.[78,79,80,66] However, most noncarbon materials are known to have disadvantages such as small porosity, low surface area, and poor electronic conduction. Therefore, these noncarbon materials must be effectively combined with each other to form new noncarbon-based composites for the bifunctional catalysis of both ORR and OER in RMABs. Apart from oxide- and metal-based composites, other noncarbon-based composites have also been studied and developed in the exploration of high-performance carbon-free bifunctional catalysts.

Metallic layered double oxides (LDO) have been reported as a family of ionic lamellar compounds in which the multiple valences of the cations in the host structure can provide donor-acceptor chemisorption sites, the increased number of active edge sites,[8] and the improved electronic conductivity[82] resulted from the strongly layered orientation, making contributions to both ORR[81] and OER[83] catalytic activities. Furthermore, the intercalation of electrolyte bis (trifluoromethane) sulfonimide (TFSI) anions into interlayers of LDO can not only improve the compatibility with the interface of LDO and electrolyte but also increase the interlayer spacing of the LDO, benefiting ion diffusion, electrolyte permeation, and then electrochemical performance.[82,84,85] Combining the advantages of LDO and the intercalation of TFSI anions, Xu et al.[86] designed and prepared a hierarchically porous flower-like cobalt-titanium layered double oxide on nickel foam with intercalated TFSI anions for RLOB air-electrodes. The hierarchical porous structure and high specific surface area of LDO could lead to the improvement of adsorption and catalytic properties. In their synthesis (see Figure 5.5a), CoTi LDH was first electrodeposited onto Ni foam by gradual co-precipitation of Co^{2+} and Ti^{4+}. Then, CoTi LDH was converted to porous CoTi LDO after calcination at 300°C in air. Upon dispersing the CoTi LDO nanoflakes into a LITFSI alcoholic aqueous solution, TFSI anions were intercalated into the interlayers of CoTi LDO to generate TFSI/CoTi LDH. Finally, the regenerated TFSI/CoTi LDH was heat-treated at 300°C in air to obtain CoTi LDO with intercalated TFSI anions (labeled as TFSI/CoTi LDO). Under the monitor of field emission scanning electron microscope (FESEM) and TEM (see Figure 5.5b–e), a hierarchically porous structure (porous size: 3–10 nm) with mesopores and macropores is observed in the nanoflakes (~1.0 μm) of CoTi LDO due to the formation of metal oxides and the release of interlayer species. According to the results of ICP-MS, the Co/Ti molar ratio of CoTi LDH and CoTi LDO was determined to be 2.58, and the elemental mapping analysis showed highly dispersive and homogeneous Co and Ti components in the structure of CoTi LDO. After the presence of TFSI was confirmed by FTIR and Raman spectroscopic analysis, the electrochemical properties of TFSI/CoTi LDO was evaluated as an air-electrode using a RLOB. The discharge and charge curves showed that, compared to CoTi LDO (~710 mAh g^{-1}), TFSI/CoTi LDO could deliver a much higher initial discharge capacity (~1133 mAh g^{-1}) with a round-trip efficiency of 93.1%, and that the charge-discharge overpotential of TFSI/CoTi LDO was only 0.45 V, which was much lower than that of CoTi LDO (~0.72 V) and even lower than that of nanostructured Pd electrodes (~0.5 V) in literature.[87] When the current density was increased from 100 mA g^{-1} to 400 mA g^{-1} and up to 800 mA g^{-1}, the discharge capacity of TFSI/CoTi LDO was decreased from 1133 mAh g^{-1} to 321 mAh g^{-1} and finally down to 161 mAh g^{-1}, showing that TFSI/CoTi LDO displayed a good rate performance. Under a cycling test at 100 mA g^{-1}, both the TFSI/CoTi LDO and CoTi LDO electrodes showed a relatively good cycling stability without other distinct capacity decays after the initial capacity loss. They discussed that TFSI/CoTi LDO outperforming CoTi LDO in terms of catalytic activities of both ORR and OER was because the reactive sites of TFSI/CoTi LDO could adsorb electron donor groups on organic solvent molecules and weaken the chemical bonds near the adsorbed atoms,[88] and TFSI/CoTi LDO could significantly decrease the polarization of the electrode due to the intercalation of TSFI, resulting in low overpotentials and high round-trip efficiencies. In the analysis of morphology and reversibility of discharge products, a combination of SEM and XRD revealed that the enlarged interlayer spacing could confine the growth of Li_2O_2 and act as a "transit depot" to store intermediate products

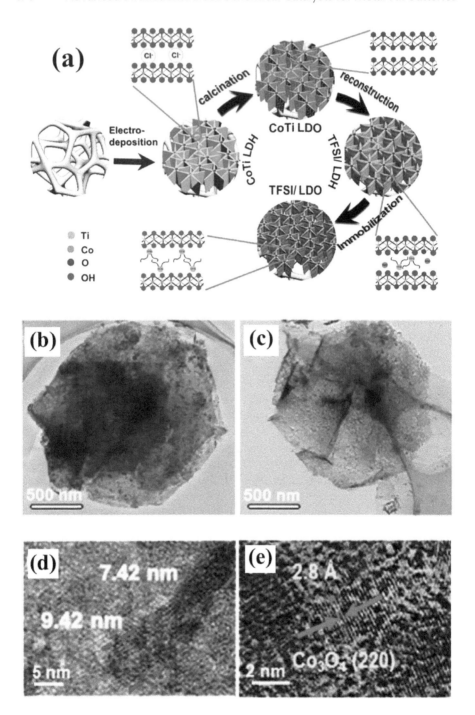

FIGURE 5.5 Schematic illustration of the preparation procedure for the carbon- and binder-free TFSI/LDO electrode. (Modified from Xu, S.-M. et al. *Adv. Funct. Mater.* **2016**, *26*, 1365–1374.[86])

for the further formation of Li_2O_2 on the edge of the TFSI/LDO nanoflakes, leading to the formation of nanoflaky Li_2O_2 species on TFSI/LDO, and the intercalation of TFSI anions into interlayers of LDO could provide more active sites and solid-liquid interfaces for catalytic reactions. After 80 cycles, the morphology analysis for the recovered electrode indicated a good structural reversibility and stability for TFSI/LDO. This design and fabrication of TFSI/LDO can provide a new strategy for the preparation of carbon- and binder-free composite cathodes for RLOBs with good compatibility at the interface between the electrode and the electrolyte.

Despite poor electronic conductivity can limit the catalytic activity of Co and Mo-based spinel oxides (i.e., $MnCo_2O_4$), these oxides have been observed in potential applications for electrocatalysis of both ORR and OER due to the multiple valence states of Co and Mn.[89–92] Cao et al.[93] fabricated a composite of $MnCo_2O_4$-polypyrrole (MCO@PPY) as a carbon-free catalyst for both ORR and OER. They discussed that, as a flexible, stable, and conductive polymer, PPY could provide a conductive network for fast electron transfer and the coupling between PPY and $MnCo_2O_4$ could promote electronic transfer from PPY to $MnCo_2O_4$, resulting in enhancement of electrocatalytic activity. In a typical experiment, $MnCo_2O_4$ was first synthesized by a co-precipitation method with $Co(NO_3)_2·6H_2O$ and $Mn(CH_3COO)_2·4H_2O$ as Co and Mn sources. The $MnCo_2O_4$ was then coated with the PPY layer using a facile chemical oxidation polymerization approach with H_2O_2 as an oxygen agent. The MCO content was 6 wt% according to TGA analysis. For comparison, a physical mixture of MCO + PPY was also prepared. A reference sample of MCO/acetylene black (MCO/AB) was also prepared with a weight ratio of MCO to AB at 1:1. After XRD confirmed the phase of $MnCo_2O_4$, SEM, TEM, and HRTEM were employed for the characterization, and the results showed that the primary aggregated $MnCo_2O_4$ had diameters ranging between 100 and 150 nm, significantly larger than the average crystallite size of 12 nm calculated from XRD. Furthermore, the edge of MCO was covered with a thin layer of amorphous PPY with a thickness of about 5 nm. The combination of N 1 s and O 1 s in the compared XPS spectra of PPY, $MnCo_2O_4$, and $MnCo_2O_4$@PPY revealed a significant coupling effect between $MnCo_2O_4$ and PPY through oxygen atoms. In the evaluation of electrocatalytic performance in O_2-saturated 0.1 M KOH solution that the LSV onset potential of MCO@PPY was higher than MCO/CB, PPY, and MCO + PPY samples, being very close to that of Pt/C. The enhanced ORR activity of MCO@PPY could be attributed to the introduction of PPY into MCO, resulting in an improvement of conductivity as well as the covalent coupling between MCO and PPY. Based on RRDE measurements, the electron transfer numbers of MCO@PPY were 3.83–3.94, suggesting a four-electron ORR mechanism. For the OER, the LSVs at a voltage range of 0.1–1 V in N_2-saturated 0.1 M KOH solution showed that both the onset potential and the current density of MCO@PPY were very close to that of benchmark 20 wt% RuO_2/C, and was also higher than the other samples of MCO/CB, PPY, and MCO + PPY. When a further comparison of stability was conducted using the chronoamperometric method at −0.35 V (vs. Ag/AgCl), MCO@PPY had only 1% loss in ORR currents after 60,000 seconds of continuous operation while MCO/CB, PPY and 20 wt% Pt/C had lost 10.0%, 33%, and 60%, respectively. When chronoamperometric measurements were run at 0.6 V (vs. Ag/AgCl), the OER current loss of MCO@PPY was 20%, which was lower than MCO/CB (~72%), PPY (80%), and 20 wt% RuO_2 (~72%) samples. All the electrochemical results demonstrated that MCO@PPY not only possessed the enhanced electrocatalytic activities for both ORR and OER, but also possessed excellent stability when used as a bifunctional catalyst.

Zhang et al.[94] combined CoFe layered double hydroxides (CoFe-LDHs) with polydopamine spheres (PDAS) to fabricate CoFe-LDHs/PDAS composites for the catalysis of ORR. They used N-containing polydopamine (PDA) for the preparation of transition metal oxides and N-containing polymer ORR catalysts. After CoFe-LDHs was in-situ grown on the surface of PDAS to CoFe-LDHs/PDAS, other reference samples such as PDA, CoFe-LDHs, acid-treated CoFe-LDHs/PDAS, and heat-treated CoFe-LDHs/PDAS were also prepared for ORR activity comparisons. Their results showed that, with a four-electron mechanism, CoFe-LDHs/PDAS presented higher ORR activities than the other four reference samples. Furthermore, CoFe-LDHs/PDAS also showed higher stability than commercial Pt/C. This research provides a novel approach for developing noncarbon-based composite catalysts for RMABs.

5.4   CHAPTER SUMMARY

Considering the instability of carbon induced by discharge intermediates and products during discharge–charge, a significant number of researchers have been devoted to the development of noncarbon materials as advanced bifunctional catalysts for RMABs. However, most noncarbon materials, such as metals and oxides, often exhibit low electronic conductivity, poor surface area, and small porosity. Therefore, the combination of different noncarbon materials have become an effective approach to form new noncarbon-based composite bifunctional catalysts for ORR and OER in RMABs.

As two important replacements of carbon-based RMAB catalysts, the metal/metal (e.g., bimetallic metals or alloy) and oxide/metal composites have been researched in the development of advanced bifunctional catalysts for ORR and OER. It is found that coupled with the optimized morphology of catalyst, the synergistic effect of different components plays a significant role in not only enhancing the fast transport of O_2 and electrolytes but also accommodating the reaction product during the discharging-charging process, resulting in the improvement of energy output, round-trip efficiency, and cyclability. Particularly, perovskite oxide acting as a special oxide have a good bifunctionality and can produce the effective interaction with other component like metal to obtain high active catalysts for both ORR and OER. Additionally, some other noncarbon components (e.g., the conductive polymer/oxide composite) have been composited together as a new strategy in the development of carbon-free composite cathode catalysts in which different noncarbon components exhibited their high compatibility with each other, thereby producing a good synergy for enhancing bifunctional catalytic activity and stability/durability.

REFERENCES

1. Liao, K.; Zhang, T.; Wang, Y.; Li, F.; Jian, Z.; Yu, H.; Zhou, H. Nanoporous Ru as a carbon- and binder-free cathode for Li-O_2 batteries. *Chem. Sus. Chem.* **2015**, *8*, 1429–1434.

2. Laoire, C. O.; Mukerjee, S.; Abraham, K. M.; Plichta, E. J.; Hendrickson, M. A. Influence of non-aqueous solvents on the electrochemistry of oxygen in the rechargeable lithium-air battery. *J. Phys. Chem. C* **2010**, *114*, 9178–9186.

3. Laoire, C. O.; Mukerjee, S.; Abraham, K. M.; Plichta, E. J.; Hendrickson, M. A. Elucidating the mechanism of oxygen reduction for lithium-air battery applications. *J. Phys. Chem. C* **2009**, *113*, 20127–20134.

4. Lu, Y.-C.; Gasteiger, H. A.; Shao-Horn, Y. Method development to evaluate the oxygen reduction activity of high-surface-area catalysts for Li-air batteries. *Electrochem. Solid-State Lett.* **2011**, *14*, A70–A74.

5. Allen, C. J.; Mukerjee, S.; Plichta, E. J.; Hendrickson, M. A.; Abraham, K. M. Oxygen electrode rechargeability in an ionic liquid for the Li-air battery. *J. Phys. Chem. Lett.* **2011**, *2*, 2420–2424.

6. Xu, J.-J.; Wang, Z.-L.; Xu, D.; Men, F.-Z.; Zhang, X.-B. 3D ordered macroporous $LaFeO_3$ as efficient electrocatalyst for $Li-O_2$ batteries with enhanced rate capability and cyclic performance. *Energy Environ. Sci.* **2014**, *7*, 2213–2219.

7. Zhu, D.; Zhang, L.; Song, M.; Wang, X.; Chen, Y. An *in situ* formed Pd nanolayer as a bifunctional catalyst for Li-air batteries in ambient or simulated air. *Chem. Commun.* **2013**, *49*, 9573–9575.

8. Cui, Y.; Wen, Z.; Liu, Y. A free-standing-type design for cathodes of rechargeable $Li-O_2$ batteries. *Energy Environ. Sci.* **2011**, *4*, 4727–4734.

9. Liu, X.; Wang, D.; Shi, S. Exploration on the possibility of Ni foam as current collector in rechargeable lithium-air batteries. *Electrochim. Acta.* **2013**, *87*, 865–871.

10. Wu, X.; Chen, F.; Jin, Y.; Zhang, N.; Johnston, R. L. Silver-copper nanoalloy catalyst layer for bifunctional air electrodes in alkaline media. *ACS Appl. Mater. Interfaces* **2015**, *7*, 17782–17791.

11. Jin, Y.; Chen, F. Facile preparation of Ag-Cu bifunctional electrocatalysts for zinc-air batteries. *Electrochim. Acta.* **2015**, *158*, 437–445.

12. Ma, W.; Chen, F.; Zhang, N.; Wu, X. Oxygen reduction reaction on Cu-doped Ag cluster for fuel–cell cathode. *J. Mol. Model.* **2014**, *20*, 2454–4561.

13. Guo, X.; Han, J.; Liu, P.; Chen, L.; Ito, Y.; Jian, Z.; Jin, T. et al. Hierarchical nanoporosity enhanced reversible capacity of bicontinuous nanoporous metal based $Li-O_2$ battery. *Sci. Rep.* **2016**, *6*, 33466.

14. Zhang, L.; Chen, L.; Liu, H.; Hou, Y.; Hirata, A.; Fujita, T.; Chen, M. Effect of residual silver on surface-enhanced Raman scattering of dealloyed nanoporous gold. *J. Phys. Chem. C.* **2011**, *115*, 19583–19587.

15. Gorlin, Y.; Jaramillo, T. F. A bifunctional nonprecious metal catalyst for oxygen reduction and water oxidation. *J. Am. Chem. Soc.* **2010**, *132*, 13612–13614.

16. Gupta, S.; Kellogg, W.; Xu, H.; Liu, X.; Cho, J.; Wu, G. Bifunctional perovskite oxide catalysts for oxygen reduction and evolution in alkaline media. *Chem. Asian J.* **2016**, *11*, 10–21.

17. Wu, G.; Nelson, M. A.; Mack, N. H.; Ma, S.; Sekhar, P.; Garzon, F. H.; Zelenay, P. Titanium dioxide-supported non-precious metal oxygen reduction electrocatalyst. *Chem. Commun.* **2010**, *46*, 7489–7491.

18. Li, Y.; Liu, C.; Liu, Y.; Feng, B.; Li, L.; Pan, H.; Kellogg, W.; Higgins, D.; Wu, G. Sn-doped TiO_2 modified carbon to support Pt anode catalysts for direct methanol fuel cells. *J. Power Sources.* **2015**, *286*, 354–361.

19. Chen, C.-F.; King, G.; Dickerson, R. M.; Papin, P. A.; Gupta, S.; Kellogg, W. R.; Wu, G. Oxygen-deficient $BaTiO_{3-x}$ perovskite as an efficient bifunctional oxygen electrocatalyst. *Nano Energy.* **2015**, *13*, 423–432.

20. Grimaud, A.; May, K. J.; Carlton, C. E.; Lee, Y.-L.; Risch, M.; Hong, W. T.; Zhou, J.; Shao-Horn, Y. Double perovskites as a family of highly active catalysts for oxygen evolution in alkaline solution. *Nat. Commun.* **2013**, *4*, 2439.

21. Zhu, Y.; Zhou, W.; Chen, Z.-G.; Chen, Y.; Su, C.; Tadé, M. O.; Shao, Z. $SrNb_{0.1}Co_{0.7}Fe_{0.2}O_{3-\delta}$ perovskite as a next-generation electrocatalyst for oxygen evolution in alkaline solution. *Angew. Chem. Int. Ed.* **2015**, *54*, 3897–3901.

22. Du, J.; Zhang, T.; Cheng, F.; Chu, W.; Wu, Z.; Chen, J. Nonstoichiometric perovskite $CaMnO_{3-\delta}$ for oxygen electrocatalysis with high activity. *Inorg. Chem.* **2014**, *53*, 9106–9114.

23. Malkhandi, S.; Yang, B.; Manohar, A. K.; Manivannan, A.; Surya Prakash, G. K.; Narayanna, S. R. Electrocatalytic properties of nanocrystalline calcium-doped lanthanum cobalt oxide for bifunctional oxygen electrodes. *J. Phys. Chem. Lett.* **2012**, *3*, 967–972.

24. Bie, S.; Zhu, Y.; Su, J.; Jin, C.; Liu, S.; Yang, R.; Wu, J. One-pot fabrication of yolk-shell structured $La_{0.9}Sr_{0.1}CoO_3$ perovskite microspheres with enhanced catalytic activities for oxygen reduction and evolution reactions. *J. Mater. Chem. A* **2015**, *3*, 22448–22453.

25. Zhang, D.; Song, Y.; Du, Z.; Wang, L.; Li, Y.; Goodenough, J. B. Active $LaNi_{1-x}Fe_xO_3$ bifunctional catalysts for air cathodes in alkaline media. *J. Mater. Chem. A* **2015**, *3*, 9421–9426.

26. Lopez, K.; Park, G.; Sun, H.-J.; An, J.-C.; Eom, S.; Shim, J. Electrochemical characterization of $LaMO_3$ (M = Co, Mn, Fe, and Ni) and partially substituted $LaNi_xM_{1-x}O_3$ (x = 0.25 or 0.5) for oxygen reduction and evolution in alkaline solution. *J. Appl. Electrochem.* **2015**, *45*, 313–323.

27. Jin, C.; Cao, X.; Lu, F.; Yang, Z.; Yang, R. Electrochemical study of $Ba_{0.5}Sr_{0.5}Co_{0.8}Fe_{0.2}O_3$ perovskite as bifunctional catalyst in alkaline media. *Int. J. Hydrogen Energ.* **2013**, *38*, 10389–10393.

28. Jung, J.; Park, S.; Kim, M.-G.; Cho, J. Tunable internal and surface structure of the bifunctional oxygen perovskite catalysts. *Adv. Energy Mater.* **2015**, *5*, 1501560.

29. Petrie, J. R.; Cooper, V. R.; Freeland, J. W.; Meyer, T. L.; Zhang, Z.; Lutterman, D. A.; Lee, H. N. Enhanced bifunctional oxygen catalysis in strained $LaNiO_3$ perovskites. *J. Am. Chem. Soc.* **2016**, *138*, 2488–2491.

30. Sickafus, K. E.; Wills, J. M. Grimes, N. W. Structure of spinel. *J. Am. Ceram. Soc.* **1999**, *82*, 3279–3292.

31. Bragg, W. H. The structure of magnetite and the spinels. *Nature* **1915**, *95*, 561–561.

32. Cheng, F.; Shen, J.; Peng, B.; Pan, Y.; Tao, Z.; Chen, J. Rapid room-temperature synthesis of nanocrystalline spinels as oxygen reduction and evolution electrocatalysts. *Nature Chem.* **2011**, *3*, 79–84.

33. Pletcher, D.; Li, X.; Price, S. W. T.; Rusell, A. E.; Sönmez, T.; Thompson, S. J. Comparison of the spinels Co_3O_4 and $NiCo_2O_4$ as bifunctional oxygen catalysts in alkaline media. *Electrochim. Acta.* **2016**, *188*, 286–293.

34. Meng, Y.; Song, W.; Huang, H.; Ren, Z.; Chen, S.-Y.; Suib, S. L. Structure-property relationship of bifunctional MnO_2 nanostructures: Highly efficient, ultra-stable electrochemical water oxidation and oxygen reduction reaction catalysts identified in alkaline media. *J. Am. Chem. Soc.* **2014**, *136*, 11452–11464.

35. Yu, J.; Sunarso, J.; Zhu, Y.; Xu, X.; Ran, R.; Zhou, W.; Shao, Z. Activity and stability of Ruddlesden-Popper-type $La_{n+1}Ni_nO_{3n+1}$ (n = 1, 2, 3 and ∞) electrocatalysts for oxygen reduction and evolution reactions in alkaline media. *Chem. Eur. J.* **2016**, *22*, 2719–2727.

36. Amow, G.; Skinner, S. J. Recent developments in Ruddlesden-Popper nickelate systems for solid oxide fuel cell cathodes. *J. Solid State Electrochem.* **2006**, *10*, 538–546.

37. Amow, G.; Davidson, I. J.; Skinner, S. J. A comparative study of the Ruddlesden-Popper series, $La_{n+1}Ni_nO_{3n+1}$ n = 1, 2 and 3, for solid-oxide fuel-cell cathode applications. *Solid State Ionics.* **2006**, *177*, 1205–1210.

38. Burriel, M.; Garcia, G.; Rossell, M. D.; Figueras, A.; Tendeloo, G. V.; Santiso, J. Enhanced high-temperature electronic transport properties in nanostructured epitaxial thin films of the $La_{n+1}Ni_nO_{3n+1}$ Ruddlesden-Popper series (n = 1, 2, 3, ∞). *Chem. Mater.* **2007**, *19*, 4056–4062.

39. Wang, D.; Chen, X.; Evans, D. G.; Yang, W. Well-dispersed Co_3O_4/Co_2MnO_4 nanocomposites as a synergistic bifunctional catalyst for oxygen reduction and oxygen evolution reactions. *Nanoscale* **2013**, *5*, 5312–5315.

40. Han, X.; Cheng, F.; Zhang, T.; Yang, J.; Hu, Y.; Chen, J. Hydrogenated uniform Pt clusters supported on porous $CaMnO_3$ as a bifunctional electrocatalyst for enhanced oxygen reduction and evolution. *Adv. Mater.* **2014**, *26*, 2047–2051.

41. Park, S. A.; Lee, E. K.; Song, H.; Kim, Y.T. Bifunctional enhancement of oxygen reduction reaction activity on Ag catalysts due to water activation on $LaMnO_3$ supports in alkaline media. *Sci. Rep.* **2015**, *5*,13552.

42. Yang, W.; Salim, J.; Li, S.; Sun, C.; Chen, L.; Goodenough, J. B.; Kim, Y. Perovskite $Sr_{0.95}Ce_{0.05}CoO_{3-\delta}$ loaded with copper nanoparticles as a bifunctional catalyst for lithium-air batteries. *J. Mater. Chem.* **2012**, *22*, 18902–18907.

43. Wang, Y.-J. Wilkinson, D. P.; Neburchilov, V.; Song, C.; Guest, A.; Zhang, J. Ta and Nb co-doped TiO_2 and its carbon-hybrid materials for supporting Pt-Pd alloy electrocatalysts for PEM fuel cell oxygen reduction reaction. *J. Mater. Chem. A* **2014**, *2*, 12681–12685.

44. Wang, Y.-J.; Zhao, N.; Fang, B.; Li, H.; Bi, X. T.; Wang, H. Effect of different solvent ratio (ethylene glycol/water) on the preparation of Pt/C catalyst and its activity toward oxygen reduction reaction. *RSC Adv.* **2015**, *5*, 56570–56577.

45. Wang, Y.-J.; Zhao, N.; Fang, B.; Li, H.; Bi, X. T.; Wang, H. A highly efficient PtCo/C electrocatalyst for the oxygen reduction reaction. *RSC Adv.* **2016**, *6*, 34484–34491.

46. Awaludin, Z.; Suzuki, M.; Masud, J.; Okajima, T.; Ohsaka, T. Enhanced electrocatalysis of oxygen reduction on $Pt/TaO_x/GC$. *J. Phys. Chem. C* **2011**, *115*, 25557–25567.

47. Beak, S.; Jung, D.; Nahm, K. S.; Kim, P. Preparation of highly dispersed Pt on TiO_2-modified carbon for the application to oxygen reduction reaction. *Catal. Lett.* **2010**, *134*, 288–294.

48. Murthi, V. S.; Urian, R. C.; Mukerjee, S. Oxygen reduction kinetics in low and medium temperature acid environment: Correlation of water activation and surface properties in supported Pt and Pt alloy electrocatalysts. *J. Phys. Chem. B* **2004**, *108*, 11011–11023.

49. Amin, H. M. A.; Baltruschat, H.; Wittmaier, D.; Friedrich, K. A. A highly efficient bifunctional catalyst for alkaline air-electrodes based on a Ag and Co_3O_4 hybrid: RRDE and online DEMS insight. *Electrochim. Acta.* **2015**, *151*, 332–339.

50. Lambert, T. N.; Vigil, J. A.; White, S. E.; David, D. J.; Limmer, S. J.; Burton, P. D.; Coker, E. N.; Beechem, T. E.; Brumbach, M. T. Electrodeposited $Ni_xCo_{3-x}O_4$ nanostructured films as bifunctional oxygen electrocatalysts. *Chem. Commun.* **2015**, *51*, 9511–9514.

51. Wu, F.; Zhang, X.; Zhao, T.; Chen, R.; Ye, Y.; Xie, M.; Li, L. Hierarchical mesoporous/macroporous Co_3O_4 ultrathin nanosheets as free-standing catalysts for rechargeable lithium-oxygen batteries. *J. Mater. Chem. A* **2015**, *3*, 17620–17626.

52. Lee, C. K.; Park, Y. J. Carbon and binder-free air electrodes composed of Co_3O_4 nanofibers for Li-Air batteries with enhanced cyclic performance. *Nanoscale Res. Lett.* **2015**, *10*, 319.

53. Kalubarme, R. S.; Jadhav, H. S.; Ngo, D. T.; Park, G. E.; Fisher, J. G.; Choi, Y.; Ryu, W. H.; Par, C. J. Simple synthesis of highly catalytic carbon-free $MnCo_2O_4$@Ni as an oxygen electrode for rechargeable Li-O_2 batteries with long-term stability. *Sci. Rep.* **2015**, *5*, 13266.

54. Lin, X.; Shang, Y.; Huang, T.; Yu, A. Carbon-free (Co, Mn)3O4 nanowires@Ni electrodes for lithium-oxygen batteries. *Nanoscale* **2014**, *6*, 9043–9049.

55. Price, S. W. T.; Thompson, S. J.; Li, X.; Gorman, S. F.; Pletcher, D.; Russell, A. E.; Walsh, F. C.; Wills, R. G. A. The fabrication of a bifunctional oxygen electrode without carbon components for alkaline secondary batteries. *J. Power Sources.* **2014**, *259*, 43–49.

56. Li, X.; Pletcher, D.; Russell, A. E.; Walsh, F. C.; Wills, R. G. A.; Gorman, S. F.; Price, S. W. T.; Thompson, S. J. A novel bifunctional oxygen GDE for alkaline secondary batteries. *Electrochim. Commun.* **2013**, *34*, 228–230.

57. Lee, D. U.; Choi, J.-Y.; Feng, K.; Park, H. W.; Chen, Z. Advanced extremely durable 3D bifunctional air electrodes for rechargeable zinc-air batteries. *Adv. Energy Mater.* **2014**, *4*, 1301389.

58. Li, L.; Manthiram, A. Decoupled bifunctional air electrodes for high-performance hybrid lithium-air batteries. *Nano Energy.* **2014**, *9*, 94–100.

59. Li, L.; Liu, C.; He, G.; Fan, D.; Manthiram, A. Hierarchical pore-in-pore and wire-in-wire catalysts for rechargeable Zn- and Li-air batteries with ultra-long cycle life and high cell efficiency. *Energy Environ. Sci.* **2015**, *8*, 3274–3282.

60. Baltruschat, H. Differential electrochemical mass spectrometry. *J. Am. Soc. Mass. Spectrom.* **2004**, *15*, 1693–1706.

61. Goodenough, J. B.; Kim, Y. Challenges for rechargeable Li batteries. *Chem. Mater.* **2010**, *22*, 587–603.

62. Tarascon, J.-M. Key challenges in future Li-battery research. *Phil. Trans. R. Soc. A* **2010**, *368*, 3227–3241.

63. Lee, H.; Kim, Y.-J.; Lee, D. J.; Song, J.; Lee, Y. M.; Kim, H.-T.; Park, J.-K. Directly grown Co_3O_4 nanowire arrays on Ni-foam: Structural effects of carbon-free and binder-free cathodes for lithium-oxygen batteries. *J. Mater. Chem. A* **2014**, *2*, 11891–11898.

64. Liu, Q. C.; Xu, J. J.; Chang, Z. W.; Zhang, X. B. Direct electrodeposition of cobalt oxide nanosheets on carbon paper as free-standing cathode for Li-O_2 battery. *J. Mater. Chem. A* **2014**, *2*, 6081–6085.

65. Liu, J.; Younesi, R.; Guestafsson, T.; Edström, K.; Zhu, J. Pt/α-MnO_2 nanotube: A highly active electrocatalyst for Li-O_2 battery. *Nano Energy* **2014**, *10*, 19–27.

66. Zhang, T.; Zhou, H. S. A reversible long-life lithium-air battery in ambient air. *Nat. Commun.* **2013**, *4*, 1817–1824.

67. Peng, Z.; Freunberger, S. A.; Chen, Y.; Bruce, P. G. A reversible and higher-rate Li-O_2 battery. *Science* **2012**, *337*, 563–566.

68. Yang, Y.; Fei, H.; Ruan, G.; Li, L.; Wang, G.; Kim, N. D.; Tour, J. M. Carbon-free electrocatalyst for oxygen reduction and oxygen evolution reactions. *ACS Appl. Mater. Interfaces* **2015**, *7*, 20607–20611.

69. Wu, B.; Zhang, H.; Zhou, W.; Wang, M.; Li, X.; Zhang, H. Carbon-free CoO mesoporous nanowire array cathode for high-performance aprotic Li-O_2 batteries. *ACS Appl. Mater. Interfaces* **2015**, *7*, 23182–23189.

70. Yang, W.; Salim, J.; Ma, C.; Ma, Z.; Sun, C.; Li, J.; Chen, L.; Kim, Y. Flowerlike Co_3O_4 microspheres loaded with copper nanoparticle as an efficient bifunctional catalyst for lithium-air batteries. *Electrochem. Commun.* **2013**, *28*, 13–16.

71. Leng, L.; Zeng, X.; Song, H.; Shu, T.; Wang, H.; Liao, S. Pd nanoparticles decorating flower-like Co_3O_4 nanowire clusters to form an efficient, carbon/binder-free cathode for Li-O_2 batteries. *J. Mater. Chem. A* **2015**, *3*, 15626–15632.

72. Zou, M.; Du, M.; Zhu, H.; Xu, C.; Fu, Y. Green synthesis of alloy site nanotubes supported Ag nanoparticles for photocatalytic decomposition of methylene blue. *J. Phys. D: Appl. Phys.* **2012**, *45*, 325302.

73. Wang, J. J.; Lv, A. F.; Wang, Y. Q.; Cui, B.; Yan, H. J.; Hu, J. S.; Hu, W. P.; Guo, Y. G.; Wan, L. J. Integrated prototype nanodevices via SnO_2 nanoparticles decorated SnSe nanosheets. *Sci. Rep.* **2013**, *3*, 2613.

74. Riaz, A.; Jung, K. N.; Chang, W.; Lee, S. B.; Lim, T. H.; Park, S. J.; Song, R. H.; Yoon, S.; Shin, K. H.; Lee, J. W. Carbon-free cobalt oxide cathodes with tunable nanoarchitectures for rechargeable lithium-oxygen batteries. *Chem. Commun.* **2013**, *49*, 5984–5986.

75. Cao, J.; Liu, S.; Xie, J.; Zhang, S.; Cao, G.; Zhao, X. Tips-bundled Pt/Co$_3$O$_4$ nanowires with directed peripheral growth of Li$_2$O$_2$ as efficient binder/carbon-free catalytic cathode for lithium-oxygen battery. *ACS Catal.* **2015**, *5*, 241–245.

76. Li, F.; Tang, D. M.; Jian, Z.; Liu, D.; Golberg, D.; Yamada, A.; Zhou, H. Li-O$_2$ battery based on highly efficient Sb-doped tin oxide supported Ru nanoparticles. *Adv. Mater.* **2014**, *26*, 4659–4664.

77. Li, F.; Tang, D. M.; Chen, Y.; Golberg, D.; Kitaura, H.; Zhang, T.; Yamada, A.; Zhou, H. Ru/ITO: A carbon-free cathode for non-aqueous Li-O$_2$ battery. *Nano Lett.* **2013**, *13*, 4702–4707.

78. Ottakam Thotiyl, M. M.; Freunberger, S. A.; Peng, Z. Q.; Bruce, P. G. The carbon electrode in non-aqueous Li-O$_2$ cells. *J. Am. Chem. Soc.* **2013**, *135*, 494–500.

79. Itkis, D. M.; Semenenko, D. A.; Kataev, E. Y.; Belova, A. I.; Neudachina, V. S.; Sirotina, A. P.; Hävecker, M. et al. Reactivity of carbon in lithium-oxygen battery positive electrodes. *Nano Lett.* **2013**, *13*, 4697–4701.

80. Ross, P. N.; Sokol, H. The corrosion of carbon black anodes in alkaline electrolyte. I. acetylene black and the effect of cobalt catalyzation. *J. Electrochem. Soc.* **1984**, *131*, 1742–1750.

81. Prabu, M.; Ketpang, K.; Shanmugam, S. Hierarchical nanostructured NiCo$_2$O$_4$ as an efficient bifunctional non-precious metal catalyst for rechargeable zinc-air batteries. *Nanoscale* **2014**, *6*, 3173–3181.

82. Yuan, C.; Wu, H. B.; Xie, Y.; Lou, X. W. Mixed transition-metal oxides: Design, synthesis, and energy-related applications. *Angew. Chem. Int. Ed.* **2014**, *53*, 1488–1504.

83. Chen, H.; Hu, L.; Chen, M.; Yan, Y.; Wu, L. Nickel-cobalt layered double hydroxide nanosheets for high-performance supercapacitor electrode materials. *Adv. Funct. Mater.* **2014**, *24*, 934–942.

84. Song, F.; Hu, X. Exfoliation of layered double hydroxides for enhanced oxygen evolution catalysis. *Nat. Commun.* **2014**, *5*, 4477.

85. Wang, L.; Dong, Z. H.; Wang, Z. G.; Zhang, F. X.; Jin, J. Layered α-CoOH2 nanocones as electrode materials for pseudocapacitors: Understanding the effect of interlayer space on electrochemical activity. *Adv. Funct. Mater.* **2013**, *23*, 2758–2764.

86. Xu, S.-M.; Zhu, Q.-C.; Long, J.; Wang, H.-H.; Xie, X.-F.; Wang, K.-X.; Chen, J.-S. Low-overpotential Li-O$_2$ batteries based on TFSI intercalated Co-Ti layered double oxides. *Adv. Funct. Mater.* **2016**, *26*, 1365–1374.

87. Lu, J.; Lei, Y.; Lau, K. C.; Luo, X.; Du, P.; Wen, J.; Assary, R. S. et al. A nanostructured cathode architecture for low charge overpotential in lithium-oxygen batteries. *Nat. Commun.* **2013**, *4*, 2383.

88. Yin, W.; Shen, Y.; Zou, F.; Hu, X.; Chi, B.; Huang, Y. Metal-organic framework derived ZnO/ZnFe$_2$O$_4$/C nanocages as stable cathode material for reversible lithium-oxygen batteries. *ACS Appl. Mater. Interfaces* **2015**, *7*, 4947–4954.

89. Lee, D. U.; Kim, B. J.; Chen, Z. One-pot synthesis of a mesoporous NiCo$_2$O$_4$ nanoplatelet and graphene hybrid and its oxygen reduction and evolution activities as an efficient bi-functional electrocatalyst. *J. Mater. Chem. A* **2013**, *1*, 4754–4762.

90. Wang, L.; Zhao, X.; Lu, Y.; Xu, M.; Zhang, D.; Ruoff, R. S.; Stevenson, K. J.; Goodenough, J. B. CoMn$_2$O$_4$ spinel nanoparticles grown on graphene as bifunctional catalyst for lithium-air batteries. *J. Electrochem. Soc.* **2011**, *158*, A1379–A1382.

91. Ge, X.; Liu, Y.; Thomas Goh, F. W.; Andy Hor, T. S.; Zong, Y.; Xiao, P.; Zhang, Z. et al. Dual-phase spinel MnCo$_2$O$_4$ and spinel MnCo$_2$O$_4$/Nanocarbon hybrids for electrocatalytic oxygen reduction and evolution. *ACS Appl. Mater. Interfaces* **2014**, *6*, 12684–12691.

92. Bian, W.; Yang, Z.; Strasser, P., Yang, R. A CoFe$_2$O$_4$/graphene nanohybrid as an efficient bi-functional electrocatalyst for oxygen reduction and oxygen evolution. *J. Power Sources* **2014**, *250*, 196–203.
93. Cao, X.; Yan, W.; Jin, C.; Tian, J.; Ke, K.; Yang, R. Surface modification of MnCo$_2$O$_4$ with conducting polypyrrole as a highly active bifunctional electrocatalyst for oxygen reduction and oxygen evolution reaction. *Electrochim. Acta.* **2015**, *180*, 788–794.
94. Zhang, X.; Wang, Y.; Dong, S.; Li, M. Dual-site polydopamine spheres/CoFe layered double hydroxides for electrocatalytic oxygen reduction reaction. *Electrochim. Acta.* **2015**, *170*, 248–255.

6 Performance Comparison and Optimization of Bifunctional Electrocatalysts for Rechargeable Metal-Air Batteries

6.1 INTRODUCTION

Regardless of the type of rechargeable metal-air batteries (RMABs), the air-electrode (or cathode) must be continuously supplied by active materials (oxygen from air or pure oxygen). This feature leads to higher theoretical energy densities of MABs than other metal-based batteries if the mass of oxygen is not accounted in the calculation of energy density. Air-electrodes have also been found to play a critical role in the performance of RMABs. The air-electrode, typically composited of a gas diffusion layer and an active catalytic layer, has several essential functions. One of the key features is the porous structure of the air-electrode composed of inter- and intra-connected carbon fibers, promoting the diffusion of gases and the deposition of discharge products. This structure provides good electrical conductivity that facilitates the transport of electrons and high catalytic activities and stability towards both the oxygen reduction reaction (ORR) when the battery is in discharge and the oxygen evolution reaction (OER) when the battery is in charge.[1–3]

The core component of an air-electrode is the bifunctional catalyst. Besides the catalyst's contribution to improve the mass transport of reactants and products due to its porous structure, it can also speed up both the ORR and OER kinetics. Inevitably, the physicochemical properties of such a bifunctional catalyst exert a significant effect on the efficiency of an air-electrode.

6.2 REQUIREMENTS FOR BIFUNCTIONAL ELECTROCATALYSTS IN RMAB APPLICATION

To push RMAB technology towards large-scale commercial deployment, several important requirements are set for bifunctional electrocatalysts as follows:

1. *High catalytic activities for both ORR and OER*: With bifunctional activities towards both ORR and OER, electrocatalyst is the key in determining the power density, energy density, and energy efficiency of RMABs. Currently, a large portion of voltage losses during charge and discharge of RMABs resulted from the sluggish kinetics of ORR and OER, especially from the OER process when charging. These two electrochemical reactions often proceed via multi-step mechanisms through the sequential breaking and forming of strong chemical bonds such as O-H bonds as well as the adsorption of electro-generated species on the catalyst surface. For an efficient electrode process, the electrocatalyst should be able to reduce overpotentials and, thus, the activation barriers of both processes. The respective mechanisms are strongly related to the type of catalyst materials. In general, noble metal-based electrocatalysts, such as platinum and its alloys, demonstrate highly effective ORR performance but relatively low OER performance. Although they are effective, their high-cost and scarcity limit their large-scale usage. Alternatively, transition metal oxides such as spinel and perovskite oxides supported on carbon materials have shown promising bifunctional catalytic activities in RMABs in alkaline electrolytes.

2. *Suitable porous structures, high surface areas, and small particle sizes with uniform distribution*: A highly porous bifunctional catalyst layer consisting of connected channels can facilitate gas diffusion (into the active layer), thus improving the ORR/OER processes and the transportation of electrolytes to the catalytic active centers. Along with high surface areas, the size of the pores is also important in generating void volumes to accommodate discharge products. The morphology of a catalyst has incredible influences on cycling performances and battery stability. Moreover, high surface areas can efficiently alleviate the blocking of gas diffusion channels due to the aggregation of discharge products. Furthermore, the high surface area of the catalyst layer can facilitate the dispersion of catalytically active sites and their exposure towards electrolytes and reactants. Regarding the effects of catalyst particle sizes, it is already proven that small particles with uniform distribution play a vital role in both catalytic processes and in the fabrication of air-electrodes.[4,5]

3. *Long-term stability*: To maintain long-term operation of RMABs, it is required that all cell components are chemically stable in both non-aqueous and/or aqueous solutions (alkaline or acidic). The stability of RMABs is required and correlates with (i) the stability of the electrolyte and electrode and (ii) the thermodynamic behavior of the discharge products (e.g., the lithium oxide or lithium peroxide in RLABs).[6,7] The electrochemical stability is mainly controlled by both ORR and OER, which determine the battery performance during charge and discharge, reflected by the

charge-discharge cycle life and coulombic efficiency.[7] Therefore, not only does the chemical and thermodynamic stability affect the efficiency of RMABs, electrochemical stability is also vital in determining the overall battery performance and operation.

4. *Low material cost, easy fabrication, and safe operation*: The air-electrode, in particular its bifunctional catalyst, is a major component that contributes to the overall cost and performance of RMABs.[1,8] Finding cost-effective and high-performance air-electrode catalysts is critical in achieving success of large-scale commercialization. In this respect, the most promising substitutions of noble metal catalysts are nonnoble metal and metal oxides. Ease of fabrication and scale-up of air-electrode catalysts are also important aspects to consider for large-scale commercialization, especially for the mass production required in automotive applications.[9,10]

Generally, bifunctional electrocatalysts can be selected for air-electrode in RMABs based on the interplay between the electrocatalytic activity, physicochemical and electrochemical stability, corrosion resistance, material and fabrication cost, and long-term safe operation. These properties can be tailored by optimizing the composition and structure of the catalyst. To some extent, the proposed compositions offer both desired electronic properties and surface properties, hence acceptable catalytic activity, while the structure or morphology gives the surface of the catalyst the distribution and density of active sites for both ORR and OER.[11]

6.3   COMPARISON OF BATTERY PERFORMANCE AND BIFUNCTIONALITY

The physicochemical structure and property of bifunctional catalysts play important roles in overall battery performance. This performance is assessed by the evaluation of several electrochemical parameters, including maximum discharge capacity, number of charge-discharge cycles (i.e., cycle number), overpotentials for initial ORR and OER, and/or energy efficiency. To further evaluate the influence of a catalyst on RMABs, air-electrode with noble metal, non-noble metal, or carbon-based bifunctional catalyst is listed and compared in Tables 6.1, 6.2, and 6.3.[12–151] In general, compared to pure single-component catalysts (e.g., Pt-C), composites can demonstrate better electrochemical performances. This draws attention to the fact that compositing varied materials to obtain composite catalyst-based air-electrodes are a suitable strategy in improving the battery performance of RMABs.

Noble metal-based air-electrode catalysts, especially Pt-based bifunctional catalysts, have been tested and used in the pioneering research of various metal-air batteries due to their high catalytic activities. Since then, Pt-C has become the baseline sample in the validation of new alternative electrocatalysts.[152] As demonstrated in Table 6.1, by pairing Pt with other components (e.g., carbon, other metal, and/or oxide), Pt can give the improved RMAB performances. This battery improvement normally results from synergistic effects, leading to increased bifunctional activities for both ORR and OER. As revealed by Lim et al.,[101] air-electrodes made from Pt/CNT shows a stable charge-discharge lifetime for over 130 cycles with 1000 mA h g^{-1}

TABLE 6.1

Electrochemical Performance for Typical Noble Metal-Based Bifunctional Catalysts as Air-Electrode Materials in MABs

No.	Bifunctional Catalyst as Cathode Materials	Maximum Capacity/ mAh g⁻¹ (Current Density/ᵃmA g⁻¹ or ᵇmA cm⁻²)	Cycle Number (Current Density/ᵃmA g⁻¹ or mA cm⁻², Upper-limit Capacity/mAh g⁻¹)	Potential Range for Cycle Testing (V)	Potential Difference for ORR/OER (V) (Initial ORR/OER Overpotentialsᶜ/V vs. Li/Li⁺)	Coulombic Efficiency (%, After Cycle Number)	Electrolyte Type	MAB Type	Year, References
1	Pt/CaMnO₃	4300 (50ᵃ)	30 (100ᵃ, 2000)	2.0–4.6	–	–	1 M LiTFSI/TEGDME	LOB	2014, 17
2	Au/α-MnO₂/KB	5759 (100ᵃ)	60 (100ᵃ, 1000)	2.2–4.4	0.79 (0.23/1.02)	–	1 M LiCF₃SO₃ in TEGDME	LOB	2015, 153
3	10 wt% Pt/GNS	1200 (70ᵇ)	20 (100ᵃ, 720)	2.0–4.8	–	–	1 M LiPF₆ in EC:DMCᵈ (1:1v/v)	LAB	2013, 36
4	20 wt% Pt₅₁Au₄₉/C	1329 (0.12ᵇ)	1 (0.12ᵇ, 240)	2.0–4.5	–	–	1 M LiPF₆ in EC:DMCᵈ (1:1v/v)	LOB	2012, 37
5	20 wt% PtCo₂/C	3040 (100ᵃ)	5 (100ᵃ, –)	2.0–4.6	1.21 (–/–)	–	1 M LiClO₄ in PC:DMEᵉ (1:2v/v)	LAB	2013, 38
6	44 wt% Pt₅₀Ru₅₀/C	700 (0.2ᵇ)	40 (0.2ᵇ, 700)	2.3–4.5	–	–	1 M LiPF₆ in TEGDME	LAB	2013, 39
7	36.5 wt% Ir@DHG	–	150 (2000ᵃ, 1000)	1.5–4.5	–	–	0.1 M LiClO4 in TEGDME:DMSO (1:2v/v)	RLOB	2015, 40
8	5 wt% Ru/CBC	–	25 (200ᵃ, 500)	2.0–4.5	–	–	LiCF₃SO₃:TEGDME (4:1 mol/mol)	LOB	2015, 41

(Continued)

TABLE 6.1 (Continued)
Electrochemical Performance for Typical Noble Metal-Based Bifunctional Catalysts as Air-Electrode Materials in MABs

No.	Bifunctional Catalyst as Cathode Materials	Maximum Capacity/ mAh g⁻¹ (Current Density/ᵃmA g⁻¹ or ᵇmA cm⁻²)	Cycle Number (Current Density/ᵃmA g⁻¹ or mA cm⁻², Upper-limit Capacity/mAh g⁻¹)	Potential Range for Cycle Testing (V)	Potential Difference for ORR/OER (V) (Initial ORR/OER Overpotentialsᶜ/V vs. Li/Li⁺)	Coulombic Efficiency (%, After Cycle Number)	Electrolyte Type	MAB Type	Year, References
9	40.12 wt% Pt/C	616 (70ᵃ)	10 (70ᵃ, –)	2.0–4.3	–	–	1 M LiPF₆ in PC	RLAB	2011, 42
	40.03 wt% Pd/C	855 (70ᵃ)	10 (70ᵃ, –)		–	–			
	40.15 wt% Ru/C	577 (70ᵃ)	10 (70ᵃ, –)		–	–			
10	4.78 wt% Pd2.5A/N-CNFs	3647 (20ᵃ)	88 (20ᵃ, 200)	2.25–4.4	–	–	5 wt% LiClO₄ in TEGDME	LOB	2016, 77
	7.06 wt% Pd2.5/N-CNFs,	7828 (20ᵃ)	73 (20ᵃ, 200)			–			
	10.65 wt% Pd5/N-CNFs	4941 (20ᵃ)	61 (20ᵃ, 200)			–			
11	Ru/Fe-Co-N-RGO	23905 (200ᵃ)	300 (200ᵃ, 600)	2.2–4.2	–	–	1 M LiTFSI/TEGDME	LOB	2015, 79
12	0.3 wt% Ru/Ni form	3720 (200ᵃ)	100 (200ᵃ, 1000)	2.75–3.75	–	–	0.5 M LiClO₄ in TEGDME	LOB	2015, 89
13	Pd/Ni/Super P	1441 (0.1ᵇ)	50 (0.1ᵇ, 1000)	2.0–4.5	–	–	0.65 M LiTFSI in TEGDME	LAB	2013, 90
14	Ag₅₀Cu₅₀/Ni	896.40 (20ᵇ)	200 (20ᵇ, –)	1.0–2.2	–	–	6 M KOH added with 0.1 M Zn(CH₃COO)₂	ZAB	2015, 91
15	AgCu/Ni foam	572 (20ᵇ)	100 (20ᵇ, –)	1.0–2.2	–	–	6 M KOH	ZAB	2015, 92

(Continued)

TABLE 6.1 (Continued)
Electrochemical Performance for Typical Noble Metal-Based Bifunctional Catalysts as Air-Electrode Materials in MABs

No.	Bifunctional Catalyst as Cathode Materials	Maximum Capacity/mAh g⁻¹ (Current Density/ᵃmA g⁻¹ or ᵇmA cm⁻²)	Cycle Number (Current Density/mA cm⁻², Upper-limit Capacity/mAh g⁻¹)	Potential Range for Cycle Testing (V)	Potential Difference for ORR/OER (V) (Initial ORR/OER Overpotentialsᶜ/V vs. Li/Li⁺)	Coulombic Efficiency (%, After Cycle Number)	Electrolyte Type	MAB Type	Year, References
16	Dealloyed Au₁₅Ag₈₅	2400 (500 mA g⁻¹)	140 (2000 mA g⁻¹, 1500 mAh g⁻¹)	2.0–4.5	–	–	0.1 M LiClO₄ in DMSO	LOB	2016, 93
17	Pd/Co₃O₄/Ni foam	1842.7 (0.05ᵇ)	70 (0.1ᵇ, 300)	2.0–4.2			1 M LiN(CF₃SO₂)₂ in TEGDME	LOB	2015, 99
18	66 wt% Pt/CNT	–/–	130 (2000ᵃ, 1000)	2.0–4.7	–	–	1 M LiPF₆ in TEGDME	LOB	2013, 101
19	Pt/CNTs/Ni-foam	4050 (20ᵃ)	80 (400ᵃ, 1500)	2.0–4.2	1.1 (–/–)		1 M LiTFSI in TEGDME	LOB	2014, 102
20	2 wt% Pt/Co₃O₄/C	>3000 (–)	40 (200ᵃ, –)	2.0–4.2			1 M LiTFSI in TEGDME	LOB	2014, 103
21	5 wt% Pt/IrO₂-CNTs	306 (0.2ᵇ)	20 (0.2ᵇ, –)	3.68–4.45			0.01 M H₂SO₄	LAB	2013, 104
22	Pt₅₀Ir₅₀/C-TiO₂	4375 (–)	35 (–, 500)	2.0–4.3	1.0 (–/–)		1 M LiTFSI in TEGDME	LOB	2016, 105
	Pt/C-TiO₂	3856 (–)	33 (–, 500)	2.0–4.3	1.12 (–/–)				
	Ir/C-TiO₂	3600 (–)	30 (–, 500)	2.0–4.3	0.99 (–/–)				
23	Pd(111)	–	50 (200ᵃ, 1000)	2.0–4.2	2.22 (0.56/1.66)		1 M LiCF₃SO₃ in TEGDME	LOB	2014, 106
	PdCu(111)	–			1.45 (0.37/1.08)				
	PdCu(110)	–			2.62 (0.66/1.96)				
	PdCu(100)	–			1.72 (0.43/1.29)				
	Cu(111)	–			3.49 (0.88/2.61)				

(Continued)

TABLE 6.1 (Continued)
Electrochemical Performance for Typical Noble Metal-Based Bifunctional Catalysts as Air-Electrode Materials in MABs

No.	Bifunctional Catalyst as Cathode Materials	Maximum Capacity/ mAh g^{-1} (Current Density/^{a}mA g^{-1} or ^{b}mA cm^{-2})	Cycle Number (Current Density/^{a}mA g^{-1} or mA cm^{-2}, Upper-limit Capacity/mAh g^{-1})	Potential Range for Cycle Testing (V)	Potential Difference for ORR/OER (V) (Initial ORR/OER Overpotentials^c/V vs. Li/Li$^+$)	Coulombic Efficiency (%, After Cycle Number)	Electrolyte Type	MAB Type	Year, References
24	5–10 wt% Ir/RGO	9529 (0.5^b)	30 (0.5^b, –)	1.5–4.5	–	–	1 M LiPF$_6$ in DMSO	LOB	2015, 107
25	Ag/β-MnO$_2$	873 (0.02^b)	10 (0.5^b, –)	2.0–4.0	0.8 (–/–)	–	1 M LiTFSI/TEGDME	LOB	2015, 108
26	1.0 wt% Ag/C/MnO$_2$	1400 (0.34^b)	5 (0.34^b, 1100)	2.0–4.5	–	–	1 M LiClO$_4$ in PC	LAB	2016, 109
27	Ag@CeO$_2$/KB	3415 (100^a)	50 (200^a, 500)	2.0–4.7	–	–	1 M LiCF$_3$SO$_3$ in TEGDME	LAB	2015, 110
28	2.43 wt% Pd/α-MnO$_2$/VC	8526 (0.1^b)	35 (0.1^b, 500)	2.0–4.3	1.00 (–/–)	–	1 M LiTFSI/TEGDME	LOB	2015, 111
29	Ag/MnO$_2$/graphite	–	270 (,–)	1.0–3.0	–	–	6 M KOH	ZAB	2013, 112
30	Au-Pd/MnO$_x$/Super P	3200 (500^a)	5 (200^a, –)	2.0–4.5	–	–	1 M LiCF$_3$SO$_3$ in TEGDME	LOB	2016, 113

^{a} mA g^{-1}.
^{b} mA cm^{-2}.
^{c} Initial overpotentials are obtained from the cyclic tests.
^{d} Dimethyl carbonate.
^{e} 1,2-dimethoxyethane.

TABLE 6.2

Electrochemical Performance for Typical Non-noble Metal-Based Bifunctional Catalysts as Air-Electrode Materials in MABs

No.	Bifunctional Catalyst as Cathode Materials	Maximum Capacity/mAh g⁻¹ (Current Density/ᵃmA g⁻¹ or ᵇmA cm⁻²)	Cycle Number (Current Density/ mA g⁻¹ or mA cm⁻², Upper-Limit Capacity/mAh g⁻¹)	Potential Ange for Cycle Testing (V)	Potential Difference for ORR/OER (V) (Initial ORR/OER Overpotentials^c/V vs. Li/Li⁺)	Energy Efficiency (%)	Electrolyte Type	MAB Type	Year, References
1	Ta/CNTs	4300 (200ᵃ)	65 (200ᵃ, 1000)	2.0–4.5	–	–	1 M LiTFSI/TEGDME	RLOB	2016, 18
2	N-CNTs@Ni	1814 (0.05ᵇ)	8 (0.05ᵇ, 500)	2.0–4.3	–	–	1 M LiPF$_6$ in DMSO	LOB	2013, 154
3	CoCu/graphene	14821 (200ᵃ)	122 (200ᵃ, 1000)	2.5–4.5	–	–	1 M LiTFSI/TEGDME	RLOB	2015, 43
4	FeCo/CNTs	3600 (250ᵃ)	50 (100ᵃ, 1000)	2.4–4.5	–	–	1 M LiTFSI/TEGDME	RLOB	2016, 155
5	CoNi/CNFs	8635 (200ᵃ)	60 (200ᵃ, 1000)	2.0–4.5	–(0.22/0.70)	–	0.5 M LiTFSI/TEGDME	RLOB	2016, 44
6	Fe-Fe$_3$C/CNFs	6250 (200ᵃ)	41 (300ᵃ, 600)	2.0–4.3	(–0.20/1.12)	–	1 M LiTFSI/TEGDME	RLOB	2014, 64
7	FeNi$_3$@GR@ Fe-NiOOH	–	100 (1ᵇ, –)	0.9–2.1		–	6 M KOH	RZAB	2016, 65
8	Co/C/NiFe LDH/AB	–	300 (40ᵇ, –)	1.05–2.05		51.2	6 M KOH	RZAB	2015, 66
9	Co-N/C	–	500 (2ᵇ, –)	0.4–1.4		–	6 M KOH	ZAB	2016, 78
10	Fe-Co-N-rGO	21825 (200ᵃ)	78 (200ᵃ, 600)	2.2–4.2		54.5	1 M LiTFSI/TEGDME	LOB	2015, 79
11	Fe-N/C	731 (100ᵇ)	–	–		–	6 M KOH	ZAB	2016, 80
12	Fe@N-C	–	100 (10ᵇ, –)	1.0–2.0		–	6 M KOH	ZAB	2015, 81
13	Co-N/MWCNTs	–	100 (0.5ᵃ, –)	1.0–2.2		–	6 M KOH	ZAB	2016, 82
14	Co-N/CNTs	–	350 (5ᵇ, –)	0.8–2.4		–	6 M KOH	ZAB	2016, 83
15	Fe-N/C	12441 (100ᵃ)	60 (200ᵃ, 800)	2.0–4.25		71.2	1 M LiCF$_3$SO$_3$ in TEGDME	LOB	2016, 84
16	Fe/Fe$_3$C/N-G	7150 (0.1ᵇ)	30 (0.1ᵇ, 800)	2.0–4.4		–	1 M LiTFSI/TEGDME	LOB	2016, 86
17	CoO@Co/N-C	–	55 (100ᵃ, 500)	2.0–4.2		67	1 M LiClO$_4$/DMSO	RLOB	2016, 87
18	Co@CoO/Co-N-C	–	100 (10ᵇ, –)	0.8–2.2		–	6 M KOH	RZAB	2016, 88

(Continued)

TABLE 6.2 (Continued)
Electrochemical Performance for Typical Non-noble Metal-Based Bifunctional Catalysts as Air-Electrode Materials in MABs

No.	Bifunctional Catalyst as Cathode Materials	Maximum Capacity/mAh g^{-1} (Current Density/^{a}mA g^{-1} or ^{b}mA cm^{-2})	Cycle Number (Current Density/mA g^{-1} or mA cm^{-2}, Upper-Limit Capacity/mAh g^{-1})	Potential Ange for Cycle Testing (V)	Potential Difference for ORR/OER (V) (Initial ORR/OER Overpotentials^c/V vs. Li/Li$^-$)	Energy Efficiency (%)	Electrolyte Type	MAB Type	Year, References
19	Co$_3$O$_4$/Ni form	11882 (100^a)	80 (200^a, 500)	2.0–4.5	–	–	1 M LiTFSI/TEGDME	RLOB	2015, 94
20	Co$_3$O$_4$/Ni mesh	8500 (400^a)	120 (400^a, 800)	2.0–4.35	–	–	1 M LiTFSI/TEGDME	LOB	2015, 95
21	MnCo$_2$O$_4$/Ni foam	10520 (100^a)	119 (500^a, 1000)	2.0–4.2	–	–	1 M LiTFSI/TEGDME	LOB	2015, 96
22	(Co,Mn)$_3$O$_4$/Ni foam	3605 (0.05^b)	50 (0.05^b, 500)	2.0–4.2	–	–	1 M LiTFSI/TEGDME	LOB	2014, 97
23	CoO/Ni form	4888 (20^a)	50 (100^a, 500)	2.0–4.3	–	–	0.1 M LiTFSI/TEGDME	LOB	2015, 98
24	TFSI/CoTi LDO/Ni form	1133 (100^a)	80 (100^a, 500)	2.25–4.2	–	–	1 M LiTFSI/Tetraglyme	LOB	2016, 100
25	Co/Ni form	11042 (100^a)	160 (100^a, 1000)	2.2–4.2	–	–	0.5 M LiClO$_4$/DMSO	LOB	2016, 114
26	Co-N-rGO/IrO$_2$	11,731 (200^a)	200 (200^a, 600)	2.0–4.5	–	65	1 M LiTFSI/TEGDME	LOB	2016, 115
27	Co-N/C	5000 (300^a)	80 (200^a, 600)	2.0–4.3	–	–	1 M LiTFSI/TEGDME	LOB	2014, 116
28	Sr$_2$CrMoO$_{6-x}$/Ni form/SP	2360 (75^a)	30 (75^a, 600)	2.0–4.5	–	–	0.1 M LiTFSI/TEGDME	RLAB	2014, 117
29	NCNT/ CoO-NiO-NiCo	594 (7^b)	100 (20^b, –)	0.8–2.4	–	–	6 M KOH	RZAB	2015, 118
30	IrO$_2$@Ti	–	20 (0.1^b, –)	1.5–3.0	0.98 (–/–)	62.3	0.5 M LiOH	ZAB	2016, 119
31	Co$_3$O$_4$/Ni/C	14830 (400^a)	48 (100^a, 2000)	2.0–4.3	–	75.1	0.1 M LiClO$_4$/DME	LOB	2014, 120
32	RuO$_x$/TiN/Ti mesh	–	300 (50^a, 500)	2.0–4.2	–	–	1 M LiTFSI/TEGDME	RLOB	2015, 121

^{a} mA g^{-1}.
^{b} mA cm^{-2}.
^{c} Initial overpotentials are obtained from the cyclic tests.

TABLE 6.3

Electrochemical Performance for Typical Carbon-Based Metal-Free Bifunctional Catalysts as Air-Electrode Materials in MABs

No.	Bifunctional Catalyst as Cathode Materials	Maximum Capacity/mAh g^{-1} (Current Density/^{a}mA g^{-1} or ^{b}mA cm^{-2})	Cycle Number (Current Density/ mA g^{-1} or mA cm^{-2}, Upper-Limit Capacity/mAh g^{-1})	Potential Range for Cycle Testing (V)	Potential Difference for ORR/OER (V) (Initial ORR/OER Overpotentials^c/V vs. Li/Li$^+$)	Energy Efficiency (%)	Electrolyte Type	MAB Type	Year, References
1	N-CNT/Co$_3$O$_4$ VC/Co$_3$O$_4$ N-CNT	—	240 (20^b, —) 22, (20^b, —) 40 (20^b, —)	1.0–2.2 0.5–3.0 0.75–2.75	0.802 (−/−) — —	—	6 M KOH + 0.2 M Zinc acetate	RZAB	2015, 12
2	Porous carbon	1767 (100^a)	25 (250^a, 500)	2.3–4.0	—	—	1 M LiTFSI/TEGDME	RLOB	2015, 13
3	C/NiCo$_2$O$_4$	580 (20^b)	50 (20^b, —)	1.0–2.0	0.7 (−/−)	—	6 M KOH	RZAB	2014, 14
4	N-G/Carbon cloth	873 (20^a)	150 (2^b, —)	1.0–2.5	—	—	6 M KOH + 0.2 M ZnCl$_2$	ZAB	2016, 15
5	KB/RM-TIT^d KB/RM-FIT^e KB	3250 (100^a) 2700 (100^a) 5950 (100^a)	121 (400^a, 1000) 105 (400^a, 1000) 24 (400^a, 1000)	2.0–4.8	—	—	1 M LITFSI/TEGDME	RLOB	2016, 16
6	C microspheres/ AB	12081 (200^a)	60 (200^a, 1000)	2.0–4.5	—	—	1 M LiTFSI/TEGDME	RLOB	2015, 19
7	CNT@RuO$_2$	4350 (100^a)	100 (500^a, 300)	2.3–4.0	—	—	LITFSI/tri(ethylene)glycol dimethyl ether (1:5 mol/mol)	RLOB	2014, 20
8	C sheets	—	160 (2^b, —)	0.6–2.2	—	—	0.6 M KOH + 0.2 M ZnCl$_2$	RZAB	2015, 21
9	O,N-CNW N-CNW	—	30 (0.5^b, —) 30 (0.5^b, —)	2.0–3.2	0.92 (−/−) 1.0 (−/−)	—	0.5 M LiOH + 1 M LiNO$_3$	HLAB	2014, 23
10	N-C	5749 (200a)	66 (200^a, 500)	1.0–4.5	—	—	0.5 NaCF$_3$SO$_3$/TEGDME	Na-O$_2$ batteries	2016, 24
11	N-C	1750 (100^a)	15 (250^a, 1000)	2.0–4.0	—	—	0.5 M LiTFSI/TEGDME	LOB	2015, 25

(Continued)

Performance Comparison and Optimization of Bifunctional Electrocatalysts 193

TABLE 6.3 (Continued)
Electrochemical Performance for Typical Carbon-Based Metal-Free Bifunctional Catalysts as Air-Electrode Materials in MABs

No.	Bifunctional Catalyst as Cathode Materials	Maximum Capacity/mAh g⁻¹ (Current Density/ᵃmA g⁻¹ or ᵇmA cm⁻²)	Cycle Number (Current Density/ mA g⁻¹ or mA cm⁻², Upper-Limit Capacity/mAh g⁻¹)	Potential Range for Cycle Testing (V)	Potential Difference for ORR/OER (V) (Initial ORR/OER Overpotentialsᶜ/V vs. Li/Li⁺)	Energy Efficiency (%)	Electrolyte Type	MAB Type	Year, References
12	N,S-C	4222 (200ᵃ)	100 (200ᵃ, 600)	2.2–4.4	—	—	1 M LiTFSI/TEGDME	LOB	2016, 27
	rGO	4702 (200ᵃ)	21 (200ᵃ, 600)						
	Super P	2199 (200ᵃ)	16 (200ᵃ, 600)						
	XC-72R	1743 (200ᵃ)	23 (200ᵃ, 600)						
13	N-G aerogels	10081 (200ᵃ)	72 (300ᵃ, 1000)	2.0–4.5	—	—	1 M LiTFSI/TEGDME	LOB	2015, 28
14	N-Graphene	7300 (50ᵃ)	22 (100ᵃ, 500)	2.0–4.5	—	—	1 M LiCF₃SO₃/TEGDME	RLOB	2016, 29
	Graphene	2250 (50ᵃ)	9 (100ᵃ, 500)						
15	N-MWCNTs	2927 (0.05ᵇ)	13 (0.1ᵇ, 1000)	2.0–4.2	—	—	1 M LiTFSI/TEGDME	LOB	2015, 30
	MWCNTs	2390 (0.05ᵇ)	13 (0.1ᵇ, 1000)						
16	N-CNFs	660 (5ᵇ)	500 (10b, –)	0.8–2.0	0.73 (–/–)	—	6.0 M KOH + 0.2 M zinc acetate	ZAB	2016, 31
17	N-CNFs/C paper	–	200 (250ᵃ, 500)	2.2–4.4	–	—	1 M LiTFSI/TEGDME	LOB	2014, 32
18	N-C	8040 (0.1ᵇ)	30 (0.1ᵇ, 480)	2.2–4.5	0.68	—	1 M LiClO₄/DMSO	LOB	2016, 33
19	B-C	10200 (0.15ᵇ)	125 (0.3ᵇ, –)	1.5–5.0	—	—	1 M NaSO₃CF₃/TEGDME	Na-OB	2016, 34
	C	7455 (0.15ᵇ)	56 (0.3ᵇ, –)						
	SP	160 (0.15ᵇ)	6 (0.3ᵇ, –)						
20	N,P-C	735 (2ᵇ)	180 (2ᵇ, –)	1.0–3.0	—	—	6 M KOH	ZAB	2015, 35
21	α-MnO₂@GNS	2413 (50ᵃ)	47 (100ᵃ, 1000)	2.2–4.2	—	—	1 M LiClO₄/DMSO	RLOB	2014, 45
22	MnOₓ/CP	–	500 (15ᵇ, –)	1.0–2.4	—	—	6 M KOH + 20 g L⁻¹ ZnCl₂	RZAB	2015, 46

(Continued)

TABLE 6.3 (Continued)
Electrochemical Performance for Typical Carbon-Based Metal-Free Bifunctional Catalysts as Air-Electrode Materials in MABs

No.	Bifunctional Catalyst as Cathode Materials	Maximum Capacity/mAh g⁻¹ (Current Density/ᵃmA g⁻¹ or ᵇmA cm⁻²)	Cycle Number (Current Density/mA g⁻¹ or mA cm⁻², Upper-Limit Capacity/mAh g⁻¹)	Potential Range for Cycle Testing (V)	Potential Difference for ORR/OER (V) (Initial ORR/OER Overpotentialsᶜ/V vs. Li/Li⁺)	Energy Efficiency (%)	Electrolyte Type	MAB Type	Year, References
23	MnO$_2$/GNS	11235 (75^a)	30 (215^a, 1500)	2.3–4.2	1.1 (–/–)	73	1 M LiTFSI/DME	RLAB	2012, 47
24	CoO/SP	5637 (200^a)	50 (200^a, 1000)	2.0–4.5	–	–	1 M LiTFSI/TEGDME	RLOB	2016, 48
25	CoO/CNF	3882.5 (0.2^b)	50 (0.2^b, 1000)	2.0–4.2	–	–	1 M LiTFSI/TEGDME	RLOB	2014, 49
26	RuO$_2$/CNT	1150 (0.4^b)	50 (0.4^b, 244)	2.0–4.4	–	65.4	1 M LiTFSI/TEGDME	LOB	2015, 50
27	G/Zr-CeO$_2$	3254 (0.2^b)	15 (1^b, 500)	2.0–4.5	–	–	1 M LiTFSI/TEGDME	LOB	2014, 52
28	TiN/C	7100 (100^a)	35 (200^a, 1000)	2.0–4.5	–	–	LiCF$_3$SO$_3$/TEGDME (1:4, mol/mol)	LOB	2013, 53
29	WC/C	7000 (100^a)	36 (100^a, 1000)	2.0–4.5	–	–	1 M LiTFSI/TEGDME	LOB	2015, 54
30	B$_4$C/CNT	16000 (0.2^b)	120 (0.4^b, 1000)	2.5–4.4	–	–	LiCF$_3$SO$_3$/TEGDME (1:4, mol/mol)	LOB	2015, 55
31	CoS$_2$/rGO	2200 (200^a)	20 (200^a, 500)	2.0–4.5	–	–	1 M LiClO$_4$/DMSO (1:4, mol/mol)	LOB	2016, 56
32	β-FeOOH/C aerogels	10230 (0.1^b)	60 (0.1^b, 800),	2.0–4.4	–	–	1 M LiTFSI/TEGDME	RLOB	2015, 58
	β-FeOOH/Super P	6050 (0.1^b)	42 (0.1^b, 800)						
33	FeOOH/MWCNT	6000 (200^a)	20 (200^a, 550)	2.0–4.3	–	–	1 M LiTFSI/TEGDME	RLOB	2015, 59
34	VC/soil	7640 (0.2^b)	100 (0.2^b, 1000)	2.3–4.2	–	–	1 M LiTFSI/TEGDME	RLOB	2015, 60
35	Polyimide/CNT	11000 (500^a)	137 (500^a, 1500)	2.0–4.35	–	–	1 M LiTFSI/TEGDME	LAB	2015, 61
36	FePc/GNS	865.6 (0.5^b)	30 (0.5^b, –)	2.0–4.8	0.67 (–/–)	–	1 M LiClO$_4$/ED/DEC + 1 M LiNO3/0.5 M LiOH	LAB	2013, 62
	FePc/CNTs	632.4 (0.5^b)	30 (0.5^b, –)		0.85 (–/–)				
	FePc/AB	795.4 (0.5^b)	30 (0.5^b, –)		1.05 (–/–)				

(Continued)

TABLE 6.3 (Continued)

Electrochemical Performance for Typical Carbon-Based Metal-Free Bifunctional Catalysts as Air-Electrode Materials in MABs

No.	Bifunctional Catalyst as Cathode Materials	Maximum Capacity/mAh g⁻¹ (Current Density/ᵃmA g⁻¹ or ᵇmA cm⁻²)	Cycle Number (Current Density/mA g⁻¹ or mA cm⁻², Upper-Limit Capacity/mAh g⁻¹)	Potential Range for Cycle Testing (V)	Potential Difference for ORR/OER (V) (Initial ORR/OER Overpotentialsᶜ/V vs. Li/Li⁺)	Energy Efficiency (%)	Electrolyte Type	MAB Type	Year, References
37	MnO_2-$LaNiO_3$/CNT	—	75 (20ᵇ, —)	0.8–2.4	—	—	6 M KOH + 0.4 M ZnO	RZAB	2014, 156
38	N-CNT/CNF	710 (0.5ᵇ)	21 (0.5ᵇ, —)	2.5–4.5	1.35 (—/—)	—	1 M $LiPF_6$ in EC/Ethyl methyl carbonate (3:1 v/v) + 0.5 M $LiNO_3$/0.5 M LiOH	LAB	2013, 67
39	N,S-G/C	11431 (100ᵃ)	100 (50ᵃ, 500) 38 (100ᵃ, 1000)	2.5–4.3	—	—	1 M LiTFSI/TEGDME20	LOB	2015, 68
40	$MnCo_2O_4$@rGO KB	11092.1 (200ᵃ) 6014.6 (200ᵃ)	35 (200ᵃ, 1000) 13 (200ᵃ, 1000)	2.3–4.3	1.36 (—/—), —	—	1 M LiPF6/TEGDME	LOB	2015, 69
41	N-CNT/Co-$LaMO_3$	—	60 (18ᵇ, —)	1.0–2.2	—	—	6 M KOH	RZAB	2015, 70
42	N-C/$LaTi_{0.65}Fe_{0.35}O_{3-\delta}$	450 (5ᵃ)	60 (5ᵃ, —)	1.0–2.0	—	—	6 M KOH	RZAB	2015, 71
43	N-CNT/$La_{0.5}Sr_{0.5}Co_{0.8}O_3$	—	100 (0.5ᵇ, —)	2.5–4.0	1.56 (—/—)	—	1 M LiPF6 in EC/DMC (1:1 v/v) + 1 M $LiNO_3$/0.5 M LiOH	RLAB	2015, 72
44	CNT/$CoFe_2O_4$	3670 (200ᵃ)	35 (200ᵃ, 430)	2.0–4.3	—	—	1 M LiTFSI/TEGDME	RLOB	2014, 73
45	$CoMn_2O_4$/rGO $CoMn_2O_4$/N-rGO	610 (20ᵇ) 460 (20ᵇ)	200 (20ᵇ, —) 200 (20ᵇ, —)	0.8–2.4	0.7 (—/—), 0.95 (—/—)	—	6 M KOH	ZAB	2014, 74
46	P-C/$MnCo_2O_4$	13150 (200ᵃ)	200 (200ᵃ, 1000)	2.2–4.4	1.25 (—/—)	—	1 M LiTFSI/TEGDME	RLOB	2015, 75

(Continued)

TABLE 6.3 (Continued)

Electrochemical Performance for Typical Carbon-Based Metal-Free Bifunctional Catalysts as Air-Electrode Materials in MABs

No.	Bifunctional Catalyst as Cathode Materials	Maximum Capacity/mAh g⁻¹ (Current Density/ᵃmA g⁻¹ or ᵇmA cm⁻²)	Cycle Number (Current Density/ mA g⁻¹ or mA cm⁻², Upper-Limit Capacity/mAh g⁻¹)	Potential Range for Cycle Testing (V)	Potential Difference for ORR/OER (Initial ORR/OER Overpotentials^c/V vs. Li/Li⁺)	Energy Efficiency (%)	Electrolyte Type	MAB Type	Year, References
47	Co_3O_4/N-CNF	—	135 (1ᵇ, —) (2ᵇ, —)	1.2–2.4 1.2–2.1	0.7 (—/—) 0.7 (—/—)	61.4 65.4	6 M KOH + 0.2 M $ZnCl_2$	RZAB	2015, 76
48	N-C/MoN	1490 (0.04ᵇ)	9 (0.08ᵇ, 1100)	2.6–4.15	—	—	$LiPF_6$ in EC:DMC (1:1, v/v)	RLOB	2011, 85
49	KB	—	20 (0.001ᵇ, 1000)	2.0–5.0	—	—	0.9 M LiTFSI/TEGDME	RLOB	2015, 22
50	N-C		30 (2ᵇ, —)	1.0–2.2	0.85 (—/—)	—	6 M KOH + 2 wt% ZnO	ZAB	2015, 26
51	N-G P-G	—	40 (0.5ᵇ, —) 40 (0.5ᵇ, —)	1.8–3.4	0.88 (—/—) 1.15 (—/—)	—	1 M $LiNO_3$ + 0.5 M LiOH	HLAB	2014, 122
52	*Cubic*-CoMn₂O/C	6400 (200ᵃ)	60 (200ᵃ, 1000)	2.4–4.1	—	—	1 M LiTFSI/TEGDME	LAB	2015, 123
53	*Cubic*-CoMn₂O/C	500 (10ᵇ)	155 (10ᵇ, —)	1.0–3.0	—	—	0.1 M KOH	ZAB	2015, 123
54	$ZnCo_2O_4$/SWCNT	3667 (0.1ᵇ)	50 (0.1ᵇ, 800)	2.0–4.5	—	—	1 M lithium trifluoromethanesulfonate/tetraglyme	RLOB	2015, 124
55	Co_3O_4/GNFs Co_3O_4/rGO	10500 (200ᵃ) 5000 (200ᵃ)	80 (200ᵃ, 1000) 40 (200ᵃ, 1000)	2.35–4.35	—	—	1 M LiTFSI/TEGDME	RLOB	2013, 125
56	Co_3O_4/G	—	50 (160ᵃ, —)	2.5–4.5	—	—	1 M $LiPF_6$/EC:DEC (v/v = 1:1)	RLOB	2014, 126
57	Co_3O_4/rGO	4150 (0.1ᵇ)	39 (0.1ᵇ, 500)	2.0–4.3	1.0 (—/—)	—	LITFSI/TEGDME	LAB	2015, 127
58	MnO_2/CP	1000 (0.1ᵇ) 600	90 (0.1ᵇ, —) 100 (0.3ᵇ, —)	2.2–4.4	—	—	1 M $LiCF_3SO_3$/TEGDME	RLOB	2014, 128
59	$RuO_2·0.64H_2O$/rGO	5000 (500ᵃ)	35 (200ᵃ, 2000)	2.0–4.3	—	—	$LiCF_3SO_3$/TEGDME (1:4 mol/mol)	LOB	2013, 51

(Continued)

TABLE 6.3 (Continued)
Electrochemical Performance for Typical Carbon-Based Metal-Free Bifunctional Catalysts as Air-Electrode Materials in MABs

No.	Bifunctional Catalyst as Cathode Materials	Maximum Capacity/mAh g^{-1} (Current Density/^{a}mA g^{-1} or ^{b}mA cm^{-2})	Cycle Number (Current Density/ mA g^{-1} or mA cm^{-2}, Upper-Limit Capacity/mAh g^{-1})	Potential Range for Cycle Testing (V)	Potential Difference for ORR/OER (V) (Initial ORR/OER Overpotentials^c/V vs. Li/Li$^+$)	Energy Efficiency (%)	Electrolyte Type	MAB Type	Year, References
60	MnO$_2$/VC/CP MnO$_2$/VC/CP/ Fe$_{0.1}$Ni$_{0.9}$Co$_2$O$_4$	—	100 (10^b, –) 100 (10^b, –)	0.5–3.0	0.96 (–/–) 0.7 (–/–)	–	6 M KOH + 0.2 M Zn acetate	RZAB	2016, 129
61	N-CNT/CNT	1887 (25^a)	50 (640^a, –)	2.0–4.5	–	–	0.5 M LiCF$_3$SO$_3$/TEGDME	Na-OB	2015, 130
62	SP/LaNiO$_3$	7076 (50^a)	155 (100^a, 500)	2.0–4.6	–	–	1 M LITFSI/TEGDME	RLOB	2016, 131
63	SP/CaMnO$_3$	7000 (200^a)	80 (200^a, 1000)	2.0–4.0	–	–	1 M NaSO$_3$CF$_3$/TEGDME	NaOB	2015, 132
64	SP/La$_2$NiO$_4$	14310.9 (0.16^b)	25 (0.16^b, 1000)	2.0–4.6	–	–	1 M LITFSI/TEGDME	RLOB	2016, 133
65	SP/ La$_{0.75}$Sr$_{0.25}$MnO$_3$	—	124 (0.15^b, 1000)	2.2–4.4	–	–	1 M LITFSI/TEGDME	RLOB	2013, 134
66	AB/ La$_{0.8}$Sr$_{0.2}$MnO$_3$	6890 (200^a)	30 (200^a, 1000)	2.2–4.4	–	–	1 M LITFSI/TEGDME	LAB	2015, 135
67	AB/ La$_{0.6}$Ni$_{0.4}$CoO$_{3-\delta}$			2.0–4.3			1 M LITFSI/TEGDME	LAB	2015, 136
68	KB/ LaNi$_x$Co$_{1-x}$O$_{3-\delta}$	7720 (0.1^b)	49 (0.15^b, 1000)	2.3–4.25	–	–	1 M LITFSI/TEGDME	RLOB	2014, 137
69	KB/La$_{0.6}$Sr$_{0.4}$Co$_{0.2}$ Fe$_{0.8}$O$_3$	13979 (100^a)	20 (200^a, 500)	2.2–4.5	0.91 (–/–)	–	1 M LITFSI/DMSO	LAB	2016, 138
70	KB/ La$_x$(Ba$_{0.5}$Sr$_{0.5}$)$_{1-x}$ Co$_{0.8}$Fe$_{0.2}$O$_{3-\delta}$	—	100 (10.5^b, –)	1.0–2.0	0.75 (–/–)	–	6 M KOH	RZAB	2016, 139
71	KB/CuCo$_2$O$_4$	7962 (50^a)	50 (100^a, 1000)	2.2–4.2	1.2 (–/–)	–	1 M LITFSI/TEGDME	RLAB	2014, 140

(Continued)

TABLE 6.3 (Continued)
Electrochemical Performance for Typical Carbon-Based Metal-Free Bifunctional Catalysts as Air-Electrode Materials in MABs

No.	Bifunctional Catalyst as Cathode Materials	Maximum Capacity/mAh g⁻¹ (Current Density/ᵃmA g⁻¹ or ᵇmA cm⁻²)	Cycle Number (Current Density/ mA g⁻¹ or mA cm⁻², Upper-Limit Capacity/mAh g⁻¹)	Potential Range for Cycle Testing (V)	Potential Difference for ORR/OER (V; Initial ORR/OER Overpotentials^c/V vs. Li/Li⁺)	Energy Efficiency (%)	Electrolyte Type	MAB Type	Year, References
72	AB/NiCo$_2$O$_4$	5842 (100ᵃ)	110 (200ᵃ, 1000)	2.2–4.5	–	–	1 M LITFSI/TEGDME	RLOB	2015, 141
73	SP/NiCo$_2$O$_4$	8658 (100ᵃ)	130 (200ᵃ, 1000)	2.0–4.6	1.31 (–/–)	–	1 M LITFSI/TEGDME	RLOB	2015, 142
74	Carbon fabric/NiCo$_2$O$_4$	29280 (0.1ᵇ)	150 (0.1ᵇ, –)	2.0–4.5	–	–	1 M LITFSI/Tetraglyme	LOB	2016, 143
75	KB/Mn-Ru oxides	6500 (0.1ᵇ)	50 (0.1ᵇ, 1100)	2.0–4.4	–	–	1 M LICF$_3$SO$_3$/TEGDME	RLOB	2014, 144
76	KB/Co$_3$S$_4$	5917 (0.1ᵇ)	25 (0.1ᵇ, 500)	2.0–4.3	0.921 (–/–)	–	1 M LITFSI/TEGDME	RLOB	2015, 145
77	CB/Co-Mn oxide	–	30 (5ᵇ, –)	0.8–2.5	0.77 (–/–)	–	6 M KOH + 2% ZnO	RZAB	2016, 146
78	SP/Co$_3$(PO$_4$)$_2$	–	50 (0.05ᵇ, 500)	2.5–3.5	–	91	0.1 M NaOH	NaOB	2016, 147
79	SP/RuO$_2$@rGO	–	50 (200ᵃ, 1000)	2.0–4.2	–	–	PMS/TiO$_2$ GPE	Polymer LOB	2016, 148
80	KB/α-MnO$_2$/RuO$_2$	5520 (0.1ᵇ)	50 (0.1ᵇ, 500)	2.5–4.5	1.0 (–/–)	–	LITFSI/TEGDME (1:1)	RLOB	2015, 149
81	CB/MnO$_2$/Co$_3$O$_4$	4024 (100ᵃ)	150 (100ᵃ, 500)	2.0–4.5	–	–	1 M LITFSI/TEGDME	RLOB	2016, 150
82	VC/g-C$_3$N$_4$-LaNiO$_3$	5500 (50ᵃ)	65 (250ᵃ, 500)	2.6–4.7	–	–	1 M LICF$_3$SO$_3$/TEGDME	LOB	2016, 151

ᵃ mA g⁻¹.
ᵇ mA cm⁻².
ᶜ Initial overpotentials are obtained from the cyclic tests.
ᵈ RuO$_2$ tube in Mn$_2$O$_3$ tube.
ᵉ RuO$_2$ fiber in Mn$_2$O$_3$ tube.

at a high current rate of 2000 mA g^{-1} in a RLOB. In addition, composites such as Pt/CNTs/Ni,[102] Pt/CaMnO$_3$,[17] and Pt$_{50}$Ir$_{50}$/C-TiO$_2$,[105] can also deliver higher discharge capacities of 4000 mAh g^{-1}. Like Pt, other noble metals also show enhanced MAB performances. As reported by Zeng et al.,[79] a quarterly Ru/Fe-Co-N-rGO composite catalyst displays a very high discharge capacity of 23,905 mAh g^{-1} at 200 mA g^{-1} when used as the air-electrode in a RLOB. Significant improvements in electrochemical features of RMABs are observed when noble metals are combined with various carbon allotropes. These synergistic effects lead to improved performance of the battery and are strongly influenced by the type, structure, and composition of the carbon material in noble metal-based composite catalysts. Overall, the high surface area and tailored porous structure of carbon are highly desired features that can result in better performance of air-electrodes of RMABs. One of the exceptions reported to date is a carbon-free, dealloyed Au$_{15}$Ag$_{85}$ cathode catalyst.[123] This catalyst shows a stable capacity of 1500 mAh g^{-1} with over 140 cycles at a current density of 2000 mA g^{-1}, indicating that air-electrodes without carbon can also demonstrate acceptable performance for RLOBs.

Due to the high price and limited resources of noble metals, non-noble metals have gained increasing attention in the exploration of high-performance bifunctional catalysts. Current developments in the synthesis methods of non-noble metal-derived composite catalysts can provide decent structures and properties, thus improving the performance of RMABs. Based on electrochemical features compared in Table 6.2, it is evidenced that at the same current density (ca. 200 mA g^{-1}), three different types of composite catalysts, CoCu/graphene,[43] Co-N-rGO/IrO$_2$,[115] and Fe-Co-N-rGO,[79] show high discharge capacities over 10,000 mAh g^{-1} and high cycle performance (i.e., cycle number above 70), suggesting that non-noble metal-derived air-electrode catalysts are strong competitors to noble-metal based bifunctional air-electrode catalysts. Carbon and doped-carbon both can contribute to the improved electrochemical features of the air-electrodes due to additional synergistic effects (e.g., electronic effects). For example, Co$_3$O$_4$/Ni/C composite catalysts[120] show even better ORR and OER activities than noble metal-based catalysts. It is speculated that this catalyst's superior capacity of 14,830 mAh g^{-1} at a high current density of 400 mA g^{-1} is related to its unique electronic structure and the improved porosity that results from the innovative synthesis methods (bimolecular strategy). Although initial activities of these catalysts are superior, their cycle lifetimes have to be improved. Interestingly, without the incorporation of carbon materials, four noncarbon and nonnoble metal-based catalysts, Co/Ni foam,[114] MnCo$_2$O$_4$/Nifoam,[96] Co$_3$O$_4$/Ni mesh,[95] and RuO$_x$/TiN/Ti mesh,[121] show very stable capacities over 100 RLOB charge-discharge cycles. This stability improvement, as compared to carbon containing air-electrodes, is related to their resistance to corrosion at operating voltages of RMABs.

Another combination of non-noble and carbon materials, Co/C/NiFe LDH/AB,[66] displays an excellent cycling stability over 300 cycles at a high current density of 40 mA cm^{-2} with 2 hours per cycle, under operations of RZAB. The harsh battery testing conditions leads to a final energy efficiency of ~51.2%. Another composite catalyst, Fe and N co-doped carbon (i.e., Fe-N-C),[80] in a RZAB demonstrates a maximum capacity of 731 mAh g^{-1} at 100 mA cm^{-2}, which corresponds to a high energy density above 800 Wh kg^{-1}. It has been proven that Fe and N co-doped carbons can outperform state-of-the-art Pt-based catalysts in RZAB.

Compared to metal-based catalysts, metal-free catalysts composed of carbon and modified carbon (i.e., non-metal-doped carbon) represent a large group of advanced bifunctional air-electrode components in RMABs. Being cost-effective and possessing high electronic conduction and large surface areas, as shown in Table 6.3, such carbon-based and metal-free catalysts have been intensely studied as air-electrode catalysts for various RMABs (i.e., RZAB, RLOB, RLAB, NaOB, etc.). In addition, binary, ternary, and quaternary carbon composites as air-electrode catalysts for RLOBs show superior cycling stability that is better than that of pure carbon catalysts under low (e.g., <100 mA g^{-1}) and high current densities (above 200 mA g^{-1}). For example, N-CNFs/C-paper[32] and P-C/MnCo$_2$O$_4$[75] catalysts can run 200 cycles with satisfactory reversibility under specific capacities of 500 and 1000 mAh g^{-1}, respectively, which corresponds to a current density of 250 and 200 mA g^{-1}. Particularly, P-C/MnCo$_2$O$_4$ catalyst shows a good rechargeability and delivers a maximum discharge capacity of 13,150 mAh g^{-1} at 200 mA g^{-1}.[75] This indicates that the combination of spinel oxides and the doping of carbon are the promising ways in the fabrication of highly efficient air-electrodes for RLABs and RLOBs. For carbon-based metal-free catalysts in RZABs (see Table 6.3) operated at a high current density of 20 mA cm^{-2}, CoMn$_2$O$_4$/rGO,[74] CoMn$_2$O$_4$/N-rGO,[74] and N-CNT/Co$_3$O$_4$[12] catalysts display stable charge and discharge voltages (over 200 cycles), demonstrating their better cycle stability in RZAB as compared to other listed catalysts. Interestingly, at 20 mA cm^{-2}, air-electrode made from N-G/C composite catalyst shows the highest specific capacity of 873 mAh g^{-1}, as well as a peak power density of 65 mW cm^{-2}.[15] This battery is also stable for over 150 charged-discharged cycles. This N-G/C catalyst derived from N-doped 3D graphene nanoribbon networks is the promising low-cost and scalable bifunctional electrocatalyst, not only for RZAB systems but also for a variety of electrochemical and catalytic applications. Additionally, SP/CaMnO$_3$ catalysts[132] in a rechargeable sodium-oxygen battery (NaOB) shows a charge-discharge stability for over 80 cycles (with a high current density of 200 mA g^{-1} and a limited capacity of 1000 mAh g^{-1}) and a high discharge capacity of 7000 mAh g^{-1}. These results are better than those obtained by the catalysts listed in Table 6.3, suggesting that the combination of carbon and oxides is a sound strategy in the development of abundant, low-cost, and efficient bifunctional catalysts for rechargeable NaOBs.

Based on the electrochemical performances of catalysts for various RMABs listed in Tables 6.1, 6.2, and 6.3, it can be concluded that all catalysts containing carbon show less cycle lifetime due to carbon corrosion. On the other hand, noncarbon-based metal-free bifunctional catalysts in RMABs (e.g., oxides, carbides, nitrides, etc.) show disadvantages such as low electronic conductivity, low surface area, and insufficient porosity. These limitations could lead to inadequate bifunctional activities and poor battery performances. Interestingly, other than a few studies focusing on hybrid electrolytes for RLAB, only one article touches upon polymer LOBs[148] in which SP/RuO$_2$@rGO catalyst is used for the air-electrode, while poly(methyl methacrylate-styrene)/TiO$_2$-based (PMS/TiO$_2$) gel polymer electrolyte (GPE) including LiClO$_4$ acts as an electrolyte. As demonstrated, this polymer LOB can run for over 50 cycles at a current density of 200 mA g^{-1} and a fixed capacity of 1000 mAh g^{-1} (proven charge-discharge cycle stability). This indicates that more effort is needed in the improvement of cycling stability and new formulations, leading to performance improvement in the development of solid electrolytes for polymer RLOB.

6.4 OPTIMIZATION OF ELECTROCATALYSTS IN RECHARGEABLE METAL-AIR BATTERIES

The optimization of composite catalyst materials is related to the improvement of bifunctional activity and minimization of electrode overpotentials, especially during the charge-discharge process. To obtain high battery performance, a trade-off strategy should focus on the selection of optimal catalyst formulation and structure, as well as its integration into the remaining air-electrode components. Several research efforts should be made in the synthesis, assembly, and testing of air-electrodes, including:

1. *Catalyst morphological design*: Because catalyst activity and stability are closely related to their morphology and structure, the synthesis procedures are critical in the development of both carbon- and noncarbon-based bifunctional composite electrocatalysts for RMABs. Unique fabrication strategies (e.g., using a template material to create desired porosity) and optimal synthesis conditions (e.g., temperature, time, starting material, solution, pH, and/or component ratio) can lead to products with high electrochemical properties that lead to high battery performance. One of the most challenging tasks for catalyst synthesis is to expand the active surface areas of the catalyst materials. If the synthesis is successful, the catalyst and catalyst layer should have a large surface area and be composed of particles with uniform size and shape, containing multiple redox active centers that are well distributed onto the substrate surface. The uniform distribution seems very important because it can prevent the aggregation of catalyst particles during electrochemical cycling, facilitate the access of electrolyte and reactants to the catalytic active centers, and then give high catalytic activity and durability of the air-electrode. The utilization of high-surface-area materials along with a high dispersion of active sites is vital to the success and further improvement of RMAB air-electrodes.

2. *Composition control*: The optimized formulation of multicomponent bifunctional catalysts has become a critical factor in improving catalytic activity and stability. The adjustment of mass ratio between redox active and non-active constituents, the grafting or combination of several active materials, as well as the selection of catalyst supports can all result in the desired catalytic performance of the composite catalyst in RMABs. Another aspect of composition optimization is to multiply ORR and OER catalytic active centers on the catalyst surface. One of the ways to achieve this is through N-doping into various carbon-based catalysts. This procedure can improve the electronic conductivity, thus the electron transport to/from active sites leading to improvement in both ORR and OER kinetics and then better battery performance.

3. *Optimization of supporting materials and/or active materials*: The type, chemical/physical structure/composition, and conductivity of the support material are vital to the development of well-performing bifunctional air-electrode catalysts. Although primary catalytic functions are taken by the catalyst itself, supporting materials can participate in this process via additional electronic effects or by facilitating electron transfer and mass transport for both ORR and OER processes. Another important function for supporting materials is to facilitate the even distribution of catalyst particles, leading to an expanded catalytic interface that could be fully utilized in the electrochemical process. The support material morphology should be adjusted to the reaction conditions and not suppress the redox processes in any way. Thus, the supporting materials should be optimized in terms of physical and/or chemical properties, such as surface area, porous structure, electronic structure, and conductivity, as well as their interaction with the catalyst. Similar requirements apply to the active materials in which their structure and morphology (e.g., size, shape, electronic and/or composition) should be optimized to form high-performance composite catalysts for RMAB air-electrodes.

4. *Electrochemical evaluation*: The electrochemical properties of the developed composite catalysts can be validated using a RDE half-cell testing and an assembled RMAB operation. In both experimental procedures, the activity and stability towards both ORR and OER in discharge-charge profiles can be justified. In addition, electrochemical impedance spectroscopy provides deeper understandings of the kinetics and mass transport processes, as well as the dynamics of the degradation process. A practical evaluation of RMAB performances can also be assessed using the potential difference between the ORR (potential at -3 mA cm^{-2}, or half-wave potential) and the OER (potential at 10 mA cm^{-2}).[17,157,158] Table 6.4 shows the assessment of bifunctional activities towards ORR and OER for the most promising composite catalysts. This table also shows the low potential difference, indicating the high bifunctionality of the catalysts.[63,88,159–188]

6.5 CHALLENGES AND POSSIBLE RESEARCH DIRECTIONS FOR DEVELOPING BIFUNCTIONAL ELECTROCATALYSTS OF RECHARGEABLE METAL-AIR BATTERIES

6.5.1 CHALLENGES

In the last two decades, large efforts have been given to the development and exploration of next-generation bifunctional catalysts for RMAB air-electrodes, however, several major technological challenges still exist, including: (1) low electrocatalytic activities for both ORR and OER at the air-electrodes which affect energy, power densities, and efficiency of MABs; (2) insufficient stability/durability of bifunctional air-electrode catalysts due to a low resistance to electrochemical corrosion, resulting in degradation of RMAB performance; (3) insufficient synthesis strategies closely related to the design and fabrication of high-performance bifunctional catalysts;

TABLE 6.4

Assessment of Bifunctional Activities toward Both ORR and OER for Typical Electrocatalysts in 0.1 M KOH Solution at an Electrode Rotation Rate of 1600 rpm

No.	Bifunctional Catalyst	E_{ORR} (V) (at −3 mA cm^{-2})	$E_{ORR,1/2}$ (V) (Half-Wave Potential for ORR)	E_{OER} (at 10 mA cm^{-2})	ΔE (V) ($E_{OER} - E_{ORR}$)	Year, References
1	20 wt%Pt/C	0.86 vs. RHE	—	2.02 vs. RHE	1.16[a]	2010, 159
2	20 wtRu/C	0.61 vs. RHE	—	1.62 vs. RHE	1.01[a]	2010, 159
3	20 wt%Ir/C	0.69 vs. RHE	—	1.61 vs. RHE	0.92[a]	2010, 159
4	MnO$_x$	0.73 vs. RHE	—	1.77 vs. RHE	1.04[a]	2010, 159
5	Co$_3$O$_4$/N-rGO	0.85 vs. RHE	—	1.54 vs. RHE	0.69[a]	2011, 160
6	NiCo$_2$O$_4$/G	0.56 vs. RHE	—	1.69 vs. RHE	1.13[a]	2013, 161
7	Co$_3$O$_4$/N-csCNT-GNR[b]	0.79 vs. RHE	—	1.59 vs. RHE	0.80[a]	2015, 162
8	CoS$_2$(400)/N,S-GO	0.79 vs. RHE	—	1.61 vs. RHE	0.82[a]	2015, 163
9	CoS$_2$(500)/N,S-GO	0.76 vs. RHE	—	1.62 vs. RHE	0.86[a]	2015, 163
10	CoS$_2$(600)/N,S-GO	0.75 vs. RHE	—	1.63 vs. RHE	0.88[a]	2015, 163
11	H-Pt/CaMnO$_3$	—	—	—	1.01[a]	2014, 17
12	N,S-CNS	−0.20 vs. Ag/AgCl	—	0.68 vs. Ag/AgCl	0.88[a]	2016, 164
13	LaCoO$_3$/NC	0.64 vs. RHE	—	1.64 vs. RHE	1.0[a]	2014, 165
14	LaNi$_{0.75}$Fe$_{0.25}$O$_3$/NC	0.67 vs. RHE	—	1.68 vs. RHE	1.01[a]	2014, 165
15	LaNiO$_3$/NC	0.64 vs. RHE	—	1.66 vs. RHE	1.02[a]	2014, 165
16	P-CN/CFP	0.72 vs. RHE	—	1.63 vs. RHE	0.91[a]	2015, 166
17	Pt/9LiCoO$_2$	0.81 vs. RHE	—	1.67 vs. RHE	0.86[a]	2016, 167
18	Pt/19LiCoO$_2$	0.80 vs. RHE	—	1.67 vs. RHE	0.87[a]	2016, 167
19	Pt/49LiCoO$_2$	0.79 vs. RHE	—	1.70 vs. RHE	0.91[a]	2016, 167
20	LiCoO$_2$	0.56 vs. RHE	—	1.71 vs. RHE	1.15[a]	2016, 167
21	CoMn$_2$O$_4$/PDDA-CNTs	−0.133 vs. Ag/AgCl	—	0.716 vs. Ag/AgCl	0.849[a]	2013, 63

(Continued)

TABLE 6.4 (Continued)
Assessment of Bifunctional Activities toward Both ORR and OER for Typical Electrocatalysts in 0.1 M KOH Solution at an Electrode Rotation Rate of 1600 rpm

No.	Bifunctional Catalyst	E_{ORR} (V) (at −3 mA cm^{-2})	$E_{ORR,1/2}$ (V) (Half-Wave Potential for ORR)	E_{OER} (at 10 mA cm^{-2})	ΔE (V) ($E_{OER} - E_{ORR}$)	Year, References
22	N,P-G/CNS	0.86 vs. RHE	—	1.57 vs. RHE	0.71[a]	2015, 168
23	LaNiO$_3$/N-C	0.64 vs. RHE	—	1.66 vs. RHE	1.02[a]	2013, 169
24	La(Co$_{0.55}$Mn$_{0.45}$)$_{0.99}$O$_{3-x}$/N-rGO	−0.242 vs. Ag/AgCl	—	0.718 vs. Ag/AgCl	0.96[a]	2015, 170
25	Ba$_{0.5}$Sr$_{0.5}$Co$_{0.8}$Fe$_{0.2}$O$_{3-\delta}$/N-C	0.74 vs. RHE	—	1.58 vs. RHE	0.84[a]	2016, 171
26	MnCoFeO$_4$/N-rGO	0.78 vs. RHE	—	1.71 vs. RHE	0.93[a]	2014, 172
27	MnO$_x$/S-C	0.81 vs. RHE	—	1.62 vs. RHE	0.81[a]	2015, 173
28	CoO@N,S-CNF	0.721 vs. RHE	—	1.549 vs. RHE	0.828[a]	2016, 174
29	Co/N-C-800	0.74 vs. RHE	—	1.60 vs. RHE	0.86[a]	2014, 175
30	Co/N-CNTs	0.84 vs. RHE	—	1.62 vs. RHE	0.78[a]	2016, 176
31	Fe/N-CNTs	0.81 vs. RHE	—	1.75 vs. RHE	0.94[a]	2016, 176
32	Ni/N-CNTs	0.73 vs. RHE	—	1.82 vs. RHE	1.09[a]	2016, 176
33	MWCNTs	0.43 vs. RHE	—	1.88 vs. RHE	1.45[a]	2016, 176
34	Co/CoO@Co-N-C-700	−0.23 vs. Ag/AgCl	—	0.76 vs. Ag/AgCl	0.99[a]	2016, 88
35	Co/CoO@Co-N-C-800	−0.17 vs. Ag/AgCl	—	0.64 vs. Ag/AgCl	0.81[a]	2016, 88
36	Co@Co-N-C-900	−0.22 vs. Ag/AgCl	—	0.63 vs. Ag/AgCl	0.85[a]	2016, 88
37	Co@Co-N-C-1000	−0.23 vs. Ag/AgCl	—	0.71 vs. Ag/AgCl	0.94[a]	2016, 88
38	Fe,N-CNF	−0.14 vs. Ag/AgCl	—	1.21 vs. Ag/AgCl	1.35[a]	2015, 88
39	N-G/SWCNT	0.70 vs. RHE	—	1.63 vs. RHE	0.93[a]	2014, 177
40	Co$_3$O$_4$/2.7Co$_2$MnO$_4$	0.68 vs. RHE	—	1.77 vs. RHE	1.09[a]	2013, 178
41	Ni$_{0.4}$Co$_{2.56}$O$_4$	0.79 vs. RHE	—	1.75 vs. RHE	0.96[a]	2015, 179
42	Ni$_{0.6}$Co$_{2.4}$O$_4$	0.78 vs. RHE	—	1.76 vs. RHE	0.98[a]	2015, 180

(Continued)

TABLE 6.4 (Continued)

Assessment of Bifunctional Activities toward Both ORR and OER for Typical Electrocatalysts in 0.1 M KOH Solution at an Electrode Rotation Rate of 1600 rpm

No.	Bifunctional Catalyst	E_{ORR} (V) (at -3 mA cm^{-2})	$E_{ORR,1/2}$ (V) (Half-Wave Potential for ORR)	E_{OER} (at 10 mA cm^{-2})	ΔE (V) ($E_{OER} - E_{ORR}$)	Year, References
43	$Ni_{0.9}Co_{2.1}O_4$	0.75 vs. RHE	—	1.76 vs. RHE	1.01[a]	2015, 180
44	$NiCo_2O_4$	0.78 vs. RHE	—	1.62 vs. RHE	0.84[a]	2014, 14
45	$NiCo_2O_4/C$	—	0.718 vs. RHE	1.67 vs. RHE	0.96[c]	2016, 181
46	$CoMn_2O_4/rGO$	—	0.89 vs. Hg/Hg$_2$Cl$_2$	1.54 vs. Hg/Hg$_2$Cl$_2$	0.65[c]	2015, 182
47	Co_3O_4/C	—	0.83 vs. Ag/AgCl	1.75 vs. Ag/AgCl	0.92[c]	2014, 183
48	Co/N-C	0.83 vs. RHE	—	1.69 vs. RHE	0.86[a]	2016, 184
49	$Co@Co_3O_4/N$-C-1	0.80 vs. RHE	—	1.65 vs. RHE	0.85[a]	2016, 184
50	$Co@Co_3O_4/N$-C-2	0.74 vs. RHE	—	1.64 vs. RHE	0.90[a]	2016, 184
51	$CoMn_2O_4$	—	0.76 vs. RHE	1.82 vs. RHE	1.06[c]	2015, 185
52	NiCoFe-LDH	—	—	—	1.26[a]	2015, 186
53	O- NiCoFe-LDH	—	—	—	1.05[a]	2015, 186

[a] (E_{OER}, 10 mA cm^{-2} − E_{ORR}, 10 mA cm^{-2}).

[b] N-doped, core-shell structured carbon nanotube-graphene nanoribbon.

[c] (E_{OER}, 10 mA cm^{-2} − $E_{ORR,1/2}$).

(4) low cost-effectiveness, which is a key step to the commercialization of bifunctional catalysts and their RMABs, dependent on the selection, preparation, and processing of the materials, as well as the nanotechnology used; and (5) immature capabilities for bifunctional catalyst optimization and scale-up production in the application of MAB air-electrodes. Moreover, due to unknown mechanisms and not-yet-optimized conditions in practical applications, RMAB assembly technology still needs further improvement.

6.5.2   FUTURE RESEARCH DIRECTIONS

In the research of advanced bifunctional catalysts for RMABs, many novel bifunctional composite catalysts have been developed and reported in published literature. However, the commercialization requirements of RMABs possess major technological challenges, as described above, for the development of high-performance bifunctional composite catalysts. To overcome these technical challenges, future research directions can be proposed for the synthesis and fabrication of highly active bifunctional composite catalysts for RMAB air-electrodes as follows:

1. Further improving the bifunctional catalytic activity and stability/ durability of both ORR and OER by developing credible methodologies to obtain optimized porous structures, including surface area, pore size, and distribution. In such morphology-controlled synthesis strategies, the recommended pores are mesopores and macropores, not micropores. It is known that at pore sizes below 2 nm, micropores will not work continuously due to pore clogging during charge-discharge, while mesopores and macropores are large enough to accommodate RMAB discharge product and at the same time facilitate the transport of reactants.
2. Establishing a more advanced fundamental understanding of catalytic mechanisms for bifunctional composite catalysts in different types of RMABs using both experimental and theoretical studies. It is necessary to further build a close relationship between the bifunctional catalytic mechanism of ORR and OER and the electronic structure and composition of catalysts using a combination of molecular/atomic modeling and experimental characterization. This will be beneficial to improve the catalytic activity of both ORR and OER in RMABs. Moreover, to solve corrosion issues, understanding and controlling the degradation mechanisms and failure modes of different composite catalysts through a combination of experimental and theoretical approaches is necessary.
3. Developing innovating synthesis approaches with low-cost and/or green materials effective high-performance and cost-effective bifunctional composite catalysts for RMAB air-electrodes. For example, biomasses consisting of natural polymers (e.g., cellulose, lignin, and hemicellulose) have been used as starting materials to produce biomass-derived carbons and their composites for advanced bifunctional catalysts. Such biomass precursors are also sources of doping elements, including various non-metal and/or metal heteroatom dopants, such as N, S, O, K, Ca, and/or

Mg. Particularly, biomass-derived carbon catalysts combined with further chemical doping or structural modification, which then pair with other catalytic components can result in even better catalysts.

4. Preventing and eliminating unwanted side reactions arising from the electrolyte and air-electrodes (particularly carbon-containing ones). These side reactions can produce unwanted products, leading to a loss of capacity during the cycle of charge-discharge. Therefore, effective strategies and approaches need to be explored and introduced to enhance the chemical/electrochemical stability of the air-electrodes. For instance, surface coating can be used to protect the surface of the electrode and thereby maintain its stability during RMAB operation. For example, to eliminate side reactions in RLABs, the electrode catalysts with low OER overpotential should be used to avoid electrolyte decomposition and evolution of CO_2 at high charge potential of around 4 V. In this sense, the selection and use of electrolytes are also very important and will influence the stability of bifunctional composite catalysts used for RMAB air-electrodes.

6.6   CHAPTER SUMMARY

Metal-air batteries have attracted considerable research interest because they have high theoretical specific energies with the potential to meet the ever-increasing demands of electrical energy storage and conversion in many emerging applications, such as electric vehicles. Recently, developments related to the selectivity, preparation, and processing of the electrode materials continuously provides progresses in increasing the energy efficiency, power density and cycle-life for RMABs. Moreover, RMABs can be crafted in diverse sizes and geometries, providing easy utilization in residential applications and businesses. However, the sluggish kinetics of the ORR and OER processes, the limited stability/durability, and low-rate capability are major hurdles in the commercialization of RMABs. To improve the performance of RMABs and to realize their commercial potential, tremendous efforts have been made in the development and exploration of high performance bifunctional composite air-electrode catalysts. Relying on different nanotechnologies, three important catalogues of bifunctional composite electrocatalysts including carbon-based, modified carbon-based, and noncarbon-based bifunctional catalysts have been created and studied for the relationships between their advanced nanostructure and chemical/electrochemical/electrocatalytic performances induced by their deep interaction and synergetic effect of the catalyst components. The advanced nanostructures of these composite electrocatalysts are usually related to their tailored design and optimized geometries.

In this chapter, several different bifunctional composite catalysts (i.e., carbon-based, doped carbon-based, and noncarbon-based bifunctional composite catalysts) for RMAB air-electrodes are discussed in terms of their bifunctionality, performance comparison, and catalyst optimization with respect to the RMAB performances. The roles of material selection, synthesis method, structural characterization, and electrochemical validation as well as resistance to carbon corrosion in achieving high-performing air-electrode catalysts are emphasized. The challenges in developing RMAB air-electrode catalysts are reviewed and analyzed, and the future research directions to overcome the challenges are also proposed to facilitate the further research and development of the bifunctional electrocatalysts for RMABs.

REFERENCES

1. Zhang, X.; Wang, X.-G.; Xie, Z.; Zhou, Z. Recent progress in rechargeable alkali metal-air batteries. *Green Energy Environ* **2016**, *1*, 4–17.
2. Song, K.; Agyeman, D. A.; Jung, J.; Ru, M. R.; Yang, J.; Kang, Y.-M. A review of the design strategies for tailored cathode catalyst materials in rechargeable Li-O$_2$ batteries. *Isr. J. Chem.* **2015**, *55*, 458–471.
3. Feng, N.; He, P.; Zhou, H. Critical challenges in rechargeable aprotic Li-O$_2$ batteries. *Adv. Energy Mater.* **2016**, *6*, 1502303.
4. Bevara, V.; Andrei, P. Changing the cathode microstructure to improve the capacity of Li-air batteries: Theoretical predictions. *J. Electrochem. Soc.* **2014**, *161*, A2068–A2079.
5. Pei, P.; Wang, K.; Ma, Z. Technologies for extending zinc-air battery's cycle life: A review. *Appl. Energ.* **2014**, *128*, 315–324.
6. Chen, F. Y.; Chen, J. Metal-air batteries: From oxygen reduction electrochemistry to cathode catalysts. *Chem. Soc. Rev.* **2012**, *41*, 2172–2192.
7. Richards, W. D.; Miara, L. J.; Wang, Y.; Kim, J. C.; Ceder, G. Interface stability in solid-state batteries. *Chem. Mater.* **2016**, *28*, 266–273.
8. Lee, D. U.; Xu, P.; Cano, Z. P.; Kashkooli, A. G.; Park, M. G.; Chen, Z. Recent progress and perspectives on bi-functional oxygen electrocatalysts for advanced rechargeable metal-air batteries. *J. Mater. Chem. A* **2016**, *4*, 7107–7134.
9. Cao, X.; Yan, W.; Jin, C.; Tian, J.; Ke, K.; Yang, R. Surface modification of MnCo$_2$O$_4$ with conducting polypyrrole as a highly active bifunctional electrocatalyst for oxygen reduction and oxygen evolution reaction. *Electrochim. Acta* **2015**, *180*, 788–794.
10. Yong, J. Y.; Ramachandaramurthy, V. K.; Tan, K. M.; Mithulananthan, N. A review on the state-of-the-art technologies of electric vehicle its impacts and prospects. *Renew Sust. Energ. Rev.* **2015**, *49*, 365–385.
11. Shen, P. K.; Wang, C.-Y.; Jiang, S. P.; Sun, X.; Zhang, J. *Electrochemical energy: Advanced materials and technologies.* Boca Raton, FL: CRC Press, **2016**.
12. Lee, D. U.; Park, M. G.; Park, H. W.; Seo, M. H.; Wang, X.; Chen, Z. Highly active and durable nanocrystal-decorated bifunctional electrocatalyst for rechargeable zinc–air batteries. *ChemSusChem* **2015**, *8*, 3129–3138.
13. Sakaushi, K.; Yang, S. J.; Fellinger, T.-P.; Antonietti, M. Impact of large-scale mesor- and macropore structures in adenosine-derived affordable noble carbon on efficient reversible oxygen electrocatalytic redox reactions. *J. Mater. Chem. A* **2015**, *3*, 11720–11724.
14. Prabu, M.; Ketpang, K.; Shanmugam, S. Hierarchical nanostructured NiCo$_2$O$_4$ as an efficient bifunctional non-precious metal catalyst for rechargeable zinc-air batteries. *Nanoscale* **2014**, *6*, 3173–3181.
15. Yang, H. B.; Miao, J.; Hung, S.-F.; Chen, J.; Tao, H. B.; Wang, X.; Zhang, L. et al. Identification of catalytic sites for oxygen reduction and oxygen evolution in N-doped graphene materials: Development of highly efficient metal-free bifunctional electrocatalyst. *Sci. Adv.* **2016**, *2*, e1501122.
16. Yoon, K. R.; Lee, G. Y.; Jung, J.-W.; Kim, N.-H.; Kim, S. O.; Kim, I.-D. One-dimensional RuO$_2$/Mn$_2$O$_3$ hollow architectures as efficient bifunctional catalysts for lithium-oxygen batteries. *Nano Lett.* **2016**, *16*, 2076–2083.
17. Han, X.; Cheng, F.; Zhang, T.; Yang, J.; Hu, Y.; Chen, J. Hydrogenated uniform Pt clusters supported on porous CaMnO$_3$ as a bifunctional electrocatalyst for enhanced oxygen reduction and evolution. *Adv. Mater.* **2014**, *26*, 2047–2051.

18. Yu, R.; Fan, W.; Guo, X.; Dong, S. Highly ordered and ultra-long carbon nanotube arrays as air cathodes for high-energy-efficiency Li-oxygen batteries. *J. Power Sources* **2016**, *306*, 402–407.
19. Meng, W.; Liu, S.; Wen, L.; Qin, X. Carbon microspheres air electrode for rechargeable Li-O$_2$ batteries. *RSC Adv.* **2015**, *5*, 52206–52209.
20. Jian, Z.; Liu, P.; Li, F.; He, P.; Guo, X.; Chen, M.; Zhou, H. Core-shell-structured CNT@RuO$_2$ composite as a high-performance cathode catalyst for rechargeable Li-O$_2$ batteries. *Angew. Chem., Int. Ed.* **2014**, *53*, 442–446.
21. Li, B.; Geng, D.; Lee, X. S.; Ge, X.; Chai, J.; Wang, Z.; Zhang, J.; Liu, Z.; Andy Hor, T. S.; Zong, Y. Eggplant-derived microporous carbon sheets: Towards mass production of efficient bifunctional oxygen electrocatalysts at low cost for rechargeable Zn–air batteries. *Chem. Commun.* **2015**, *51*, 8841–8844.
22. Luo, G.; Huang, S.-T.; Zhao, N.; Cui, Z.-H.; Guo, X.-X. A super high discharge capacity induced by a synergetic effect between high-surface-area carbons and a carbon paper current collector in a lithium–oxygen battery. *Chin. Phys. B* **2015**, *24*, 088102.
23. Li, L.; Manthiram, A. O- and N-doped carbon nanowebs as metal-free catalysts for hybrid Li-air batteries. *Adv. Energy Mater.* **2014**, *4*, 1301795.
24. Ma, J.-L.; Zhang, X.-B. Optimized nitrogen-doped carbon with a hierarchically porous structure as a highly efficient cathode for Na-O$_2$ batteries. *J. Mater. Chem. A* **2016**, *4*, 10008–10013.
25. Sakaushi, K.; Fellinger, T.-P.; Antonietti, M. Bifunctional metal-free catalysis of mesoporous noble carbons for oxygen reduction and evolution reactions. *ChemSusChem* **2015**, *8*, 1156–1160.
26. Hadidi, L.; Davari, E.; Lqbal, M.; Purkait, T. K.; Lvey, D. G.; Veinot, J. G. C. Spherical nitrogen-doped hollow mesoporous carbon as an efficient bifunctional electrocatalyst for Zn-air batteries. *Nanoscale* **2015**, *7*, 20547–20556.
27. Zeng, X.; Leng, L.; Liu, F.; Wang, G.; Dong, Y.; Du, L.; Liu, L.; Liao, S. Enhanced Li-O$_2$ battery performance, using graphene-like nori-derived carbon as the cathode and adding LiI in the electrolyte as a promoter. *Electrochim. Acta* **2016**, *200*, 231–238.
28. Zhao, C.; Yu, C.; Liu, S.; Yang, J.; Fan, X.; Huang, H.; Qiu, J. 3D porous N-doped graphene frameworks made of interconnected nanocages for ultrahigh-rate and long-life Li-O$_2$ batteries. *Adv. Funct. Mater.* **2015**, *25*, 6913–6920.
29. He, M.; Zhang, P.; Liu, L.; Liu, B.; Xu, S. Hierarchical porous nitrogen doped three-dimensional graphene as a free-standing cathode for rechargeable lithium-oxygen batteries. *Electrochim. Acta* **2016**, *191*, 90–97.
30. Zhang, Z.; Peng, B.; Chen, W.; Lai, Y.; Li, J. Nitrogen-doped carbon nanotubes with hydrazine treatment as cathode materials for lithium-oxygen batteries. *J. Solid State Electrochem.* **2015**, *19*, 195–200.
31. Liu, Q.; Wang, Y.; Dai, L.; Yao, J. Scalable fabrication of nanoporous carbon fiber films as bifunctional catalytic electrodes for flexible Zn-air batteries. *Adv. Mater.* **2016**, *28*, 3000–3006.
32. Shui, J.; Du, F.; Xue, C.; Li, Q.; Dai, L. Vertically aligned N-doped coral-like carbon fiber arrays as efficient air electrodes for high-performance non-aqueous Li-O$_2$ batteries. *ACS Nano* **2014**, *8*, 3015–3022.
33. Liu, J.; Wang, Z.; Zhu, J. Binder-free nitrogen-doped carbon paper electrodes derived from polypyrrole/cellulose composite for Li-O$_2$ batteries. *J. Power Sources* **2016**, *306*, 559–566.
34. Shu, C.; Lin, Y.; Zhang, B.; Hamid, S. B. A.; Su, D. Mesoporous boron-doped onion-like carbon as long-life oxygen electrode for sodium-oxygen batteries. *J. Mater. Chem. A* **2016**, *4*, 6610–6619.
35. Zhang, J.; Zhao, Z.; Xia, Z.; Dai, L. A metal-free bifunctional electrocatalyst for oxygen reduction and oxygen evolution reactions. *Nat. Nanotechn.* **2015**, *10*, 444–452.

36. Wang, L.; Ara, M.; Wadumesthrige, K.; Salley, S.; Simon Ng, K. Y. Graphene nanosheet supported bifunctional catalyst for high cycle life Li-air batteries. *J. Power Sources* **2013**, *234*, 8–15.

37. Yin, J.; Fang, B.; Luo, J.; Wanjala, B.; Mott, D.; Loukrakpam, R.; Shan Ng, M. et al. Nanoscale alloying effect of gold-platinum nanoparticles as cathode catalysts on the performance of a rechargeable lithium-oxygen battery. *Nanotechnology* **2012**, *23*, 305404.

38. Su, D.; Kim, H.-S.; Kim, W.-S.; Wang, G. A study of Pt_xCo_y alloy nanoparticles as cathode catalysts for lithium-air batteries with improved catalytic activity. *J. Power Sources* **2013**, *244*, 488–493.

39. Ko, B. K.; Kim, M. K.; Kim, S. H.; Lee, M. A.; Shim, S. E.; Baeck, S.-H. Synthesis and electrocatalytic properties of various metals supported on carbon for lithium-air battery. *J. Mol. Catal. A-Chem.* **2013**, *379*, 9–14.

40. Zhou, W.; Cheng, Y.; Yang, X.; Wu, B.; Nie, H.; Zhang, H.; Zhang, H. Iridium incorporated into deoxygenated hierarchical graphene as a high-performance cathode for rechargeable $Li-O_2$ batteries. *J. Mater. Chem. A* **2015**, *3*, 14556–14561.

41. Tong, S.; Zheng, M.; Lu, Y.; Lin, Z.; Zhang, X.; He, P.; Zhou, H. Binder-free carbonized bacterial cellulose-supported ruthenium nanoparticles for $Li-O_2$ batteries. *Chem. Commun.* **2015**, *51*, 7302–7304.

42. Cheng, H.; Scott, K. Selection of oxygen reduction catalyst for rechargeable lithium-air batteries-metal or oxide? *Appl. Catal. B Environ.* **2011**, *108–109*, 140–151.

43. Chen, Y.; Zhang, Q.; Zhang, Z.; Zhou, X.; Zhong, Y.; Yang, M.; Xie, Z.; Wei, J.; Zhou, Z. Two better than one: Cobalt-copper bimetallic yolk-shell nanoparticles supported on graphene as excellent cathode catalyst for $Li-O_2$ batteries. *J. Mater. Chem. A* **2015**, *3*, 17874–17879.

44. Huang, J.; Zhang, B.; Xie, Y. Y.; Lye, W. W. K.; Xu, Z.-L.; Abouali, S.; Garakani, M. A. et al. Electrospun graphitic carbon nanofibers with in-situ encapsulated Co-Ni nanoparticles as freestanding electrodes for $Li-O_2$ batteries. *Carbon* **2016**, *100*, 329–336.

45. Cao, Y.; Zheng, M.-S.; Cai, S.; Lin, X.; Yang, C.; Hu, W.; Dong, Q.-F. Carbon embedded $\alpha-MnO_2$@graphene nanosheet composite: A bifunctional catalyst for high performance lithium oxygen batteries. *J. Mater. Chem. A* **2014**, *2*, 18736–18741.

46. Sumboja, A.; Ge, X.; Thomas Goh, F. W.; Li, B.; Geng, D.; Andy Hor, T. S.; Zong, Y.; Liu, Z. Manganese oxide catalyst grown on carbon paper as an air cathode for high-performance rechargeable zinc-air batteries. *ChemPlusChem* **2015**, *80*, 1341–1346.

47. Yang, Y.; Shi, M.; Li, Y.-S.; Fu, Z.-W. MnO_2-graphene composite air electrode for rechargeable Li-air batteries. *J. Electrochem. Soc.* **2012**, *159*, A1917–A1921.

48. Gao, R.; Li, Z.; Zhang, X.; Zhang, J.; Hu, Z.; Liu, X. Carbon-dotted defective CoO with oxygen vacancies: A synergetic deign of bifunctional cathode catalyst for $Li-O_2$ batteries. *ACS Catal.* **2016**, *6*, 400–406.

49. Huang, B.-W.; Li, L.; He, Y.-J.; Liao, X.-Z.; He, Y.-S.; Zhang, W.; Ma, Z.-F. Enhanced electrochemical performance of nanofibrous CoO/CNF cathode catalyst for $Li-O_2$ batteries. *Electrochim. Acta.* **2014**, *137*, 183–189.

50. Tan, P.; Shyy, W.; Zhao, T. S.; Zhu, X. B.; Wei, Z. H. A RuO_2 nanoparticle-decorated buckypaper cathode for non-aqueous lithium-oxygen batteries. *J. Mater. Chem. A* **2015**, *3*, 19042–19049.

51. Jung, H.-G.; Jeong, Y. S.; Park, J.-B.; Sun, Y.-K.; Scrosati, B.; Lee, Y. J. Ruthenium-based electrocatalysts supported on reduced graphene oxide for lithium-air batteries. *ACS Nano* **2013**, *7*, 3532–3539.

52. Ahn, C.-H.; Kalubarme, R. S.; Kim, Y.-H.; Jung, K.-N.; Shin, K.-H.; Par, C.-J. Graphene/doped ceria nano-blend for catalytic oxygen reduction in non-aqueous lithium-oxygen batteries. *Electrochim. Acta.* **2014**, *117*, 18–25.

53. Park, J.; Jun, Y.-S.; Lee, W.; Gerbec, J. A.; See, K. A.; Stucky, G. D. Bimodal mesoporous titanium nitride/carbon microfibers as efficient and stable electrocatalysts for Li-O_2 batteries. *Chem. Mater.* **2013**, *25*, 3779–3781.

54. Koo, B. S.; Lee, J. K.; Yoon, W. Y. Improved electrochemical performances of lithium-oxygen batteries with tungsten carbide-coated cathode. *Japan. J. Appl. Phys.* **2015**, *54*, 047101.

55. Luo, W.-B.; Chou, S.-L.; Wang, J.-Z.; Liu, H.-K. A B4C nanowire and carbon nanotube composite as a novel bifunctional electrocatalyst for high energy lithium oxygen batteries. *J. Mater. Chem. A* **2015**, *3*, 18395–18399.

56. Lyu, Z.; Zhang, J.; Wang, L.; Yuan, K.; Luan, Y.; Xiao, P.; Chen, W. CoS_2 nanoparticles-graphene hybrid as a cathode catalyst for aprotic Li-O_2 batteries. *RSC Adv.* **2016**, *6*, 31739–31743.

57. Wu, J.; Dou, S.; Shen, A.; Wang, X.; Ma, Z.; Ouyang, C.; Wang, S. One-step hydrothermal synthesis of $NiCo_2S_4$-rGO as an efficient electrocatalyst for the oxygen reduction reaction. *J. Mater. Chem. A* **2014**, *2*, 20990–20995.

58. Chen, W.; Lai, Y.; Zhang, Z.; Gan, Y.; Jiang, S.; Li, J. β-FeOOH decorated highly porous carbon aerogels composites as a cathode material for rechargeable Li-O_2 batteries. *J. Mater. Chem. A* **2015**, *3*, 6447–6454.

59. Li, J.; Wen, W.; Zhou, M.; Guan, L.; Huang, Z. MWNT-supported bifunctional catalyst of β-FeOOH nanospindles for enhanced rechargeable Li-O_2 batteries. *J. Alloy. Compd.* **2015**, *639*, 428–434.

60. Hu, X.; Fu, X.; Chen, J. A soil/Vulcan XC-72 hybrid as a highly-effective catalytic cathode for rechargeable Li-O_2 batteries. *Inorg. Chem. Front.* **2015**, *2*, 1006–1010.

61. Lee, C. K.; Park, Y. J. Polyimide-wrapped carbon nanotube electrodes for long cycle Li-air batteries. *Chem. Commun.* **2015**, *51*, 1210–1213.

62. Yoo, E.; Zhou, H. Fe phthalocyanine supported by graphene nanosheet as catalyst in Li-air battery with the hybrid electrolyte. *J. Power Sources* **2013**, *244*, 429–434.

63. Zhai, X.; Yang, W.; Li, M.; Lv, G.; Liu, J.; Zhang, X. Noncovalent hybrid of $CoMn_2O_4$ spinel nanocrystals and poly(diallyl dimethylammonium chloride) functionalized carbon nanotubes as efficient electrocatalyst for oxygen reduction reaction. *Carbon* **2013**, *65*, 277–286.

64. Li, J.; Zou, M.; Chen, L.; Huang, Z.; Guan, L. An efficient bifunctional catalyst of Fe/Fe_3C carbon nanofibers for rechargeable Li-O_2 batteries. *J. Mater. Chem. A* **2014**, *2*, 10634–10638.

65. Wang, X.; Liu, X.; Tong, C.-J.; Yuan, X.; Dong, W.; Lin, T.; Liu, L.-M.; Huang, F. An electron injection promoted highly efficient electrocatalyst of $FeNi_3$@GR@Fe-NiOOH for oxygen evolution and rechargeable metal-air batteries. *J. Mater. Chem. A* **2016**, *4*, 7762–7771.

66. Wang, Q.; Zhou, D.; Yu, H.; Zhang, Z.; Bao, X.; Zhang, F.; Zhou, M. NiFe layered double-hydroxide and cobalt-carbon composite as a high-performance electrocatalyst for bifunctional oxygen electrode. *J. Electrochem. Soc.* **2015**, *162*, A2362–A2366.

67. Li, Y.; Huang, Z.; Huang, K.; Carnahan, D.; Xing, Y. Hybrid Li-air battery cathodes with sparse carbon nanotube arrays directly grown on carbon fiber papers. *Energy Environ. Sci.* **2013**, *6*, 3339–3345.

68. Kim, J.-H.; Kannan, A. G.; Woo, H.-S.; Jin, D.-G.; Kim, W.; Ryu, K.; Kim, D.-W. A bi-functional metal-free catalyst composed of dual-doped graphene and mesoporous carbon for rechargeable lithium-oxygen batteries. *J. Mater. Chem. A* **2015**, *3*, 18456–18465.

69. Kim, J. G.; Kim, Y.; Noh, Y.; Kim, W. B. $MnCo_2O_4$ nanowires anchored on reduced graphene oxide sheets as effective bifunctional catalysts for Li-O_2 battery cathodes. *ChemSusChem*, **2015**, *22*, 1752–1760.

70. Lee, D. U.; Par, M. G.; Park, H. W.; Seo, M. H.; Ismayilov, V.; Ahmed, R.; Chen, Z. Highly active Co-doped LaMnO$_3$ perovskite oxide and N-doped carbon nanotube hybrid bi-functional catalyst for rechargeable zinc-air batteries. *Electrochem. Commun.* **2015**, *60*, 38–41.
71. Prabu, M.; Ramakrishnan, P.; Ganesan, P.; Manthiram, A.; Shanmugam, S. LaTi$_{0.65}$Fe$_{0.35}$O$_{3-\delta}$ nanoparticle-decorated nitrogen-doped carbon nanorods as an advanced hierarchical air electrode for rechargeable metal-air batteries. *Nano Energy* **2015**, *15*, 92–103.
72. Park, H. W.; Lee, D. U.; Park, M. G.; Ahmed, R.; Seo, M. H.; Nazar, L. F.; Chen, Z. Perovskite-nitrogen-doped carbon nanotube composite as bifunctional catalyst for rechargeable lithium-air batteries. *ChemSusChem* **2015**, *8*, 1058–1065.
73. Li, J.; Zou, M.; Wen, W.; Zhao, Y.; Lin, Y.; Chen, L.; Lai, H.; Guan, L.; Huang, Z. Spinel MFe$_2$O$_4$ (M=Co, Ni) nanoparticles coated on multi-walled carbon nanotubes as electrocatalysts for Li-O$_2$ batteries. *J. Mater. Chem. A* **2014**, *2*, 10257–10262.
74. Prabu, M.; Ramakrishnan, P.; Nara, H.; Momma, T.; Osaka, T. Shanmugam, S. Zinc-air battery: Understanding the structure and morphology changes of graphene-supported CoMn$_2$O$_4$ bifunctional catalysts under practical rechargeable conditions. *ACS Appl. Mater. Interfaces* **2014**, *6*, 16545–16555.
75. Cao, X.; Wu, J.; Jin, C.; Tian, J.; Strasser, P.; Yang, R. MnCo$_2$O$_4$ anchored on P-doped hierarchical porous carbon as an electrocatalyst for high-performance rechargeable Li-O$_2$ batteries. *ACS Catal.* **2015**, *5*, 4890–4896.
76. Li, B.; Ge, X.; Goh, T.; Andy Hor, T. S.; Geng, D.; Du, G.; Liu, Z.; Zhang, J.; Liu, X.; Zong, Y. Co$_3$O$_4$ nanoparticles decorated carbon nanofiber mat as binder-free air-cathode for high performance rechargeable zinc-air batteries. *Nanoscale* **2015**, *7*, 1830–1838.
77. Crespiera, S. M.; Amantia, D.; Knipping, E.; Aucher, C.; Aubouy, L.; Amici, J.; Zeng, J.; Francia, C.; Bodoardo, S. Electrospun Pd-doped mesoporous carbon nanofibers as catalysts for rechargeable Li-O$_2$ batteries. *RSC Adv.* **2016**, *6*, 57335–57345.
78. Li, B.; Chen, Y.; Ge, X.; Chai, J.; Zhang, X.; Andy Hor, T. S.; Du, G.; Liu, Z.; Zhang, H.; Zong, Y. Mussel-inspired one-pot synthesis of transition metal and nitrogen co-doped carbon (M/N-C) as efficient oxygen catalysts for Zn-air batteries. *Nanoscale* **2016**, *8*, 5067–5075.
79. Zeng, X.; You, C.; Leng, L.; Dang, D.; Qiao, X.; Li, X.; Li, Y.; Liao, S.; Adzic, R. R. Ruthenium nanoparticles mounted on multielement co-doped graphene: An ultra-high-efficiency cathode catalyst for Li-O$_2$ batteries. *J. Mater. Chem. A* **2015**, *3*, 11224–11231.
80. Wu, M.; Tang, Q.; Dong, F.; Wang, Y.; Li, D.; Guo, Q.; Liu, Y.; Qiao, J. The design of Fe, N-doped hierarchically porous carbons as highly active and durable electrocatalysts for a Zn-air battery. *Phys. Chem. Chem. Phys.* **2016**, *18*, 18665–18669.
81. Wang, J.; Wu, H.; Gao, D.; Miao, S.; Wang, G.; Bao, X. High-density iron nanoparticles encapsulated within nitrogen-doped carbon nanoshell as efficient oxygen electrocatalyst for zinc-air battery. *Nano Energy* **2015**, *13*, 387–396.
82. Thippani, T.; Mandal, S.; Wang, G.; Ramani, V. K.; Kothandaraman, R. Probing oxygen reduction and oxygen evolution reactions bifunctional non-precious metal catalysts for metal-air batteries. *RSC Adv.* **2016**, *6*, 71122–71133.
83. Song, J.; Zhu, C.; Fu, S.; Song, Y.; Du, D.; Lin, Y. Optimization of cobalt/nitrogen embedded carbon nanotubes as an efficient bifunctional oxygen electrode for rechargeable zinc-air batteries. *J. Mater. Chem. A* **2016**, *4*, 4864–4870.
84. Guo, G.; Yao, X.; Ang, H.; Tan, H.; Zhang, Y.; Guo, Y.; Fong, E.; Yan, Q. Using elastin protein to develop highly efficient air cathodes for lithium-O$_2$ batteries. *Nanotechnology* **2016**, *27*, 045401.
85. Dong, S.; Chen, X.; Zhang, K.; Gu, L.; Zhang, L.; Zhou, X.; Li, L. et al. Molybdenum nitride based hybrid cathode for rechargeable lithium-O$_2$ batteries. *Chem. Commun.* **2011**, *47*, 11291–11293.

86. Lai, Y.; Chen, W.; Zhang, Z.; Qu, Y.; Gan, Y.; Li, J. Fe/Fe$_3$C decorated 3-D porous nitrogen-doped graphene as a cathode material for rechargeable Li-O$_2$ batteries. *Electrochim. Acta.* **2016**, *191*, 733–742.

87. Ni, W.; Liu, S.; Fei, Y.; He, Y.; Ma, X.; Lu, L.; Deng, Y. CoO@Co and N-doped mesoporous carbon composites derived from ionic liquids as cathode catalysts for rechargeable lithium-oxygen batteries. *J. Mater. Chem. A* **2016**, *4*, 7746–7753.

88. Zhang, X.; Liu, R.; Zang, Y.; Liu, G.; Wang, G.; Zhang, Y.; Zhang, H.; Zhao, H. Co/CoO nanoparticles immobilized on Co-N-doped carbon as trifunctional electrocatalysts for oxygen reduction, oxygen evolution and hydrogen evolution reactions. *Chem. Commun.* **2016**, *52*, 5946–5949.

89. Liao, K.; Zhang, T.; Wang, Y.; Li, F.; Jian, Z.; Yu, H.; Zhou, H. Nanoporous Ru as a carbon- and binder-free cathode for Li-O$_2$ batteries. *ChemSusChem* **2015**, *8*, 1429–1434.

90. Zhu, D.; Zhang, L.; Song, M.; Wang, X.; Chen, Y. An *in situ* formed Pd nanolayer as a bifunctional catalyst for Li-air batteries in ambient or simulated air. *Chem. Commun.* **2013**, *49*, 9573–9575.

91. Wu, X.; Chen, F.; Jin, Y.; Zhang, N.; Johnston, R. L. Silver-copper nanoalloy catalyst layer for bifunctional air electrodes in alkaline media. *ACS Appl. Mater. Interfaces* **2015**, *7*, 17782–17791.

92. Jin, Y.; Chen, F. Facile preparation of Ag-Cu bifunctional electrocatalysts for zinc-air batteries. *Electrochim. Acta* **2015**, *158*, 437–445.

93. Guo, X.; Han, J.; Liu, P.; Chen, L.; Ito, Y.; Jian, Z.; Jin, T. et al. Hierarchical nanoporosity enhanced reversible capacity of bicontinuous nanoporous metal-based Li-O$_2$ battery. *Sci. Rep.* **2016**, *6*, 33466.

94. Wu, F.; Zhang, X.; Zhao, T.; Chen, R.; Ye, Y.; Xie, M.; Li, L. Hierarchical mesoporous/macroporous Co$_3$O$_4$ ultrathin nanosheets as free-standing catalysts for rechargeable lithium-oxygen batteries. *J. Mater. Chem. A* **2015**, *3*, 17620–17626.

95. Lee, C. K.; Park, Y. J. Carbon and binder-free air electrodes composed of Co$_3$O$_4$ nanofibers for Li-Air batteries with enhanced cyclic performance. *Nanoscale Res. Lett.*, **2015**, *10*, 319.

96. Kalubarme, R. S.; Jadhav, H. S.; Ngo, D. T.; Park, G.-E.; Fisher, J. G.; Choi, Y.; Ryu, W.-H.; Par, C.-J. Simple synthesis of highly catalytic carbon-free MnCo$_2$O$_4$@Ni as an oxygen electrode for rechargeable Li-O$_2$ batteries with long-term stability. *Sci. Rep.* **2015**, *5*, 13266.

97. Lin, X.; Shang, Y.; Huang, T.; Yu, A. Carbon-free (Co, Mn)3O4 nanowires@Ni electrodes for lithium-oxygen batteries. *Nanoscale* **2014**, *6*, 9043–9049.

98. Wu, B.; Zhang, H.; Zhou, W.; Wang, M.; Li, X.; Zhang, H. Carbon-free CoO mesoporous nanowire array cathode for high-performance aprotic Li-O$_2$ batteries. *ACS Appl. Mater. Interfaces* **2015**, *7*, 23182–23189.

99. Leng, L.; Zeng, X.; Song, H.; Shu, T.; Wang, H.; Liao, S. Pd nanoparticles decorating flower-like Co$_3$O$_4$ nanowire clusters to form an efficient, carbon/binder-free cathode for Li-O$_2$ batteries. *J. Mater. Chem. A* **2015**, *3*, 15626–15632.

100. Xu, S.-M.; Zhu, Q.-C.; Long, J.; Wang, H.-H.; Xie, X.-F.; Wang, K.-X.; Chen, J.-S. Low-overpotential Li-O$_2$ batteries based on TFSI intercalated Co-Ti layered double oxides. *Adv. Funct. Mater.* **2016**, *26*, 1365–1374.

101. Lim, H.-D.; Song, H.; Gwon, H.; Park, K.-Y.; Kim, J.; Bae, Y.; Kim, H. et al. A new catalyst-embedded hierarchical air electrode for high-performance Li-O$_2$ batteries. *Energy Environ. Sci.* **2013**, *6*, 3570–3575.

102. Li, J.; Zhao, Y.; Zou, M.; Wu, C.; Huang, Z.; Guan, L. An effective integrated design for enhanced cathodes of Ni foam-supported Pt/carbon nanotubes for Li-O$_2$ batteries. *ACS Appl. Mater. Interfaces* **2014**, *6*, 12479–12485.

103. Zhao, G.; Lv, J.; Xu, Z.; Zhang, L.; Sun, K. Carbon and binder free rechargeable Li-O$_2$ battery cathode with Pt/Co$_3$O$_4$ flake arrays as catalyst. *J. Power Sources* **2014**, *248*, 1270–1274.

104. Huang, K.; Li, Y.; Xing, Y. Increasing round trip efficiency of hybrid Li-air battery with bifunctional catalysts. *Electrochim. Acta* **2013**, *103*, 44–49.

105. Zahoor, A.; Christy, M.; Kim, Y.; Arul, A.; Lee, Y. S.; Nahm, K. S. Carbon/titanium oxide supported bimetallic platinum/iridium nanocomposites as bifunctional electrocatalysts for lithium-air batteries. *J. Solid State Electrochem.* **2016**, *20*, 1397–1404.

106. Choi, R.; Jung, J.; Kim, G.; Song, K.; Kim, Y.-I.; Jung, S. C.; Han, Y.-K.; Song, H.; Kang, Y.-M. Ultra-low overpotential and high rate capability in Li-O$_2$ batteries through surface atom arrangement of PdCu nanocatalysts. *Energy Environ. Sci.* **2014**, *7*, 1362–1368.

107. Kumar, S.; Chinnathambi, S.; Munichandraiah, N. Ir nanoparticles-anchored reduced graphene oxide as a catalyst for oxygen electrode in Li-O$_2$ cells. *New J. Chem.* **2015**, *39*, 7066–7075.

108. Huang, Z.; Zhang, M.; Cheng, J.; Gong, Y.; Li, X.; Chi, B.; Pu, J.; Jian, L. Silver decorated beta-manganese oxide nanorods as an effective cathode electrocatalyst for rechargeable lithium-oxygen battery. *J. Alloy. Compd.* **2015**, *626*, 173–179.

109. Minguzzi, A.; Longoni, G.; Cappelletti, G.; Pargoletti, E.; Bari, C. D.; Locatelli, C.; Marelli, M.; Rondinini, S.; Vertova, A. The influence of carbonaceous matrices and electrocatalytic MnO$_2$ nanopowders on lithium-air battery performances. *Nanomaterials* **2016**, *6*, 10.

110. Liu, Y.; Wang, M.; Cao, L.-J.; Yang, M.-Y.; Cheng, S. H.-S.; Cao, C.-W.; Leung, K.-L.; Chung, C.-Y.; Lu, Z.-G. Interfacial redox reaction-directed synthesis of silver@cerium oxide core-shell nanocomposites as catalysts for rechargeable lithium-air batteries. *J. Power Sources* **2015**, *286*, 136–144.

111. Zahoor, A.; Christy, M.; Jeon, J. S.; Lee, Y. S.; Nahm, K. S. Improved lithium oxygen battery performance by addition of palladium nanoparticles on manganese oxide nanorod catalysts. *J. Solid State Electrochem.* **2015**, *19*, 1501–1509.

112. Thomas Goh, F. W.; Liu, Z.; Ge, X.; Zong, Y.; Du, G.; Andy Hor, T. S. Ag nanoparticle-modified MnO$_2$ nanorods catalyst for use as an air electrode in zinc-air battery. *Electrochim. Acta* **2013**, *114*, 598–604.

113. Lu, X.; Zhang, L.; Sun, X.; Si, W.; Yan, C.; Schmidt, O. G. Bifunctional Au-Pd decorated MnO$_x$ nanomembranes as cathode materials for Li-O$_2$ batteries. *J. Mater. Chem. A* **2016**, *4*, 4155–4160.

114. Shen, C.; Wen, Z.; Wang, F.; Wu, T.; Wu, X. Cobalt-metal-based cathode for lithium-oxygen battery with improved electrochemical performance. *ACS Catal.* **2016**, *6*, 4149–4153.

115. Zeng, X.; Dang, D.; Leng, L.; You, C.; Wang, G.; Zhu, C.; Liao, S. Doped reduced graphene oxide mounted with IrO$_2$ nanoparticles shows significantly enhanced performance as a cathode catalyst for Li-O$_2$ batteries. *Electrochim. Acta* **2016**, *192*, 431–438.

116. Zhang, Z.; Su, L.; Yang, M.; Hu, M.; Bao, J.; Wei, J.; Zhou, Z. A composite of Co nanoparticles highly dispersed on N-rich carbon substrates: An efficient electrocatalyst for Li-O$_2$ battery cathodes. *Chem. Commun.* **2014**, *50*, 776–778.

117. Ma, Z.; Yuan, X.; Li, L.; Ma, Z.-F. The double perovskite oxide Sr$_2$CrMoO$_{6-\delta}$ as an efficient electrocatalyst for rechargeable lithium air batteries. *Chem. Commun.* **2014**, *50*, 14855–14858.

118. Liu, X.; Park, M.; Kim, M. G.; Gupta, S.; Wu, G.; Cho, J. Integrating NiCo alloys with their oxides as efficient bifunctional cathode catalysts for rechargeable zinc-air batteries. *Angew. Chem. Int. Ed.* **2015**, *54*, 9654–9658.

119. Li, L.; Manthiram, A. Long-life, high-voltage acidic Zn-air batteries. *Adv. Energy Mater.* **2016**, *6*, 1502054.

120. Oh, D.; Qi, J.; Han, B.; Zhang, G.; Garney, T. J.; Ohmura, J.; Zhang, Y.; Shao-Horn, Y.; Belcher, A. M. M13 virus-directed synthesis of nanostructured metal oxides for lithium-oxygen batteries. *Nano Lett.* **2014**, *14*, 4837–4845.

121. Chang, Y.; Dong, S.; Ju, Y.; Xiao, D.; Zhou, X.; Zhang, L.; Chen, X. et al. A carbon- and binder-free nanostructured cathode for high-performance non-aqueous Li-O$_2$ battery. *Adv. Sci.* **2015**, *2*, 1500092.

122. Yoo, E.; Zhou, H. Hybrid electrolyte Li-air rechargeable batteries based on nitrogen- and phosphorus-doped graphene nanosheets. *RSC Adv.* **2014**, *4*, 13119–13122.

123. Li, C.; Han, X.; Cheng, F.; Hu, Y.; Chen, C.; Chen, J. Phase and composition controllable synthesis of cobalt manganese spinel nanoparticles towards efficient oxygen electrocatalysis. *Nat. Commun.* **2015**, *6*, 7345.

124. Liu, B.; Xu, W.; Yan, P.; Bhattacharya, P.; Cao, R.; Bowden, M. E.; Engelhard, M. H.; Wang, C.-M.; Zhang, J.-G. In situ-grown ZnCo$_2$O$_4$ on single-walled carbon nanotubes as air electrode materials for rechargeable lithium-oxygen batteries. *ChemSusChem* **2015**, *8*, 3697–3703.

125. Ryu, W.-H.; Yoon, T.-H.; Song, S. H.; Jeon, S.; Park, Y.-J.; Kim, I.-D. Bifunctional composite catalysts using Co$_3$O$_4$ nanofibers immobilized on non-oxidized graphene nanoflakes for high-capacity and long-cycle Li-O$_2$ batteries. *Nano Lett.* **2013**, *13*, 4190–4197.

126. Sun, C.; Li, F.; Ma, C.; Wang, Y.; Ren, Y.; Yang, W.; Ma, Z. et al. Graphene-Co$_3$O$_4$ nanocomposite as an efficient bifunctional catalyst for lithium-air batteries. *J. Mater. Chem. A* **2014**, *2*, 7188–7196.

127. Kumar, G. G.; Christy, M.; Jang, H.; Nahm, K. S. Cobaltite oxide nanosheets anchored graphene nanocomposite as an efficient oxygen reduction reaction (ORR) catalyst for the application of lithium-air batteries. *J. Power Sources* **2015**, *288*, 451–460.

128. Zhang, L.; Zhang, F.; Huang, G.; Wang, J.; Du, X.; Qin, Y.; Wang, L. Freestanding MnO$_2$@carbon papers air electrodes for rechargeable Li-O$_2$ batteries. *J. Power Sources* **2014**, *261*, 311–316.

129. Li, P.-C.; Chien, Y.-J.; Hu, C.-C. Novel configuration of bifunctional air electrodes for rechargeable zinc-air batteries. *J. Power Sources* **2016**, *313*, 37–45.

130. Sun, Q.; Yadegari, H.; Banis, M. N.; Liu, J.; Xiao, B.; Wang, B.; Lawes, S.; Li, X.; Li, R.; Sun, X. Self-stacked nitrogen-doped carbon nanotubes as long-life air electrode for sodium-air batteries: Elucidating the evolution of discharge product morphology. *Nano Energy* **2015**, *12*, 698–708.

131. Xu, Q.; Han, X.; Ding, F.; Zhang, L.; Sang, L.; Liu, X.; Xu, Q. A highly efficient electrocatalyst of perovskite LaNiO$_3$ for non-aqueous Li-O$_2$ batteries with superior cycle stability. *J. Alloy. Compd.* **2016**, *664*, 750–755.

132. Hu, Y.; Han, X.; Zhao, Q.; Du, J.; Cheng, F.; Chen, J. Porous perovskite calcium-manganese oxide microspheres as an efficient catalyst for rechargeable sodium-oxygen batteries. *J. Mater. Chem. A* **2015**, *3*, 3320–3324.

133. Wei, Z.; Cui, Y.; Huang, K.; Ouyang, J.; Wu, J.; Baker, A. P.; Zhang, X. Fabrication of La$_2$NiO$_4$ nanoparticles as an efficient bifunctional cathode catalyst for rechargeable lithium-oxygen batteries. *RSC Adv.* **2016**, *6*, 17430–17437.

134. Xu, J.-J.; Xu, D.; Wang, Z.-L.; Wang, H.-G.; Zhang, L.-L.; Zhang, X.-B. Synthesis of perovskite-based porous La$_{0.75}$Sr$_{0.25}$MnO$_3$ nanotubes as a highly efficient electrocatalyst for rechargeable lithium-oxygen batteries. *Angew. Chem. Int. Ed.* **2013**, *52*, 3887–3890.

135. Lu, F.; Wang, Y.; Jin, C.; Li, F.; Yang, R.; Chen, F. Microporous La$_{0.8}$Sr$_{0.2}$MnO$_3$ perovskite nanorods as efficient electrocatalyst for lithium-air battery. *J. Power Sources* **2015**, *293*, 726–733.

136. Oh, M. Y.; Jeon, J. S.; Lee, J. J.; Kim, P.; Nahm, K. S. The bifunctional electrocatalytic activity of perovskite La$_{0.6}$Ni$_{0.4}$CoO$_{3-\delta}$ for oxygen reduction and evolution reactions. *RSC Adv.* **2015**, *5*, 19190–19198.

137. Kalubarme, R. S.; Park, G.-E.; Jung, K.-N.; Shin, K.-H.; Ryu, W.-H.; Park, C.-J. LaNi$_x$Co$_{1-x}$O$_{3-\delta}$ perovskites as catalyst material for non-aqueous lithium-oxygen batteries. *J. Electrochem. Soc.* **2014**, *161*, A880–A889.

138. Cheng, J.; Zhang, M.; Jiang, Y.; Zou, L.; Gong, Y.; Chi, B.; Pu, J.; Jian, L. Perovskite $La_{0.6}Sr_{0.4}Co_{0.2}Fe_{0.8}O_3$ as an effective electrocatalyst for non-aqueous lithium air batteries. *Electrochim. Acta* **2016**, *191*, 106–115.

139. Juang, J.-I.; Risch, M.; Park, S.; Kim, M. G.; Nam, G.; Jeong, H.-Y.; Shao-Horn, Y.; Cho, J. Optimizing nanoparticle perovskite for bifunctional oxygen electrocatalysis. *Energy Environ. Sci.* **2016**, *9*, 176–183.

140. Liu, Y.; Cao, L.-J.; Cao, C.-W.; Wang, M.; Leung, K.-L.; Zeng, S.-S.; Hung, T. F.; Chung, C. Y.; Lu, Z.-G. Facile synthesis of spinel $CuCo_2O_4$ nanocrystals as high-performance cathode catalysts for rechargeable Li-air batteries. *Chem. Commun.* **2014**, *50*, 14635–14638.

141. Li, L.; Shen, L.; Nie, P.; Pang, G.; Wang, J.; Li, H.; Dong, S.; Zhang, X. Porous $NiCo_2O_4$ nanotubes as a noble-metal-free effective bifunctional catalyst for rechargeable $Li-O_2$ batteries. *J. Mater. Chem. A* **2015**, *3*, 24309–24314.

142. Peng, S.; Hu, Y.; Li, L.; Han, X.; Cheng, F.; Srinivasan, M.; Yan, Q.; Ramakrishna, S.; Chen, J. Controlled synthesis of porous spinel cobaltite core-shell microspheres as high-performance catalysts for rechargeable $Li-O_2$ batteries. *Nano Energy* **2015**, *13*, 718–726.

143. Liu, B.; Yan, P.; Xu, W.; Zheng, J.; He, Y.; Luo, L.; Bowden, M. E.; Wang, C.-M.; Zhang, J.-G. Electrochemically formed ultrafine metal oxide nanocatalysts for high-performance lithium-oxygen batteries. *Nano Lett.* **2016**, *16*, 4932–4939.

144. Guo, K.; Li, Y.; Yang, J.; Zou, Z.; Xue, X.; Li, X.; Yang, H. Nanosized Mn-Ru binary oxides as effective bifunctional cathode electrocatalysts for rechargeable $Li-O_2$ batteries. *J. Mater. Chem. A* **2014**, *2*, 1509–1514.

145. Sennu, P.; Christy, M.; Aravindan, V.; Lee, Y.-G.; Nahm, K. S.; Lee, Y.-S. Two-dimensional mesoporous cobalt sulfide nanosheets as a superior anode for a Li-ion battery and a bifunctional electrocatalyst for the $Li-O_2$ system. *Chem. Mater.* **2015**, *27*, 5726–5735.

146. Davari, E.; Johnson, A. D.; Mittal, A.; Xiong, M.; Ivey, D. G. Manganese-cobalt mixed oxide film as a bifunctional catalyst for rechargeable zinc-air batteries. *Electrochim. Acta.* **2016**, *211*, 735–743.

147. Senthilkumar, B.; Khan, Z.; Park, S.; Seo, I.; Ko, H.; Kim, Y. Exploration of cobalt phosphate as a potential catalyst for rechargeable aqueous sodium-air battery. *J. Power Sources* **2016**, *311*, 29–34.

148. Yi, J.; Wu, S.; Bai, S.; Liu, Y.; Li, N.; Zhou, H. Interfacial construction Li_2O_2 for a performance-improved polymer $Li-O_2$ battery. *J. Mater. Chem. A* **2016**, *4*, 2403–2407.

149. Jang, H.; Zahoor, A.; Jeon, J. S.; Kim, P.; Lee, Y. S.; Nahm, K. S. Sea urchin-shaped $\alpha-MnO_2/RuO_2$ mixed oxides nanostructure as promising electrocatalyst for lithium-oxygen battery. *J. Electrochem. Soc.* **2015**, *162*, A300–A307.

150. Zhao, C.; Yu, C.; Yang, J.; Liu, S.; Zhang, M.; Qiu, J. A dual component catalytic system composed of non-noble metal oxides for $Li-O_2$ batteries with enhanced cyclability. *Part. Part. Syst. Charact.* **2016**, *33*, 228–234.

151. Wu, Y.; Wang, T.; Zhang, Y.; Xin, S.; He, X.; Zhang, D.; Shui, J. Electrocatalytic performances of $g-C_3N_4-LaNiO_3$ composite as bi-functional catalysts for lithium-oxygen batteries. *Sci. Rep.* **2016**, *6*, 24314.

152. Gwirth, A. A.; Thorum, M. S. Electroreduction of dioxygen for fuel-cell applications: Materials and challenges. *Inorg. Chem.* **2010**, *49*, 3557–3566.

153. Lu, M.; Qu, J.; Yao, Q.; Xu, C.; Zhan, Y.; Xie, J.; Lee, J. Y. Exploring metal nanoclusters for lithium-oxygen batteries. *ACS Appl. Mater. Interface* **2015**, *7*, 5488–5496.

154. Lin, X.; Lu, X.; Huang, T.; Liu, Z.; Yu, A. Binder-free nitrogen-doped carbon nanotubes electrodes for lithium-oxygen batteries. *J. Power Sources* **2013**, *242*, 855–859.

155. Kwak, W.-J.; Kang, T.-G.; Sun, Y.-K.; Lee, Y. J. Iron-cobalt bimetal decorated carbon nanotubes as cost-effective cathode catalysts for $Li-O_2$ batteries. *J. Mater. Chem. A* **2016**, *4*, 7020–7026.

156. Ma, H.; Wang, B. A bifunctional electrocatalyst α-MnO$_2$-LaNiO$_3$/carbon nanotube composite for rechargeable zinc-air batteries. *RSC Adv.* **2014**, *4*, 46084–46092.

157. Chen, D.; Chen, C.; Baiyee, Z. M.; Shao, Z.; Ciucci, F. Nonstoichiometric oxides as low-cost and highly-efficient oxygen reduction/evolution catalysts for low-temperature electrochemical devices. *Chem. Rev.* **2015**, *115*, 9869–9921.

158. Wang, D.; Chen, X.; Evans, D. G.; Yang, W. Well-dispersed Co$_3$O$_4$/Co$_2$MnO$_4$ nanocomposites as a synergistic bifunctional catalyst for oxygen reduction and oxygen evolution reactions. *Nanoscale* **2013**, *5*, 5312–5315.

159. Gorlin, Y.; Jaramillo, T. F. A bifunctional nonprecious metal catalyst for oxygen reduction and water oxidation. *J. Am. Chem. Soc.* **2010**, *132*, 13612–13614.

160. Liang, Y.; Li, Y.; Wang, H.; Zhou, J.; Wang, J.; Regier, T.; Dai, H. Co$_3$O$_4$ nanocrystals on graphene as a synergistic catalyst for oxygen reduction reaction. *Nat. Mater.* **2011**, *10*, 780–786.

161. Lee, D. U.; Kim, B. J.; Chen, Z. One-pot synthesis of a mesoporous NiCo$_2$O$_4$ nanoplatelet and graphene hybrid and its oxygen reduction and evolution activities as an efficient bi-functional electrocatalyst. *J. Mater. Chem. A* **2013**, *1*, 4754–4762.

162. Lu, X.; Chan, H. M.; Sun, C.-L.; Tseng, C.-M.; Zhao, C. Interconnected core-shell carbon nanotube-graphene nanoribbon scaffolds for anchoring cobalt oxides as bifunctional electrocatalysts for oxygen evolution and reduction. *J. Mater. Chem. A* **2015**, *3*, 13371–13376.

163. Ganesan, P.; Prabu, M.; Sanetuntikul, J.; Shanmugam, S. Cobalt sulfide nanoparticles grown on nitrogen and sulfur co-doped graphene oxide: An efficient electrocatalyst for oxygen reduction and evolution reactions. *ACS Catal.* **2015**, *5*, 3625–3637.

164. Qu, K.; Zheng, Y.; Dai, S.; Qiao, S. Z. Graphene oxide-polydopamine derived N, S-co-doped carbon nanosheets as superior bifunctional electrocatalyst for oxygen reduction and evolution. *Nano Energy* **2016**, *19*, 373–381.

165. Hardi, W. G.; Mefford, J. T.; Slanac, D. A.; Patel, B. B.; Wang, X. Q.; Dai, S.; Zhao, X.; Ruoff, R. S.; Johnston, K. P.; Stevenson, K. J. Tuning the electrocatalytic activity of perovskites through active site variation and support interactions. *Chem. Mater.* **2014**, *26*, 3368–3376.

166. Ma, T. Y.; Ran, J.; Dai, S.; Jaroniec, M.; Qiao, S. Z. Phosphorus-doped graphitic carbon nitrides grown *in situ* on carbon-fiber paper: Flexible and reversible oxygen electrodes. *Angew. Chem. Int. Ed.* **2015**, *54*, 4646–4650.

167. Su, C.; Yang, T.; Zhou, W.; Wang, W.; Xu, X.; Shao, Z. Pt/C-LiCoO$_2$ composites with ultralow Pt loadings as synergistic bifunctional electrocatalyst for oxygen reduction and evolution reactions. *J. Mater. Chem. A* **2016**, *4*, 4516–4524.

168. Li, R.; Wei, Z.; Gou, X. Nitrogen and phosphorus dual-doped graphene/carbon nanosheets as bifunctional electrocatalysts for oxygen reduction and evolution. *ACS Catal.* **2015**, *5*, 4133–4142.

169. Hardin, W. G.; Slanac, D. A.; Wang, X.; Dai, S.; Johnston, K. P.; Stevenson, K. J. Highly active, nonprecious metal perovskite electrocatalysts for bifunctional metal-air battery electrodes. *J. Phys. Chem. Lett.* **2013**, *4*, 1254–1259.

170. Ge, X.; Thomas Goh, F. W.; Li, B.; Anday Hor, T. S.; Zhang, J.; Xiao, P.; Wang, X.; Zong, Y.; Liu, Z. Efficient and durable oxygen reduction and evolution of a hydrothermally synthesized La(Co$_{0.55}$Mn$_{0.45}$)$_{0.99}$O$_{3-\delta}$ nanorod/graphene hybrid in alkaline media. *Nanoscale* **2015**, *7*, 9046–9054.

171. Wang, J.; Zhao, H.; Gao, Y.; Chen, D.; Chen, C.; Saccoccio, M.; Ciucci, F. Ba$_{0.5}$Sr$_{0.5}$Co$_{0.8}$Fe$_{0.2}$O$_{3-\delta}$ on N-doped mesoporous carbon derived from organic waste as a bi-functional oxygen catalyst. *Int. J. Hydrogen Energ.* **2016**, *41*, 10744–10754.

172. Zhan, Y.; Xu, C.; Lu, M.; Liu, Z.; Lee, J. Y. Mn and Co co-substituted Fe$_3$O$_4$ nanoparticles on nitrogen-doped reduced graphene oxide for oxygen electrocatalysis in alkaline solution. *J. Mater. Chem. A.* **2014**, *2*, 16217–16223.

173. Gao, Y.; Zhao, H.; Chen, D.; Chen, C.; Ciucci, F. In situ synthesis of mesoporous manganese oxide/sulfur-doped graphitized carbon as a bifunctional catalyst for oxygen evolution/reduction reactions. *Carbon* **2015**, *94*, 1028–1036.

174. Liu, T.; Guo, Y.-F.; Yan, Y.-M.; Wang, F.; Deng, C.; Rooney, D.; Sun, K.-N. CoO nanoparticles embedded in three-dimensional nitrogen/sulfur co-doped carbon nanofiber networks as a bifunctional catalyst for oxygen reduction/evolution reactions. *Carbon* **2016**, *106*, 84–92.

175. Su, Y.; Zhu, Y.; Jiang, H.; Shen, J.; Yang, X.; Zou, W.; Chen, J.; Li, C. Cobalt nanoparticles embedded in N-doped carbon as an efficient bifunctional electrocatalyst for oxygen reduction and evolution reactions. *Nanoscale* **2014**, *6*, 15080–15089.

176. Liu, Y.; Jiang, H.; Zhu, Y.; Yang, X.; Li, C. Transition metals (Fe, Co, and Ni) encapsulated in nitrogen-doped carbon nanotubes as bi-functional catalysts for oxygen electrode reactions. *J. Mater. Chem. A* **2016**, *4*, 1694–1701.

177. Wu, Z.-Y.; Xu, X.-X.; Hu, B.-C.; Liang, H.-W.; Lin, Y.; Chen, L.-F.; Yu, S.-H. Iron carbide nanoparticles encapsulated in mesoporous Fe-N-doped carbon nanofibers for efficient electrocatalysis. *Angew. Chem. Int. Ed.* **2015**, *54*, 8179–8183.

178. Tian, G.-L.; Zhao, M.-Q.; Yu, D.; Kong, X.-Y.; Huang, J.-Q.; Zhang, Q.; Wei, F. Nitrogen-doped graphene/carbon nanotube hybrids: *In situ* formation on bifunctional catalysts and their superior electrocatalytic activity for oxygen evolution/reduction reaction. *Small* **2014**, *10*, 2251–2259.

179. Wang, D.; Chen, X.; Evans, D. G.; Yang, W. Well-dispersed Co_3O_4/Co_2MnO_4 nanocomposites as a synergistic bifunctional catalyst for oxygen reduction and oxygen evolution reactions. *Nanoscale* **2013**, *5*, 5312–5315.

180. Lambert, T. N.; Vigil, J. A.; White, S. E.; David, D. J.; Limmer, S. J.; Burton, P. D.; Coker, E. N.; Beechem, T. E. Brumbach, M. T. Electrodeposited $Ni_xCo_{3-x}O_4$ nanostructured films as bifunctional oxygen electrocatalysts. *Chem. Commun.* **2015**, *51*, 9511–9514.

181. Wang, J.; Wu, Z.; Han, L.; Lin, R.; Xin, H.; Wang, D. Hollow-structured carbon-supported nickel cobaltite nanoparticles as an efficient bifunctional electrocatalyst for the oxygen reduction and evolution reactions. *ChemCatChem* **2016**, *8*, 736–742.

182. Du, J.; Chen, C.; Cheng, F.; Chen, J. Rapid synthesis and efficient electrocatalytic oxygen reduction/evolution reaction of $CoMn_2O_4$ nanodots supported on graphene. *Inorg. Chem.* **2015**, *54*, 5467–5474.

183. Zhang, C.; Antonietti, M.; Fellinger, T.P. Blood ties: Co_3O_4 decorated blood derived carbon as a superior bifunctional electrocatalyst. *Adv. Funct. Mater.* **2014**, *24*, 7655–7665.

184. Aijaz, A.; Masa, J.; Rösler, C.; Xia, W.; Weide, P.; Botz, A. J. R.; Fischer, R. A.; Schuhmann, W.; Muhler, M. Co@Co_3O_4 encapsulated in carbon nanotube-grafted nitrogen-doped carbon polyhedral as an advanced bifunctional oxygen electrode. *Angew. Chem. Int. Ed.* **2016**, *55*, 4087–4091.

185. Menezes, P. W.; Indra, A.; Sahraie, N. R.; Bergmann, A.; Strasser, P.; Driess, M. Cobalt-manganese-based spinels as multifunctional materials that unify catalytic water oxidation and oxygen reduction reactions. *ChemSusChem* **2015**, *8*, 164–171.

186. Qian, L.; Lu, Z.; Xu, T.; Wu, X.; Tian, Y.; Li, Y.; Huo, Z.; Sun, X.; Duan, X. Trinary layered double hydroxides as high-performance bifunctional materials for oxygen electrocatalysis. *Adv. Energy Mater.* **2015**, *5*, 1500245.

187. Strmcnik, D.; Kodama, K.; van der Vliet, D.; Greeley, J.; Stamenkovic, V. R.; Marković, N. M. The role of non-covalent interactions in electrocatalytic fuel-cell reactions on platinum. *Nat. Chem.* **2009**, *1*, 466–472.

188. Wiechen, M.; Zaharieva, I.; Dau, H.; Kurz, P. Layered manganese oxides for water-oxidation: Alkaline earth cations influence catalytic activity in a photosystem II-like fashion. *Chem. Sci.* **2012**, *3*, 2330–2339.

Index

T - #0424 - 071024 - C244 - 234/156/11 - PB - 9780367780500 - Gloss Lamination